Production and Properties of Starch—Current Research

Production and Properties of Starch—Current Research

Editors

Litao Tong
Lili Wang

Basel • Beijing • Wuhan • Barcelona • Belgrade • Novi Sad • Cluj • Manchester

Editors

Litao Tong
Institute of Food Science and Technology, Chinese Academy of Agricultural Sciences
Beijing
China

Lili Wang
Institute of Food Science and Technology, Chinese Academy of Agricultural Sciences
Beijing
China

Editorial Office
MDPI
St. Alban-Anlage 66
4052 Basel, Switzerland

This is a reprint of articles from the Special Issue published online in the open access journal *Molecules* (ISSN 1420-3049) (available at: https://www.mdpi.com/journal/molecules/special_issues/Starch_Production_Properties).

For citation purposes, cite each article independently as indicated on the article page online and as indicated below:

Lastname, A.A.; Lastname, B.B. Article Title. *Journal Name* **Year**, *Volume Number*, Page Range.

ISBN 978-3-7258-0513-6 (Hbk)
ISBN 978-3-7258-0514-3 (PDF)
doi.org/10.3390/books978-3-7258-0514-3

© 2024 by the authors. Articles in this book are Open Access and distributed under the Creative Commons Attribution (CC BY) license. The book as a whole is distributed by MDPI under the terms and conditions of the Creative Commons Attribution-NonCommercial-NoDerivs (CC BY-NC-ND) license.

Contents

About the Editors vii

Preface ix

Lili Wang and Litao Tong
Production and Properties of Starch: Current Research
Reprinted from: *Molecules* **2024**, *29*, 646, doi:10.3390/molecules29030646 1

Hesham Alqah, Shahzad Hussain, Mohamed Saleh Alamri, Abdellatif A. Mohamed, Akram A. Qasem, Mohamed A. Ibraheem, et al.
Effect of Germinated Sorghum Extract on the Physical and Thermal Properties of Pre-Gelatinized Cereals, Sweet Potato and Beans Starches
Reprinted from: *Molecules* **2023**, *28*, 7030, doi:10.3390/molecules28207030 5

Junio Flores Castellanos, Arsalan Khan and Joerg Fettke
Gradual Analytics of Starch-Interacting Proteins Revealed the Involvement of Starch-Phosphorylating Enzymes during Synthesis of Storage Starch in Potato (*Solanum tuberosum* L.) Tubers
Reprinted from: *Molecules* **2023**, *28*, 6219, doi:10.3390/molecules28176219 18

Mengzi Nie, Chunhong Piao, Jiaxin Li, Yue He, Huihan Xi, Zhiying Chen, et al.
Effects of Different Extraction Methods on the Gelatinization and Retrogradation Properties of Highland Barley Starch
Reprinted from: *Molecules* **2022**, *27*, 6524, doi:10.3390/molecules27196524 33

Takahiro Noda, Koji Ishiguro, Tatsuro Suzuki and Toshikazu Morishita
Physicochemical Properties and In Vitro Digestibility of Starch from a Trace-Rutinosidase Variety of Tartary Buckwheat 'Manten-Kirari'
Reprinted from: *Molecules* **2022**, *27*, 6172, doi:10.3390/molecules27196172 46

David Tochihuitl-Vázquez, Rafael Ramírez-Bon, José Martín Yáñez-Limón and Fernando Martínez-Bustos
Reactive Extrusion as a Pretreatment in Cassava (*Manihot esculenta* Crantz) and Pea (*Pisum sativum* L.) Starches to Improve Spinnability Properties for Obtaining Fibers
Reprinted from: *Molecules* **2022**, *27*, 5944, doi:10.3390/molecules27185944 54

Wanyu Qin, Huihan Xi, Aixia Wang, Xue Gong, Zhiying Chen, Yue He, et al.
Ultrasound Treatment Enhanced Semidry-Milled Rice Flour Properties and Gluten-Free Rice Bread Quality
Reprinted from: *Molecules* **2022**, *27*, 5403, doi:10.3390/molecules27175403 71

Jau-Shya Lee, Jahurul Haque Akanda, Soon Loong Fong, Chee Kiong Siew and Ai Ling Ho
Effects of Annealing on the Properties of Gamma-Irradiated Sago Starch
Reprinted from: *Molecules* **2022**, *27*, 4838, doi:10.3390/molecules27154838 86

Xue Gong, Lin Zhu, Aixia Wang, Huihan Xi, Mengzi Nie, Zhiying Chen, et al.
Understanding the Palatability, Flavor, Starch Functional Properties and Storability of Indica-Japonica Hybrid Rice
Reprinted from: *Molecules* **2022**, *27*, 4009, doi:10.3390/molecules27134009 101

Mayra Esthela González-Mendoza, Fernando Martínez-Bustos, Eduardo Castaño-Tostado and Silvia Lorena Amaya-Llano
Effect of Microwave Irradiation on Acid Hydrolysis of Faba Bean Starch: Physicochemical Changes of the Starch Granules
Reprinted from: *Molecules* **2022**, *27*, 3528, doi:10.3390/molecules27113528 118

Zhaohang Zuo, Shuting Liu, Weiqiao Pang, Baoxin Lu, Wei Sun, Naidan Zhang, et al.
Beneficial Effect of Kidney Bean Resistant Starch on Hyperlipidemia—Induced Acute Pancreatitis and Related Intestinal Barrier Damage in Rats
Reprinted from: *Molecules* **2022**, *27*, 2783, doi:10.3390/molecules27092783 134

Szilard Pesek and Radu Silaghi-Dumitrescu
The Iodine/Iodide/Starch Supramolecular Complex
Reprinted from: *Molecules* **2024**, *29*, 641, doi:10.3390/molecules29030641 148

Jiameng Liu, Wei Lu, Yantian Liang, Lili Wang, Nuo Jin, Huining Zhao, et al.
Research Progress on Hypoglycemic Mechanisms of Resistant Starch: A Review
Reprinted from: *Molecules* **2022**, *27*, 7111, doi:10.3390/molecules27207111 163

About the Editors

Litao Tong

Li-Tao Tong graduated from Kyushu University, Japan, with a Ph.D. degree in 2012, and is currently working at the Institute of Agricultural Product Processing, Chinese Academy of Agricultural Sciences (CAAS), the Western Agricultural Research Center, CAAS, and a part-time tutor at Shanxi University and Beijing Agricultural College. His main research fields are the deep processing of cereals and functional foods, and his main research directions are (1) research on key technologies of quality control and deep processing of rice-based foods and (2) the research and development of key technologies and core equipment for processing new functional cereal foods.

Lili Wang

Lili Wang graduated from the School of Food Science and Nutrition Engineering, China Agricultural University, with a Ph.D. She is a visiting scholar at the Center for Nutrition and Food Science, University of Queensland, Australia, and has been working at the Institute of Agricultural Product Processing, Chinese Academy of Agricultural Sciences (CAAS) since 2014. Her main research area is processing and the high-value utilization of nutritious and healthy grains. Her main research interests are (1) the correlation between the structure and function of food macromolecules; metabolite differentiation and precise function development; and the establishment of in vitro digestion of starch, in vitro and ex vivo nutritional, and beneficial evaluation methods of active polysaccharides. She also carries out research on (2) the quality stabilization and improvement technology of barley, oats, and other cereal products; the stability regulation of cereal beverages and creation of clean-label products; and research and development of low-GI grains and staple food products.

Preface

Starch is a natural storage polysaccharide in the plant kingdom and is the main source of carbohydrates in the human diet. Natural starches demonstrate a wide range of applications in the food, cosmetic, biomedical, and pharmaceutical industries due to their inexpensive, renewable, biodegradable, biocompatible, and non-toxic properties. However, natural starches have certain inherent defects such as water insolubility, rapid regrowth, and bioactivity defects. Meanwhile, the multiscale structure of starch (e.g., the molecular weight and degree of molecular laminar ordering) has been shown to affect its physicochemical properties, such as digestibility, thermal properties, and viscosity. Therefore, various methods have been used to modify the structure of natural starches to achieve the desired physicochemical properties and biological activity. This Special Issue focuses on the sources of starch, separation processes, modification means, and current and future application trends of starch. Additionally, we explore the development of new methods of starch production and modification that have the advantages of being highly efficient, environmentally friendly, and easy to operate. This Special Issue explores the relationship between starch structure and functionality via the study of starch micromorphology, starch multiscale structure, physicochemical properties, pasting properties, thermal properties, gel properties, and digestibility, laying the groundwork for future starch applications in food and industry. The content of this Special Issue was prepared by a team of professionals with in-depth research in the field of starch and is intended for all those involved in research and development in the field of starch. We appreciate all contributions received and wish all authors and readers good luck and more insightful findings in future studies.

Litao Tong and Lili Wang
Editors

Editorial

Production and Properties of Starch: Current Research

Lili Wang * and Litao Tong *

Institute of Food Science and Technology, Chinese Academy of Agricultural Sciences, Haidian District, Beijing 100193, China
* Correspondence: wlland2013@163.com (L.W.); tonglitao@caas.cn (L.T.)

1. Introduction

Starch is an important carbohydrate polymer found in plants and has been widely used in food and non-food industries due to its abundance, renewability, biodegradability, low cost, biocompatibility and non-toxicity [1]. The functional properties of starch are variable due to its different sources and structures [2]. Factors such as starch granule morphology, amylose to amylopectin ratio, molecular structure, degree of branching in terms of steric hindrance and, consequently, mass transfer resistance can affect the physicochemical properties and application of starch [3–5]. Native starch has been modified by various processes for desired industrial applications over the last few decades [6,7]. There has been intense interest in developing novel methods, with may present advantages of high efficiency, environmental friendliness and easier starch production and modification [8]. Although more attention is paid to the functional properties of modified starch for practical applications in the industry, structural changes are the basis of these functional changes; hence, understanding the structural alterations induced by processing techniques is a fundamental subject when considering the better utilization of starch and starch modification techniques. The current research is also geared towards the importance of starch in human nutrition and health. Different forms of starch in foods and novel food processing, with respect to its digestibility, have received tremendous research focus [9] as the microstructure of starch has a great influence on its digestibility [10]. Starch multi-scale structures, including the amylose/amylopectin ratio, fine structures of amylose and amylopectin, short-range ordered structures, helical structures, crystalline features, lamellar structures and morphology, are factors that determine enzyme binding and catalyzation with starch in the human diet [11–13]. This Special Issue introduces the latest research, using a variety of raw materials for analysis, on the physical and chemical properties of starch after modification, intrinsic structure changes after extraction and modification methods, and starch's digestive and anti digestive properties. It contains ten articles and a review; we will provide a brief overview of the contents of this Special Issue in the following paragraphs. On this matter, we clarify that it is not the purpose of this Editorial to elaborate on each of the texts, but rather to encourage the reader to explore them.

2. An Overview of Published Articles

Hesham Alqah et al. (Contribution 1) discuss how starches from different botanical sources are affected in the presence of enzymes. This study investigated the impact of α-amylase on the water holding capacity, freezable water content, sugar content and water sorption isotherm of pregelatinized starches derived from chickpea, wheat, corn, white beans and sweet potatoes. The results show that the water holding capacity and the sugar content of starch increased significantly when the annealing temperature and GSE were increased. The ordering of starches in terms of their freezable water content is as follows: chickpea starch > white beans starch > wheat starch > chickpea starch > sweet potato starch. The moisture sorption of different pregelatinized starches increased with increasing water activity at different annealing temperatures and was characterized as Type II isotherm. The

Citation: Wang, L.; Tong, L. Production and Properties of Starch: Current Research. *Molecules* **2024**, *29*, 646. https://doi.org/10.3390/molecules29030646

Received: 31 October 2023
Accepted: 6 November 2023
Published: 30 January 2024

Copyright: © 2024 by the authors. Licensee MDPI, Basel, Switzerland. This article is an open access article distributed under the terms and conditions of the Creative Commons Attribution (CC BY) license (https://creativecommons.org/licenses/by/4.0/).

equilibrium moisture content and monolayer moisture contents decreased after treatment with 0.1 mL GSE but increased after 1.0 mL GSE treatment and longer annealing time. This study demonstrated that the cost of the analyzed procedure is significantly lower in comparison to utilizing refined α-amylase extract.

The article by Nie et al. (Contribution 2) investigated the gelatinization and retrogradation properties of highland barley starch (HBS) using different extraction methods. In this paper, the effect of different extraction methods on starch was explained by obtaining HBS through three methods: alkali extraction (A-HBS), ultrasonic extraction (U-HBS) and enzyme extraction (E-HBS), which were used to explain the apparent morphology, structural changes and physicochemical properties of the extracted starch. The results showed that A-HBS- and U-HBS-treated starch had less surface damage and the E-HBS starch had a rough surface, but none of them changed the integrity. Compared to the other two extraction methods, A-HBS showed the weakest hydrogen bonding, which resulted in the highest viscosity, and could, thus, be used as a food thickener. The different extraction methods did not change the crystalline type of HBS, but E-HBS had the highest relative crystallinity value; thus, it exhibited better thermal stability, which increased the crispness of foods such as cookies. U-HBS had the highest damaged starch content, as well as the lowest gelatinization enthalpy, after 1 and 7 days of storage, respectively, providing anti-aging properties which make it suitable for foods with a short shelf life such as bread or noodles. The authors state that the work provides a theoretical basis for the development of HBS and highland barley food with different expected properties.

Takahiro Noda et al. (Contribution 3) investigated the physicochemical properties and in vitro digestibility of "Manten-Kirari" starch with trace-rutinosidase activity. The starch was analyzed by a rapid viscosity analyzer (RVA) and differential scanning calorimetry (DSC), as well as in vitro digestibility. The study finds the lowest content of amylose was observed in 'Manten-Kirari' starch (18.1%), while the highest was located in 'Kitawasesoba' starch (22.6%). The 'Manten-Kirari' starch exhibited a larger median granule size (11.41 µm) and higher values of peak viscosity (286.8 RVU) and breakdown (115.2 RVU) than the others. The onset temperature values for gelatinization were 60.5 °C for 'Kitawasesoba', 61.3 °C for 'Manten-Kirari' and 64.7 °C for 'Hokkai T8'. The 'Manten-Kirari' and 'Hokkai T8' starches were digested more slowly than the 'Kitawasesoba' starch.

Qin et al. (Contribution 4) investigated the effects of different physical treatments on the structural and functional properties of rice flour. The treatment methods in this paper include ultrasonic treatment (US), microwave treatment (MW) and hydrothermal treatment (HT). The results of this paper show that ultrasound and microwave have little effect on the apparent morphology of starch, but hydrothermal treatment produces more fragments and cracks. Ultrasound led to a more ordered arrangement within the starch, whereas hydrothermal heat disrupted the structure of the starch. The batter with added HT exhibited the highest G' and G'' values and the lowest tan δ. Furthermore, bread made from US and MW starch presented reduced hardness, cohesion and gumminess. According to the authors' conclusions, it is presented that US rice flour improves the textural properties and appearance of rice bread.

Li et al. (Contribution 5) investigated the effects of gamma irradiation and annealing (ANN) on the functional properties of sago starch. In this study, the solubility and swelling power, pasting characteristics and amylose starch content of sago starch were explored using dual treatments of gamma irradiation and annealing. The final results showed that the content of amylose starch decreased under gamma irradiation and increased under ANN, and the effect was the same under dual modification as under ANN. After irradiation, the swelling power was reduced and solubility increased; after ANN treatment, both the solubility and swelling power were reduced, while the results after double treatment showed similar results to those of gamma irradiation. At the same time, neither single nor double treatment altered the integrity of starch. Based on the authors' conclusions, gamma irradiation and ANN were able to induce some new properties of sago starch that could be used for extended applications.

Xue et al. (Contribution 6) explored the palatability, flavor, storability and starch functionality properties of different varieties of rice. In this paper, 84 varieties were selected for comparison of their various characteristics. The authors stated that the presented three YY-IJHR varieties could be identified, through their straight chain starch and protein content, as being better for cooking and consumption in relation to the N84 variety. Significant differences were seen in the pasting characteristics of the 84 rice varieties. Rice aroma components were revealed by GC-IMS, which indicated that the alcohol content of the volatile components of YY-IJHR was generally lower, whereas the content of some aldehydes and esters was higher than in N84. In terms of storage, YY-IJHR had better rice quality and storability than N84. The authors' study clearly analyzed the variability of 84 rice varieties and is extremely helpful for both agricultural cultivation as well as food production.

Zuo et al. (Contribution 7) explored the beneficial effects of resistant starch in hyperlipidemia acute pancreatitis (HLAP). In this paper, an acute pancreatitis model was established by feeding a hyperlipidemia diet to rats and subsequently evaluating the anti-HLAP effect of RS in kidney beans. The final results show that compared with original starch, the average volume particle size of the RS granules increased significantly and the specific surface reduced significantly. Reduction in a specific surface area can avoid excessive contact between RS and enzymes, thereby enhancing its resistance to enzymatic hydrolysis. The IL-6, IL-1β and TNF-α of serum in each RS group and the depth of the crypt decreased, while the height of the villi in the small intestine and the thickness of the muscle layer of rats increased. The authors showed that resistant starch has a preventive effect on intestinal damage, which is a very interesting result for the study of human health.

3. Conclusions

This compilation of articles, focused on the production and functional properties of starch, includes a wide variety of research tools that illustrate the breadth of the field of study. This can be reflected in the variety of samples and starches analyzed, different extraction and modification methods, and means of detection.

In terms of the subject, starch modification has received increasing attention in recent years, with a total of six articles describing starch modification. However, it should be noted that one of the articles is focused on extraction, using processes that may make starch extraction more eco-friendly, with increased extraction rates as well as purity. Two of these articles describe the differences between varieties and are extremely beneficial to those in the food processing industry as well as in crop cultivation. The last two articles describe the beneficial effects of resistant starch on the human body, which could be helpful when considering society's current focus on healthy eating.

In conclusion, research should not only focus on the production and properties of starch but also on exploring its structural changes, through which we can then explain the altered physicochemical and functional properties of starch. Currently, there are fewer articles on this subject in this Special Issue. Therefore, we are extremely interested in articles of this novelty, which not only provide the reader with a deeper understanding of the structural changes of starch but also allow the principles of these to be analyzed, thus enabling a more comprehensive understanding of the field of research.

Conflicts of Interest: The authors declare no conflict of interest.

List of Contributions

1. Alqah, H.; Hussain, S.; Alamri, M.S.; Mohamed, A.A.; Qasem, A.A.; Ibraheem, M.A.; Shehzad, A. Effect of Germinated Sorghum Extract on the Physical and Thermal Properties of Pre-Gelatinized Cereals, Sweet Potato and Beans Starches. *Molecules* **2023**, *28*, 7030. https://doi.org/10.3390/molecules28207030.
2. Nie, M.; Piao, C.; Li, J.; He, Y.; Xi, H.; Chen, Z.; Wang, L.; Liu, L.; Huang, Y.; Wang, F.; et al. Effects of Different Extraction Methods on the Gelatinization and Retrogradation Properties of Highland Barley Starch. *Molecules* **2022**, *27*, 6524. https://doi.org/10.3390/molecules27196524.

3. Noda, T.; Ishiguro, K.; Suzuki, T.; Morishita, T. Physicochemical Properties and In Vitro Digestibility of Starch from a Trace-Rutinosidase Variety of Tartary Buckwheat 'Manten-Kirari'. *Molecules* **2022**, *27*, 6172. https://doi.org/10.3390/molecules27196172.
4. Qin, W.; Xi, H.; Wang, A.; Gong, X.; Chen, Z.; He, Y.; Wang, L.; Liu, L.; Wang, F.; Tong, L. Ultrasound Treatment Enhanced Semidry-Milled Rice Flour Properties and Gluten-Free Rice Bread Quality. *Molecules* **2022**, *27*, 5403. https://doi.org/10.3390/molecules27175403.
5. Lee, J.S.; Akanda, J.H.; Fong, S.L.; Siew, C.K.; Ho, A.L. Effects of Annealing on the Properties of Gamma-Irradiated Sago Starch. *Molecules* **2022**, *27*, 4838. https://doi.org/10.3390/molecules27154838.
6. Gong, X.; Zhu, L.; Wang, A.; Xi, H.; Nie, M.; Chen, Z.; He, Y.; Tian, Y.; Wang, F.; Tong, L. Understanding the Palatability, Flavor, Starch Functional Properties and Storability of IndicaJaponica Hybrid Rice. *Molecules* **2022**, *27*, 4009. https://doi.org/10.3390/molecules27134009.
7. Zuo, Z.; Liu, S.; Pang, W.; Lu, B.; Sun, W.; Zhang, N.; Zhou, X.; Zhang, D.; Wang, Y. Beneficial Effect of Kidney Bean Resistant Starch on Hyperlipidemia—Induced Acute Pancreatitis and Related Intestinal Barrier Damage in Rats. *Molecules* **2022**, *27*, 2783. https://doi.org/10.3390/molecules27092783.

References

1. Bušić, A.; Marđetko, N.; Kundas, S.; Morzak, G.; Belskaya, H.; Šantek, M.I.; Komes, D.; Novak, S.; Šantek, B. Bioethanol Production from Renewable Raw Materials and Its Separation and Purification: A Review. *Food Technol. Biotechnol.* **2018**, *56*, 289–311. [CrossRef] [PubMed]
2. Xu, J.; Li, X.; Chen, J.; Dai, T.; Liu, C.; Li, T. Effect of polymeric proanthocyanidin on the physicochemical and in vitro digestive properties of different starches. *LWT* **2021**, *148*, 111713. [CrossRef]
3. Wang, N.; Li, C.; Miao, D.; Hou, H.; Dai, Y.; Zhang, Y.; Wang, B. The effect of non-thermal physical modification on the structure, properties and chemical activity of starch: A review. *Int. J. Biol. Macromol.* **2023**, *251*, 126200. [CrossRef] [PubMed]
4. Jia, R.; Cui, C.; Gao, L.; Qin, Y.; Ji, N.; Dai, L.; Wang, Y.; Xiong, L.; Shi, R.; Sun, Q. A review of starch swelling behavior: Its mechanism, determination methods, influencing factors, and influence on food quality. *Carbohydr. Polym.* **2023**, *321*, 121260. [CrossRef] [PubMed]
5. Gebre, B.A.; Zhang, C.; Li, Z.; Sui, Z.; Corke, H. Impact of starch chain length distributions on physicochemical properties and digestibility of starches. *Food Chem.* **2024**, *435*, 137641. [CrossRef] [PubMed]
6. Salimi, M.; Channab, B.-E.; El Idrissi, A.; Zahouily, M.; Motamedi, E. A comprehensive review on starch: Structure, modification, and applications in slow/controlled-release fertilizers in agriculture. *Carbohydr. Polym.* **2023**, *322*, 121326. [CrossRef] [PubMed]
7. Chen, Z.; Yang, Q.; Yang, Y.; Zhong, H. The effects of high-pressure treatment on the structure, physicochemical properties and digestive property of starch—A review. *Int. J. Biol. Macromol.* **2023**, *244*, 125376. [CrossRef] [PubMed]
8. Sun, X.; Sun, Z.; Saleh, A.S.; Lu, Y.; Zhang, X.; Ge, X.; Shen, H.; Yu, X.; Li, W. Effects of various microwave intensities collaborated with different cold plasma duration time on structural, physicochemical, and digestive properties of lotus root starch. *Food Chem.* **2023**, *405*, 134837. [CrossRef] [PubMed]
9. Jiang, H.; Zhang, W.; Cao, J.; Jiang, W. Development of biodegradable active films based on longan seed starch incorporated with banana flower bract anthocyanin extracts and applications in food freshness indication. *Int. J. Biol. Macromol.* **2023**, *251*, 126372. [CrossRef] [PubMed]
10. Lu, C.; Zhao, Z.; Huang, G.; Liu, J.; Ye, F.; Chen, J.; Ming, J.; Zhao, G.; Lei, L. The contribution of cell wall integrity to gastric emptying and in vitro starch digestibility and fermentation performance of highland barley foods. *Food Res. Int.* **2023**, *169*, 112912. [CrossRef] [PubMed]
11. Mao, S.; Ren, Y.; Ye, X.; Kong, X.; Tian, J. Regulating the physicochemical, structural characteristics and digestibility of potato starch by complexing with different phenolic acids. *Int. J. Biol. Macromol.* **2023**, *253*, 127474. [CrossRef] [PubMed]
12. Yang, Z.; Zhang, Y.; Wu, Y.; Ouyang, J. Factors influencing the starch digestibility of starchy foods: A review. *Food Chem.* **2023**, *406*, 135009. [CrossRef] [PubMed]
13. Wang, Y.; Guo, J.; Wang, C.; Li, Y.; Bai, Z.; Luo, D.; Hu, Y.; Chen, S. Effects of konjac glucomannan and freezing on thermal properties, rheology, digestibility and microstructure of starch isolated from wheat dough. *LWT* **2021**, *177*, 114588. [CrossRef]

Disclaimer/Publisher's Note: The statements, opinions and data contained in all publications are solely those of the individual author(s) and contributor(s) and not of MDPI and/or the editor(s). MDPI and/or the editor(s) disclaim responsibility for any injury to people or property resulting from any ideas, methods, instructions or products referred to in the content.

Article

Effect of Germinated Sorghum Extract on the Physical and Thermal Properties of Pre-Gelatinized Cereals, Sweet Potato and Beans Starches

Hesham Alqah [1], Shahzad Hussain [1,*], Mohamed Saleh Alamri [1], Abdellatif A. Mohamed [1], Akram A. Qasem [1], Mohamed A. Ibraheem [1] and Aamir Shehzad [2]

[1] Department of Food Science and Nutrition, King Saud University, Riyadh 11451, Saudi Arabia; heshamfrnd@gmail.com (H.A.); msalamri@ksu.edu.sa (M.S.A.); abdmohamed@ksu.edu.sa (A.A.M.); aqasem@ksu.edu.sa (A.A.Q.); mfadol@ksu.edu.sa (M.A.I.)

[2] UniLaSalle, Univ. Artois, ULR7519—Transformations & Agro-Ressources, Normandie Université, F-76130 Mont-Saint-Aignan, France; aamir.shehzad@unilasalle.fr

* Correspondence: shhussain@ksu.edu.sa

Abstract: Starches from different botanical sources are affected in the presence of enzymes. This study investigated the impact of α-amylase on several properties of pre-gelatinized starches derived from chickpea (*Cicer arietinum* L.), wheat (*Triticum aestivum* L.), corn (*Zea mays* L.), white beans (*Phaseolus vulgaris*), and sweet potatoes (*Ipomoea batatas* L.). Specifically, the water holding capacity, freezable water content, sugar content, and water sorption isotherm (adsorption and desorption) properties were examined. The source of α-amylase utilized in this study was a germinated sorghum (*Sorghum bicolor* L. Moench) extract (GSE). The starch samples were subjected to annealing at temperatures of 40, 50, and 60 °C for durations of either 30 or 60 min prior to the process of gelatinization. A significant increase in the annealing temperature and GSE resulted in a notable enhancement in both the water-holding capacity and the sugar content of the starch. The ordering of starches in terms of their freezable water content is as follows: Chickpea starch (C.P.S) > white beans starch (W.B.S) > wheat starch (W.S) > chickpea starch (C.S) > sweet potato starch (S.P.S). The Guggenheim-Anderson-de Boer (GAB) model was only employed for fitting the data, as the Brunauer–Emmett–Teller (BET) model had a low root mean square error (RMSE). The application of annealing and GSE treatment resulted in a shift of the adsorption and desorption isotherms towards greater levels of moisture content. A strong hysteresis was found in the adsorption and desorption curves, notably within the water activity range of 0.6 to 0.8. The GSE treatment and longer annealing time had an impact on the monolayer water content (m_o), as well as the C and K parameters of the GAB model, irrespective of the annealing temperature. These results can be used to evaluate the applicability of starch in the pharmaceutical and food sectors.

Keywords: starch; enzyme; sorghum (*Sorghum bicolor* L.) extract; sorption

Citation: Alqah, H.; Hussain, S.; Alamri, M.S.; Mohamed, A.A.; Qasem, A.A.; Ibraheem, M.A.; Shehzad, A. Effect of Germinated Sorghum Extract on the Physical and Thermal Properties of Pre-Gelatinized Cereals, Sweet Potato and Beans Starches. *Molecules* **2023**, *28*, 7030. https://doi.org/10.3390/molecules28207030

Academic Editors: Litao Tong and Lili Wang

Received: 10 September 2023
Revised: 7 October 2023
Accepted: 10 October 2023
Published: 11 October 2023

Copyright: © 2023 by the authors. Licensee MDPI, Basel, Switzerland. This article is an open access article distributed under the terms and conditions of the Creative Commons Attribution (CC BY) license (https:// creativecommons.org/licenses/by/ 4.0/).

1. Introduction

Starch granules are normally insoluble in cold water, and thus cannot exhibit any of the main functions, such as increased viscosity and water binding at ambient temperature. These restrictions have limited the use of native starches in many items. Instant or pre-gelatinized starch (PGS) was used to overcome such issues which is often referred to as cold gel. This form of modified starch will swell in cold water which leads to a rapid viscosity increase depending on the solid content [1]. Pre-gelatinized starches are used in food processing for thickening or water retention without heat application. These starches are commonly used in puddings, and baby food preparation, especially cereal based designated for infants under 12 months of age [2,3]. The enzymatic hydrolysis promotes the partial degradation of starch in cereals prior to ingestion and facilitates starch digestion by infants due to the limited pancreas ability to digest starch [4]. Pre-gelatinization is accomplished by exposing a thin coating of starch slurry to a heated surface, such as a hot plate or drum dryer, with twin drum dryers producing a superior product than single drum dryers. The qualities of the finished product can be influenced by the slurry content, temperature, and speed of the drum [5,6]. The starch granules may be damaged or destroyed depending on the severity of the pre-gelatinization procedure. As a result, water easily interacts with the starch components, increasing viscosity without heating [1,7]. However, some limitations of PGS including grainy texture, inadequate consistency, and weak gels have restricted its applications to some foods. These deficiencies are mainly due to the disintegration of the granules and retrogradation of the wet starch film during drying [8].

The thermodynamic relationship between water activity and moisture balance of food at a constant temperature and pressure is represented by sorption isotherms since the quality of the stored products primarily relies on its water activity, which depends on its relative humidity and storage temperature. The moisture sorption isotherms of food products are useful information regarding their stability and prediction of their shelf life. At a given temperature, the isotherm provides information on the relation between humidity and the water activities [9,10]. Many models have been proposed to explain the moisture-sorption isotherm including multi-layered (Brunauer–Emmett–Teller (BET) and Guggenheim-Anderson-de Boer (GAB) models), semi-empirical (Ferro-Fontan, Henderson, and Halsey), or empirical (Smith and Oswin models). The BET isotherm model is the most important model for understanding the multi-layer sorption isotherm, in particular, for Type II isotherm [11]. The GAB model is known as the most flexible sorption model in the literature. The American Society of Agricultural Engineers has adopted the BET and GAB models for the description of sorption isotherms. These models are used extensively in the literature [12–14]. The study by Ocieczek, et al. [15] concluded that despite identical particle size characteristics, native cassava starch notably differs from potato starch in terms of hygroscopicity as indicated by the parameters of the BET model. The moisture sorption behavior of the pea starch films exhibited an upward trend as the water activity levels increased across various temperatures (5, 15, 25, and 40 °C), conforming to a Type III isotherm. The equilibrium moisture content and monolayer moisture contents (m_o) exhibited a decrease as the storage temperature increased while maintaining a constant water activity [16].

This study aimed to evaluate the impact of crude germinated sorghum extract on the sorption isotherm, thermal characteristics, amount of freezable water, and water retention capacity of pre-gelatinized chickpea starches, corn, White beans, wheat, and sweet potato starches. The outcome of this study demonstrates that the cost of the procedure is significantly lower in comparison to utilizing a refined α-amylase extract.

2. Results and Discussion

2.1. Water Holding Capacity

Based on the activity of pure α-amylase solution, the concentration of α-amylase in the GSE was found to be 5 mg/10 mL. The percentage (%) of amylose content of the tested starches for wheat, chickpea, sweet potato, white beans, and corn, was 25.0, 24.0, 22.6, 20.9, and 20.4, respectively. The water holding capacity (WHC) is defined as the amount of water that can be absorbed per gram of sample [17]. The WHC of native, annealed, and GSE-treated pre-gelatinized (PGS) starches is presented in (Table 1). The range of the WHC was 13.0–4.51 (g/g) and 14.35–5.28 (g/g) for the native and annealed, respectively. Annealing appeared to increase WHC of chickpea starch, corn starch, and white bean starch by 12.9, 20.1, and 15.3%, respectively, but W.S and S.P.S exhibited a drop in WHC by 12.4 and 3.6%, respectively. The WHC of native and annealed starches rank as: S.P.S > C.P.S > W.B.S > W.S > C.S. According to these ranks, amylose content was not the determining factor in the WHC since high amylose content starch, such as wheat starch, did not exhibit highest WHC. Botanical origin could be considered a factor because wheat starch and corn starch, cereal-based starches, exhibited the lowest values, whereas sweet potato starch, and tuber starch, exhibited the highest values. The data presented here indicate that the amorphous region of wheat starch granules is more compact compared to sweet potato starch, and chickpea starch since it allowed limited water penetration. Starch granules bind water via hydrogen bonding. Therefore, the differences between the WHC can be attributed to the intensity of the hydrogen bonds and the accessibility of water to binding sites in the granule. Consequently, the WHC of starches is dependent on granule structure, botanical origin and type of processing (treatment), and to some extent amylose content. This is in agreement with previous reports which indicated that WHC, swelling power, and peak viscosity are correlated, but amylose content was not a major indicator of these parameters [18]. Alqah, et al. [19] reported that no correlation was found between WHC and peak viscosity of several starches. It is clear how annealing at 50 °C increased the WHC, compared to 40 and 60 °C which may indicate how this temperature affected the granule structure which could be attributed to an increase in granule porosity leading to higher WHC. The WHC has allegedly been higher in dry-heated starch compared to their native [20]. Overall, sweet potato starch behavior stood out because it was the most temperature sensitive, had the highest WHC, and ranked third with reference to amylose content. The effect of annealing on the WHC of chickpea starch was the most perceptible with or without GSE treatment compared to the other tested starches. All GSE-treated starches exhibited higher WHC, but GSE treatment appeared to have less impact on cereal-based starches (wheat and corn) (Table 1). Given that the α-amylase attack causes holes on the starch granule surface as well as the effect of higher temperatures, obviously increased the porosity of the granules and triggered higher WHC conditions which explains the increase. The ability of starch to bind and hold water is a desirable characteristic in the food industry especially when starch is used in frozen food products as stabilizers and emulsifiers because it prevents syneresis, therefore GSE treatment is a desirable process.

Table 1. The water holding capacity of pre-gelatinized, native, and modified starches.

		Chickpea Starch	Corn Starch	White Bean Starch	Wheat Starch	Sweet Potato Starch
	Native	10.12 ± 1.48 [b]	4.51 ± 0.89 [e]	6.80 ± 0.4 [c]	5.93 ± 1.14 [d]	13.00 ± 0.64 [a]
	40 °C					
30 min	No GSE	11.43 ± 0.15 [b]	5.40 ± 0.12 [d]	7.84 ± 0.19 [c]	5.28 ± 0.30 [d]	13.47 ± 0.11 [a]
	0.1 mL	15.09 ± 0.18 [a]	5.28 ± 0.13 [e]	11.10 ± 0.16 [c]	6.19 ± 0.24 [d]	13.90 ± 0.12 [b]
	1.0 mL	17.76 ± 0.23 [a]	5.19 ± 0.01 [e]	13.73 ± 0.20 [c]	6.86 ± 0.10 [d]	14.32 ± 0.15 [b]
60 min	No GSE	11.29 ± 0.10 [b]	8.39 ± 0.13 [c]	8.65 ± 0.21 [c]	6.54 ± 0.41 [d]	16.15 ± 0.21 [a]
	0.1 mL	14.60 ± 0.13 [b]	9.90 ± 0.16 [d]	12.62 ± 0.24 [c]	6.55 ± 0.14 [e]	16.82 ± 0.16 [a]
	1.0 mL	17.00 ± 0.32 [ab]	9.40 ± 0.26 [e]	13.34 ± 0.31 [c]	10.61 ± 0.10 [d]	17.81 ± 0.21 [a]
	50 °C					
30 min	No GSE	12.04 ± 0.60 [b]	6.08 ± 0.20 [e]	9.35 ± 0.30 [c]	7.83 ± 0.2 [d]	14.35 ± 0.32 [a]
	0.1 mL	14.61 ± 0.10 [b]	6.74 ± 0.10 [e]	10.52 ± 0.10 [c]	9.54 ± 0.09 [d]	16.73 ± 0.23 [a]
	1.0 mL	16.51 ± 0.60 [b]	7.54 ± 0.10 [e]	13.03 ± 0.10 [c]	9.78 ± 0.15 [d]	17.01 ± 0.14 [a]
60 min	No GSE	12.06 ± 0.20 [b]	8.22 ± 0.30 [d]	9.95 ± 0.10 [c]	8.40 ± 0.16 [d]	15.60 ± 0.18 [a]
	0.1 mL	14.63 ± 0.50 [b]	9.03 ± 0.30 [d]	10.88 ± 0.20 [c]	8.37 ± 0.21 [e]	17.11 ± 0.13 [a]
	1.0 mL	13.65 ± 0.10 [c]	9.28 ± 0.10 [e]	14.61 ± 0.10 [b]	10.46 ± 0.26 [d]	18.22 ± 0.21 [a]

Table 1. Cont.

		Chickpea Starch	Corn Starch	White Bean Starch	Wheat Starch	Sweet Potato Starch
				60 °C		
30 min	No GSE	7.18 ± 0.24 [c]	5.56 ± 0.53 [d]	8.38 ± 0.26 [b]	Gelatinized	10.39 ± 0.14 [a]
	0.1 mL	7.84 ± 0.12 [c]	5.47 ± 0.32 [d]	10.40 ± 0.61 [b]	Gelatinized	11.82 ± 0.11 [a]
	1.0 mL	8.95 ± 0.23 [c]	5.43 ± 0.41 [d]	10.53 ± 0.36 [b]	Gelatinized	12.00 ± 0.12 [a]
60 min	No GSE	8.18 ± 0.23 [c]	8.13 ± 0.12 [c]	9.22 ± 0.36 [b]	Gelatinized	14.93 ± 0.12 [a]
	0.1 mL	8.29 ± 0.23 [c]	8.80 ± 0.12 [c]	10.80 ± 0.12 [b]	Gelatinized	15.26 ± 0.23 [a]
	1.0 mL	9.29 ± 0.51 [c]	9.80 ± 0.23 [c]	11.17 ± 0.12 [b]	Gelatinized	15.42 ± 0.28 [a]

Values (means ± S.D) followed by different letters within each row are significantly different.

2.2. Sugars Content Determination

The sugar content (SC) of the native and treated starches is shown in (Table 2). Variation between the tested starches is clear, where annealed samples in GSE exhibited the highest sugar content due to the hydrolytic action of α-amylase. This variation indicates the level of susceptibility of the starch to α-amylase. Annealed samples exhibited SC lower than the native which indicated a loss of low molecular weight sugars during annealing at 60 °C for 60 min. The SC loss could be attributed to the swelling of the granules in the course of annealing which facilitates for leaching of lower molecular weight fractions. Annealing appeared to have a limited effect on the SC of chickpea starch, but GSE treatment resulted in the highest SC among the starches (Table 2). The SC rank of the native and annealed starches was: S.P.S > W.S > C.P.S > C.S > W.B.S, whereas GSE treated rank as: W.S > C.P.S > C.S > W.B.S > S.P.S. Based on this ranking, wheat starch was influenced the most with GSE by releasing the most sugar, whereas sweet potato starch released the minimum. Although native sweet potato starch contained the most SC, it was influenced more by annealing than GSE. Therefore, wheat starch was the most susceptible to GSE among the tested starches and sweet potato starch.

Table 2. Soluble sugars of pre-gelatinized native and annealed with and without GSE.

	Glucose (μg/mL)		
Starch Type	Native	Annealed	Annealed with GSE
White bean	24.93 ± 0.56 [d]	14.93 ± 0.08 [d]	82 ± 2.01 [d]
Chickpea	33.16 ± 1.11 [c]	33.10 ± 1.02 [b]	104 ± 1.21 [b]
Corn	20.94 ± 0.98 [e]	17.21 ± 1.33 [c]	92 ± 2.54 [c]
Wheat	39.49 ± 1.32 [b]	35.18 ± 3.21 [b]	116 ± 3.34 [a]
Sweet potato	96.13 ± 2.04 [a]	77.02 ± 2.27 [a]	79 ± 2.59 [d]

Values (means ± S.D) followed by different letters within each column are significantly different.

2.3. Freezable Water Determination

Modified PGS starch treated with GSE for 60 min at 40 °C was used for FW (freezable water) determination. One distinct DSC endotherm profile was obtained in all tested starches for FW. The size of the endotherm, onset, peak, and ΔH values shifted to higher or lower temperatures depending on the starch type (Table 3). These variations are reflective of the difference in the size of the peak of the melting ice. When sufficient amounts of water are present in the system appreciable amount of ice crystals form, but in limited water conditions most of the water is bound to starch with a small freezable quantity. This is dependent on the different water binding sites present in the pre-gelatinized starch (PGS) samples. These binding sites in PGS are typically hydroxyl groups and inter-glucose oxygen atoms. The interaction of these sites with water varies according to the molecular structure and compositional properties of the starches [21]. The freezable water contents (ΔH of melted ice/ΔH of water) of chickpea starch was the highest, whereas sweet potato starch exhibited the lowest value indicating more bound water (Table 3). The FW of the tested PGS rank as: C.P.S > W.B.S > W.S> C.S > S.P.S. When comparing the FW ranking to the starch amylose content, no connection was observed, which leads to the conclusion that granule structure is the main cause of the variation in the molecular structure of the PGS rather than the amylose content of the native starch. It is also true how physical modifications, such as gelatinization can be effective in separating starches based on water binding capacity. Obviously, gelatinization improves starch–water interaction. In addition, Fu, Wang, Zou, Li and Adhikari [21,22] reported that the gelatinization process breaks the weaker bonds in the amorphous region of the starch granule first, therefore increasing the hydration capacity of the starch. Researchers have observed that the completely gelatinized starch samples contain less freezable water (more bound water) than partially gelatinized samples. This is in agreement with literature reports [23]. Waxy starches exhibited less FW that common starch which indicates that amylopectin plays an important role in starch–water interaction, possibly due to the branching [24]. Although high amylose starches should exhibit high FW due to the low amylopectin, this was not true for W.S., because it has high amylose and relatively high FW. The high amylose–high FW theory is factual for chickpea starch. This could be attributed to the sensitivity of starch to the action of α–amylase. It is important to consider the action of GSE (α-amylase activity) and the sensitivity of the tested starches to α–amylase, because it changes the molecular structure of the PGS due to the enzymatic hydrolysis of native starch. The peak temperature of the melting ice showed variation, where PGS exhibited peak temperature higher than pure water except for W.S. This is consistent with the low FW of wheat starch.

Table 3. Freezable water.

Starch Type	Onset (°C)	Peak (°C)	ΔH (J/g)	Freezable Water
White bean	−3.99 ± 0.01 [d]	9.34 ± 0.21 [a]	514.2 ± 12.32 [b]	1.52 ± 0.02 [b]
Chickpea	−4.56 ± 0.02 [c]	5.12 ± 0.10 [c]	536.3 ± 4.56 [a]	1.59 ± 0.01 [a]
Corn	−4.98 ± 0.10 [b]	6.08 ± 1.02 [b]	286.1 ± 8.21 [e]	0.85 ± 0.02 [d]
Sweet potato	−8.44 ± 0.12 [a]	3.77 ± 0.21 [d]	204.3 ± 11.20 [f]	0.60 ± 0.01 [e]
Wheat	−8.15 ± 0.13 [a]	−0.17 ± 0.01 [f]	402.6 ± 3.25 [c]	1.19 ± 0.03 [c]
Pure water	2.15 ± 0.09 [d]	1.34 ± 0.05 [e]	337.4 ± 4.87 [d]	-

Values (means ± S.D) followed by different letters within each column are significantly different.

2.4. Moisture Sorption Isotherms

The desorption and adsorption isotherms profiles of the PGS demonstrate a simultaneous increase in equilibrium moisture content with increasing equilibrium relative humidity. This profile represents a sigmoidal shape, thus reflecting the dominant type II curve (Figure 1), according to the BET classification of isotherms [25]. The experimental quantitative evaluation of adsorption and desorption data were determined based on BET and GAB models, but the experimental data of BET of the present work exhibited a low coefficient of determination (R^2); therefore, the current data represent the GAB model only. The experimental moisture isotherms data were fitted to GAB and BET models using nonlinear regression analysis. This theoretical tri-parametric model is suited for food engineering and is highly suitable for almost all foodstuffs with water activity ranging from 0.1 to 0.9 [26]. The parameters of the model also provide valuable details about the condition of water in food. For example, the definition of the monolayer moisture content (m_o) is included in the GAB equation, which is linked to product stability and shelf life [27,28]. The data showed that annealing permitted greater enzyme accessibility to the amorphous and crystalline regions of starch granules, which promoted the idea of the development of a more porous structure, which in turn accelerated enzyme hydrolysis. It was, obvious how the equilibrium moisture content (EMC) of PGS tended to increase after treatment with 1.0 mL GSE for 60 min at all annealing temperatures. This could be attributed to the increase in the number of water binding sites induced by the enzyme (Tables 4 and 5). The constant C is the total heat of sorption of the first layer of water vapor bound directly to the active binding sites, whereas K represents the multilayer water molecules with respect to the bulk water rather than vapor. The C value is always positive, and K is less than unity. The values of K and C presented here showed that GAB is suitable for fitting the pre-gelatinized starch data. The fit of GAB was evaluated by calculating the percentage square root error, RMSE, against experimental isothermal data, where GAB was found to be more fit than BET [29].

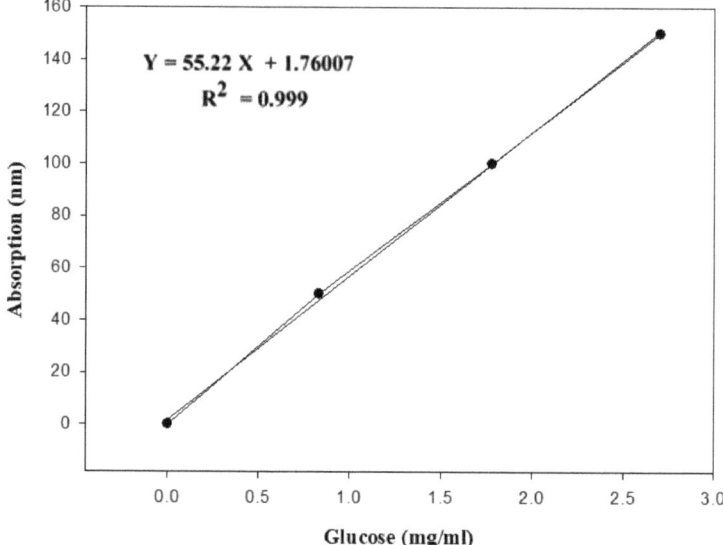

Figure 1. Glucose standard curve.

The chickpea starch isotherm data, presented in (Table 4), summarizes the estimated constants of the GAB model along with the root mean square error (RMSE) which indicates the absolute fit of the model. The coefficient of determination (R^2) is also given in the table. The low values of RMSE or the R^2 close to unity indicate that the GAB model is a good fit for the sorption isotherm data, and the projected parameters are statistically satisfactory. The monolayer moisture content (m_o) of the control (annealed without GSE) decreased at a longer annealing time within each annealing temperature (Table 4). The GSE-treated chickpea starch exhibited a reduction in m_o at 0.1 mL GSE and an increase after 1.0 mL GSE as well as a longer annealing time, regardless of annealing temperature. The same trend was observed for the C and K parameters. In particular, the downward trend of m_o with respect to increasing annealing time reflects a reduction in hygroscopicity, which goes along with longer annealing time and temperature. This may be attributed to a reduction in the total sorption capacity of the material, which may in turn reflect annealing.

Table 4. GAB parameters for moisture sorption isotherms of chickpea starch pre-gelatinization/germinated sorghum extract annealed at 40, 50, and 60 °C at 30 and 60 min (m_o g/100 g water dry basis).

		m_o	Cg	K	R^2	RMSE
40 °C						
30 min	No GSE	0.14 ± 0.01 [b]	2.47 ± 0.02 [b]	0.41 ± 0.03 [b]	0.99	0.213
	0.1 mL	0.18 ± 0.01 [a]	2.51 ± 0.03 [b]	0.45 ± 0.03 [b]	0.99	0.740
	1.0 mL	0.18 ± 0.02 [a]	3.20 ± 0.04 [a]	0.51 ± 0.02 [a]	0.99	0.860
60 min	No GSE	0.12 ± 0.03 [c]	2.05 ± 0.04 [c]	0.25 ± 0.01 [c]	0.99	0.871
	0.1 mL	0.17 ± 0.02 [b]	2.90 ± 0.04 [b]	0.46 ± 0.02 [b]	0.99	0.923
	1.0 mL	0.38 ± 0.04 [a]	3.91 ± 0.06 [a]	0.53 ± 0.04 [a]	0.99	0.914
50 °C						
30 min	No GSE	0.14 ± 0.02 [b]	2.75 ± 0.08 [c]	0.36 ± 0.07 [b]	0.99	0.741
	0.1 mL	0.16 ± 0.02 [b]	2.96 ± 0.18 [b]	0.46 ± 0.05 [a]	0.99	0.768
	1.0 mL	0.21 ± 0.01 [a]	3.74 ± 0.12 [a]	0.48 ± 0.04 [a]	0.99	0.613
60 min	No GSE	0.12 ± 0.02 [c]	2.46 ± 0.09 [b]	0.29 ± 0.03 [b]	0.99	0.417
	0.1 mL	0.16 ± 0.02 [b]	2.65 ± 0.15 [b]	0.51 ± 0.03 [a]	0.99	0.560
	1.0 mL	0.29 ± 0.02 [a]	3.62 ± 0.15 [a]	0.56 ± 0.03 [a]	0.99	0.951
60 °C						
30 min	No GSE	0.14 ± 0.04 [b]	2.08 ± 0.13 [b]	0.29 ± 0.04 [b]	0.99	0.921
	0.1 mL	0.27 ± 0.03 [a]	2.26 ± 0.17 [b]	0.33 ± 0.03 [b]	0.99	0.731
	1.0 mL	0.32 ± 0.04 [a]	2.85 ± 0.15 [a]	0.51 ± 0.05 [a]	0.99	0.860
60 min	No GSE	0.11 ± 0.03 [b]	1.72 ± 0.08 [b]	0.24 ± 0.01 [b]	0.99	1.035
	0.1 mL	0.44 ± 0.05 [a]	1.81 ± 0.10 [b]	0.24 ± 0.04 [b]	0.99	0.994
	1.0 mL	0.44 ± 0.02 [a]	3.23 ± 0.14 [a]	0.58 ± 0.02 [a]	0.99	1.004

m_o = monolayer moisture content; RMSE = root-mean-square error; C and K are GAB parameters related to monolayer and multilayer properties. Values (means ± S.D) followed by different letters under particular annealing temperature and time within each column are significantly different.

Induced physical structural changes on the starch granules during annealing and before gelatinization. These changes appeared to influence the way starch granules go through the gelatinization events, such as granule swelling rate, which has a direct effect on the molecular weight profile of the gelatinized starch. Keeping in mind, the granule structure and the entanglement between amylose and amylopectin as well as their ratio have a direct effect on the events leading to starch gelatinization. The 0.1 mL GSE treatment had little effect on the m_o within each annealing temperature, but the 1.0 mL GSE increased the m_o values indicating an increase in the hygroscopicity of the PGS. This could be ascribed to the exposure of additional hydroxyl groups which allowed for more water binding sites, due to the action of α–amylase. The comparison of annealing chickpea starch at different temperatures within 30 or 60 min is presented in Table 4. The data showed very little difference between m_o values for the chickpea starch control after annealing for 30 min regardless of annealing temperature, whereas annealing for 60 min at 60 °C showed higher m_o. Samples treated with GSE at 50 °C exhibited lower m_o after both annealing times, which indicates the effect of annealing on the starch granules structure that affected the gelatinization events leading to different molecular profiles of the PGS. Therefore, the gelatinization product which is the substrate for α-amylase will produce different molecular sizes leading to different hygroscopicity. It is also apparent how chickpea starch treated with 1.0 mL GSE and annealed at 60 °C, exhibited the highest m_o compared to the control. Therefore, GSE treatment has a direct effect on the PGS. The m_o of the sweet potato starch was dependent on annealing temperature and GSE treatment, as well. Compared to chickpea starch, the native S.P.S had higher m_o at higher annealing temperatures and time (Table 5). Higher hygroscopicity was recorded for 0.1 mL GSE at higher temperatures and a short time, but m_o exhibited a drop after 1.0 mL GSE treatment. Once again, annealing in 0.1 mL GSE for 30 and 60 min, the m_o increased at higher temperature indicating a more hygroscopic material, but 1.0 mL GSE produced a less hygroscopic material due to the low m_o. The difference between the m_o of chickpea starch and sweet potato starch could be attributed to the different amylose content which has a direct effect on the granule structure of the native starch. The activity of α-amylase appeared to be dependent on the molecular structure of the gelatinized starch. An example of PGS isotherm profiles showing the presence of hysteresis between adsorption and desorption profiles is presented in Figure 1. The hysteresis effect extended over the entire water activity range for both starches but it was most pronounced in the $0.6 < a_w < 0.8$ region. The magnitude of the hysteresis loop is larger for the samples annealed at 50 °C compared to the other temperatures. It is obvious how annealing temperature can affect the adsorption and desorption as well as the hysteresis, which indicates the ability of the PGS to lose water in a different pathway than up-taking. It is also apparent the effect of GSE treatment on the profiles is due to the action of α-amylase. Therefore, PGS absorb moisture faster, but the desorption profile indicates a stronger association between the water and the PGS which is obvious in Figure 1, where at the same water activity, the desorption profile exhibited moisture content higher than adsorption [12].

Table 5. GAB parameters for moisture sorption isotherms sweet potato starch pre-gelatinization/germinated sorghum extract annealed at 40, 50, and 60 °C at 30 and 60 min (m_o g/100 g water dry basis).

		40 °C				
		m_o	Cg	K	R^2	RMSE
30 min	No GSE	0.18 ± 0.02 [a]	3.20 ± 0.11 [b]	0.41 ± 0.02 [b]	0.99	0.864
	0.1 mL	0.18 ± 0.01 [a]	2.46 ± 0.17 [c]	0.45 ± 0.01 [a]	0.99	0.208
	1.0 mL	0.14 ± 0.03 [a]	3.54 ± 0.12 [a]	0.49 ± 0.05 [a]	0.99	0.743
60 min	No GSE	0.12 ± 0.02 [b]	3.90 ± 0.09 [a]	0.53 ± 0.04 [a]	0.99	0.603
	0.1 mL	0.17 ± 0.03 [b]	2.89 ± 0.12 [b]	0.46 ± 0.03 [a]	0.99	0.572
	1.0 mL	0.38 ± 0.02 [a]	2.05 ± 0.10 [c]	0.25 ± 0.02 [b]	0.99	0.518
		50 °C				
30 min	No GSE	0.16 ± 0.03 [b]	2.96 ± 0.10 [b]	0.47 ± 0.04 [a]	0.99	0.206
	0.1 mL	0.21 ± 0.02 [a]	2.75 ± 0.17 [b]	0.36 ± 0.03 [b]	0.99	0.385
	1.0 mL	0.14 ± 0.04 [b]	3.74 ± 0.14 [a]	0.47 ± 0.02 [a]	0.99	0.742
60 min	No GSE	0.29 ± 0.02 [a]	2.46 ± 0.12 [b]	0.28 ± 0.01 [b]	0.99	0.552
	0.1 mL	0.16 ± 0.03 [b]	2.64 ± 0.09 [b]	0.50 ± 0.04 [a]	0.99	0.044
	1.0 mL	0.12 ± 0.03 [b]	3.62 ± 0.12 [a]	0.56 ± 0.05 [a]	0.99	0.554

Table 5. Cont.

		60 °C				
30 min	No GSE	0.27 ± 0.03 [a]	2.26 ± 0.11 [b]	0.33 ± 0.04 [b]	0.99	0.920
	0.1 mL	0.32 ± 0.04 [a]	2.08 ± 0.20 [b]	0.29 ± 0.04 [b]	0.99	0.734
	1.0 mL	0.13 ± 0.03 [b]	2.85 ± 0.11 [a]	0.51 ± 0.05 [a]	0.99	0.960
60 min	No GSE	0.44 ± 0.03 [a]	1.80 ± 0.09 [b]	0.24 ± 0.02 [b]	0.99	0.687
	0.1 mL	0.46 ± 0.04 [a]	1.72 ± 0.10 [b]	0.24 ± 0.02 [b]	0.99	0.908
	1.0 mL	0.11 ± 0.04 [b]	3.22 ± 0.12 [a]	0.57 ± 0.01 [a]	0.99	0.232

m_o = monolayer moisture content; RMSE = root-mean-square error; C and K are GAB parameters related to monolayer and multilayer properties. Values (means ± S.D) followed by different letters under particular annealing temperature and time within each column are significantly different.

3. Materials and Methods

3.1. Starch Isolation

The starches used in this study were obtained from several sources. Chickpea (*Cicer arietinum* L.) starch (C.P.S), white bean (*Phaseolus vulgaris* L.) starch (W.B.S), sweet potato (*Ipomoea batatas* L.) starch (S.P.S), and wheat (*Triticum aestivum* L.) starch (W.S) were extracted from raw materials acquired from a local market in Riyadh, Saudi Arabia. Corn (*Zea mays* L.) starch (C.S) was generously supplied by ARASCO Company, also located in Riyadh, Saudi Arabia. The authors have already provided an in-depth description of the techniques employed for starch isolation in a prior paper [19]. The *Aspergillus* fungal α-amylase (EC3.2.1.1) and sulfuric acid were procured from Sigma Aldrich, a reputable supplier based in St. Louis, MI, USA. The centrifugation process for starch isolation was conducted using a Beckman Centrifuge (Beckman JXN, Brea, CA, USA).

3.2. Starch Modification

Sorghum (*Sorghum bicolor* L. Moench) seeds were germinated for four days at 24 °C and 25% moisture, air dried, and 10 g were added to 40 mL distilled water, agitated for 15 min, filtered through Whitman 40, and centrifuged for 10.0 min at 2000× g. Germinated sorghum extract (GSE) was given to the supernatant. Starch to water slurry 1:9 (w/v) (30 g + 270 mL water) and 1.0 mL or 10.0 mL GSE was added to the starch slurry, with starch slurry without GSE serving as the control. The slurry was agitated and annealed in water baths at 40, 50, and 60 °C for 30 or 60 min before being centrifuged three times with fresh water to remove the excess enzyme. After washing, the starch was air-dried with 100 mL of acetone. The dried starch was sieved through a 250 µm sieve and kept at −20 °C for later investigation. The α-amylase activity in the GSE extract was calculated by measuring the activity of a known concentration of pure amylase enzyme solution.

3.3. Pre-Gelatinized Starch

Native or GSE-treated starch slurries were prepared by adding 70 g starch (14% MC) to 200 mL distilled water. A thin layer of the slurry was spread on a heated flatbed until a dry thin layer (sheet) was formed (about 0.8 mm). The sheet was left to take room temperature, ground in a coffee grinder to pass through a 250 µm sieve and sealed in plastic bags, labeled as pre-gelatinized starches (PGS), and stored in the refrigerator for further analysis.

3.4. Water Holding Capacity

Beuchat [30] method was used to calculate the water holding capacity (WHC). In 5 mL of distilled water, 0.1 g of PGS (W0) was suspended and vortexed for 10 s. After 30 min at ambient temperature (25 ± 2 °C), the sample was centrifuged at 2000× g for 15 min, and the precipitate was weighed (W1). According to the following relationship, the WBC was determined as grams of water absorbed per gram of starch: WHC (g/g) = W1 − W0/W0.

3.5. Sugars Determination

A 10% glucose standard solution (0.2 mg glucose/2 mL) was prepared and a standard curve was constructed using 0, 50, 100, and 150 µg glucose/mL and the absorption was read at 490 nm. The phenol–sulfuric acid method as described by [31] was used for the determination of sugars with some modification. Pre-gelatinized starch (0.1 g) was added to 10 mL of double distilled water, vortexed for 60 s, and placed for 60 min at room temperature with occasional hand mixing. The sample was centrifuged for 10.0 min at 2000× g and 200 µL aliquot was transferred and diluted with double distilled water to a total of 2 mL. The diluted sample was mixed with 50 µL of 80% aqueous phenol solution and subsequently with 5 mL of concentrated sulfuric acid (98% by weight) and mixed. After 10 min, the sample was vortexed for 30 s and placed for 20 min in a water bath at room temperature for color development. Then the absorption was read at 490 nm.

3.6. Freezable Water

Freezable water (FW) of PGS was determined for starches annealed at 40 °C in 1.0 mL GSE for 60 min. The freezable water was determined using differential scanning calorimeter (DSC) analysis by scanning the samples at 10 °C/min using TA instrument DSC (TA instrument, New Castel, PA, USA). PGS sample (10–12 mg at 8% moisture content) was placed in aluminum pans and 18–20 µL distilled water was added, whereas the reference pan contained a similar weight of distilled water. After sealing, the sample was equilibrated for 2 h and cooled from 25 °C to −80 °C, equilibrated for 10 min, and heated to 50 °C. Onset and peak temperature and ΔH were determined using the software provided by TA instruments (Q2000, TA Instruments Inc., New Castle, DE, USA). The DSC was calibrated for temperature and heat flow using indium (melting point: 156.6 °C, ΔH = 28.47 J/g) and pure water (melting point: 0 °C, ΔH = 334.10 J/g). The freezable water was calculated by dividing the ΔH of the melted ice in the sample by the ΔH of the melting ice of pure water.

3.7. Sorption Isotherms Determination

PGS moisture sorption isotherms were determined gravimetrically using a sorption analyzer Q5000 SA (TA instruments, New Castile, DE, USA). The sample (10 mg) was loaded on the Q5000 autosampler using quartz pans and the relative humidity was automatically set between 10 and 90% (0.1–0.9 a_w) at 25 °C. The instrument provided the equilibrium moisture content (EMC) directly after each step. The EMC at the specific water activity is used for the determination of the GAB and BET parameters.

3.8. Sorption Isotherm Models

In order to determine the best fit corresponding to a_w at the selected temperature, GAB and BET models were applied. The GAB model is represented by the following relationship:

$$Wm = \frac{C.K.a_w.m_o}{(1 - k.a_w).(1 - k.a_w + C.K.a_w)}$$

where a_w is the water activity, Wm is the equilibrium moisture content and m_o is the monolayer moisture content. C and K are GAB constants (C is a constant related to the heat of sorption of the first layer. K is related to the heat of adsorption of the multi-layer) derived from the following polynomial equation:

$$\frac{a_w}{EMC} = \alpha(a_w)^2 + \beta + \gamma$$

where a_w is the water activity and EMC is the equilibrium moisture content. The terms α, β, and γ can be calculated by non-linear regression of the experimental EMC as a function of a_w. Therefore, the C and K values can be obtained as follows.

$$\alpha = \frac{k}{m_o}\left[\frac{1}{C} - 1\right]$$

$$\beta = \frac{1}{m_o}\left[1 - \frac{2}{C}\right]$$

$$\gamma = \frac{1}{m_o.C.K}$$

$$C = \frac{T + \sqrt{T^2 - 4T}}{2}$$

$$T = \frac{\beta^2}{-\alpha\gamma} + 4$$

$$K = \frac{1}{C.m_o.\gamma}$$

$$m_o = \left[1 - \frac{2}{C}\right] \times \frac{1}{\beta}$$

Therefore, EMC can be estimated by rearranging following equation

$$\frac{m_o}{EMC} = \frac{[1 - a_w][1 + (C-1)]a_w}{C.Ka_w}$$

where m_o is the monolayer moisture content and EMC is the equilibrium moisture content.
The BET model is represented by

$$m = \frac{C\,a_w m_o}{(1 - a_w)[1 + (C-1)]a_w}$$

After arrangement, the BET equation is as follows:

$$\frac{m_o}{EMC} = \frac{[1 - a_w][1 + (c-1)]}{C.a_w}$$

3.9. Model Validation

The GAB and BET models are fitted to a non-linear regression equation. All calculations were made using Sigma Plot version 10.0. Besides the R^2 of the non-linear regression, the goodness of the model fit was tested using the percentage root square error (RMSE), which is the absolute measure of the fit of the model [29].

$$RMSE = \sqrt{\frac{\sum_{i=1}^{N}(m^i_e - m^i_p)}{N}}$$

where m_e is the experimental EMC value and m_p is the predicted value and N is the number of experimental data.

3.10. Statistical Evaluation

All the observations were recorded in triplicate. ANOVA was used to examine the experimental data, which were expressed as mean and standard deviation. Duncan's multiple range test was used to determine whether there were significant differences between experimental mean values ($p \leq 0.05$). SAS Foundation 9.2 for Windows (SAS Institute, Inc., Cary, NC, USA) was used to analyze the data.

4. Conclusions

Sweet potato starch exhibited the highest water holding capacity and the lowest freezable water, whereas chickpea starch had the most freezable water. Therefore, it is probable to state that much more stable products may be obtained if sweet potato starch is present in starch-rich foods. The moisture sorption of different pre-gelatinized starches increased with increasing water activity at different annealing temperatures and was characterized as Type II isotherm. The equilibrium moisture content and monolayer moisture contents (m_o) decreased after treatment with 0.1 mL GSE but increased after 1.0 mL GSE-treatment and longer annealing time. The results showed that α-amylase had a significant effect on the equilibrium moisture content and monolayer moisture content (m_o) of the starches. The GAB model showed a high correlation coefficient of determination (R^2) and a low percentage square root error (RMSE), indicating the best fit of the experimental data in the whole range of water activity. These fundamental data are important in assessing the applicability of starch in food and pharmaceutical industries.

Author Contributions: Conceptualization, A.A.M. and S.H.; data curation, H.A., A.A.Q. and A.S.; formal analysis, M.A.I. and A.S.; funding acquisition, M.S.A.; methodology, M.S.A. and S.H.; project administration, S.H. and H.A.; resources, M.S.A. and S.H.; software, M.A.I. and A.A.Q.; supervision, A.A.M. and M.S.A.; validation, S.H.; visualization, A.A.Q.; writing original draft, A.A.M. and H.A.; writing—review and editing, S.H. and A.S. All authors have read and agreed to the published version of the manuscript.

Funding: The authors appreciate the support from the Researchers Supporting Project number (RSPD2023R1073), King Saud University, Riyadh, Saudi Arabia.

Institutional Review Board Statement: Not applicable.

Informed Consent Statement: Not applicable.

Data Availability Statement: Not applicable.

Conflicts of Interest: The authors declare no conflict of interest.

Sample Availability: Not applicable.

References

1. BeMiller, J.N.; Whistler, R.L. *Starch: Chemistry and Technology*; Academic Press: Cambridge, MA, USA, 2009.
2. Vanier, N.L.; Pozzada dos Santos, J.; Pinheiro Bruni, G.; Zavareze, E.D.R. Starches in Foods and Beverages. In *Handbook of Eating and Drinking*; Springer: Berlin/Heidelberg, Germany, 2020; pp. 897–913.
3. Pascari, X.; Marín, S.; Ramos, A.J.; Molino, F.; Sanchis, V. Deoxynivalenol in cereal-based baby food production process. A review. *Food Control* **2019**, *99*, 11–20. [CrossRef]
4. Fernández-Artigas, P.; Guerra-Hernández, E.; García-Villanova, B. Browning indicators in model systems and baby cereals. *J. Agric. Food Chem.* **1999**, *47*, 2872–2878. [CrossRef] [PubMed]
5. Mercier, C.; Cantarelli, C. *Pasta and Extrusion Cooked Foods: Some Technological and Nutritional Aspects: Proceedings of an International Symposium, Milan, Italy, 25–26 March 1985*; Elsevier: Amsterdam, The Netherlands, 1986.
6. Vallès-Pàmies, B.; Barclay, F.; Hill, S.E.; Mitchell, J.R.; Paterson, L.; Blanshard, J. The effects of low molecular weight additives on the viscosities of cassava starch. *Carbohydr. Polym.* **1997**, *34*, 31–38. [CrossRef]
7. Anastasiades, A.; Thanou, S.; Loulis, D.; Stapatoris, A.; Karapantsios, T. Rheological and physical characterization of pregelatinized maize starches. *J. Food Eng.* **2002**, *52*, 57–66. [CrossRef]
8. Rajagopalan, S.; Seib, P. Granular cold-water-soluble starches prepared at atmospheric pressure. *J. Cereal Sci.* **1992**, *16*, 13–28. [CrossRef]
9. Al-Muhtaseb, A.; McMinn, W.; Magee, T. Water sorption isotherms of starch powders: Part 1: Mathematical description of experimental data. *J. Food Eng.* **2004**, *61*, 297–307. [CrossRef]
10. Ayranci, E.; Duman, O. Moisture sorption isotherms of cowpea (*Vigna unguiculata* L. Walp) and its protein isolate at 10, 20 and 30 °C. *J. Food Eng.* **2005**, *70*, 83–91. [CrossRef]
11. Al-Ghouti, M.A.; Da'ana, D.A. Guidelines for the use and interpretation of adsorption isotherm models: A review. *J. Hazard. Mater.* **2020**, *393*, 122383. [CrossRef]
12. McMinn, W.; Magee, T. Principles, methods and applications of the convective drying of foodstuffs. *Food Bioprod. Process.* **1999**, *77*, 175–193. [CrossRef]
13. Pahlevanzadeh, H.; Yazdani, M. Moisture adsorption isotherms and isosteric energy for almond. *J. Food Process Eng.* **2005**, *28*, 331–345. [CrossRef]
14. Rohvein, C.; Santalla, E.; Gely, M. Note: Estimation of sorption isotherm and the heat of sorption of quinoa (*Chenopodium quinoa* Willd.) seeds. *Food Sci. Technol. Int.* **2004**, *10*, 409–413. [CrossRef]
15. Ocieczek, A.; Mesinger, D.; Toczek, H. Hygroscopic properties of three cassava (*Manihot esculenta* Crantz) starch products: Application of BET and GAB models. *Foods* **2022**, *11*, 1966. [CrossRef] [PubMed]

16. Saberi, B.; Vuong, Q.V.; Chockchaisawasdee, S.; Golding, J.B.; Scarlett, C.J.; Stathopoulos, C.E. Water sorption isotherm of pea starch edible films and prediction models. *Foods* **2015**, *5*, 1. [CrossRef]
17. Kinsella, J.E.; Melachouris, N. Functional properties of proteins in foods: A survey. *Crit. Rev. Food Sci. Nutr.* **1976**, *7*, 219–280. [CrossRef]
18. Lee, C.J.; Kim, Y.; Choi, S.J.; Moon, T.W. Slowly digestible starch from heat-moisture treated waxy potato starch: Preparation, structural characteristics, and glucose response in mice. *Food Chem.* **2012**, *133*, 1222–1229. [CrossRef]
19. Alqah, H.; Alamri, M.; Mohamed, A.; Hussain, S.; Qasem, A.; Ibraheem, M.; Ababtain, I. The Effect of Germinated Sorghum Extract on the Pasting Properties and Swelling Power of Different Annealed Starches. *Polymers* **2020**, *12*, 1602. [CrossRef] [PubMed]
20. Bae, I.Y.; Lee, H.G. Effect of dry heat treatment on physical property and in vitro starch digestibility of high amylose rice starch. *Int. J. Biol. Macromol.* **2018**, *108*, 568–575.
21. Fu, Z.-Q.; Wang, L.-J.; Zou, H.; Li, D.; Adhikari, B. Studies on the starch–water interactions between partially gelatinized corn starch and water during gelatinization. *Carbohydr. Polym.* **2014**, *101*, 727–732. [CrossRef]
22. Wootton, M.; Bamunuarachchi, A. Water binding capacity of commercial produced native and modified starches. *Starch-Stärke* **1978**, *30*, 306–309. [CrossRef]
23. Gercekaslam, K.E. Hydration level significantly impacts the freezable-and unfreezable-water contents of native and modified starches. *Food Sci. Technol.* **2020**, *41*, 426–431. [CrossRef]
24. Tananuwong, K.; Reid, D.S. Differential scanning calorimetry study of glass transition in frozen starch gels. *J. Agric. Food Chem.* **2004**, *52*, 4308–4317. [CrossRef]
25. Brunauer, S.; Deming, L.S.; Deming, W.E.; Teller, E. On a theory of the van der Waals adsorption of gases. *J. Am. Chem. Soc.* **1940**, *62*, 1723–1732. [CrossRef]
26. Saravacos, G.; Tsiourvas, D.; Tsami, E. Effect of temperature on the water adsorption isotherms of sultana raisins. *J. Food Sci.* **1986**, *51*, 381–383. [CrossRef]
27. Maroulis, Z.; Tsami, E.; Marinos-Kouris, D.; Saravacos, G. Application of the GAB model to the moisture sorption isotherms for dried fruits. *J. Food Eng.* **1988**, *7*, 63–78. [CrossRef]
28. Rosa, G.S.; Moraes, M.A.; Pinto, L.A. Moisture sorption properties of chitosan. *LWT-Food Sci. Technol.* **2010**, *43*, 415–420. [CrossRef]
29. Kaya, S.; Kahyaoglu, T. Thermodynamic properties and sorption equilibrium of pestil (grape leather). *J. Food Eng.* **2005**, *71*, 200–207. [CrossRef]
30. Beuchat, L.R. Functional and electrophoretic characteristics of succinylated peanut flour protein. *J. Agric. Food Chem.* **1977**, *25*, 258–261. [CrossRef]
31. Dubois, M.; Gilles, K.A.; Hamilton, J.K.; Rebers, P.t.; Smith, F. Colorimetric method for determination of sugars and related substances. *Anal. Chem.* **1956**, *28*, 350–356. [CrossRef]

Disclaimer/Publisher's Note: The statements, opinions and data contained in all publications are solely those of the individual author(s) and contributor(s) and not of MDPI and/or the editor(s). MDPI and/or the editor(s) disclaim responsibility for any injury to people or property resulting from any ideas, methods, instructions or products referred to in the content.

Article

Gradual Analytics of Starch-Interacting Proteins Revealed the Involvement of Starch-Phosphorylating Enzymes during Synthesis of Storage Starch in Potato (*Solanum tuberosum* L.) Tubers

Junio Flores Castellanos, Arsalan Khan and Joerg Fettke *

Biopolymer Analytics, Institute of Biochemistry and Biology, University of Potsdam, Karl-Liebknecht-Str. 24-25, Building 20, 14476 Potsdam-Golm, Germany; junio.flores.castellanos@uni-potsdam.de (J.F.C.); arsalan@uni-potsdam.de (A.K.)
* Correspondence: fettke@uni-potsdam.de

Abstract: The complete mechanism behind starch regulation has not been fully characterized. However, significant progress can be achieved through proteomic approaches. In this work, we aimed to characterize the starch-interacting proteins in potato (*Solanum tuberosum* L. cv. Desiree) tubers under variable circumstances. Starch-interacting proteins were extracted from developing tubers of wild type and transgenic lines containing antisense inhibition of glucan phosphorylases. Further, proteins were separated by SDS-PAGE and characterized through mass spectrometry. Additionally, starch-interacting proteins were analyzed in potato tubers stored at different temperatures. Most of the proteins strongly interacting with the potato starch granules corresponded to proteins involved in starch metabolism. GWD and PWD, two dikinases associated with starch degradation, were consistently found bound to the starch granules. This indicates that their activity is not only restricted to degradation but is also essential during storage starch synthesis. We confirmed the presence of protease inhibitors interacting with the potato starch surface as previously revealed by other authors. Starch interacting protein profiles of transgenic tubers appeared differently from wild type when tubers were stored under different temperatures, indicating a differential expression in response to changing environmental conditions.

Keywords: starch; starch-interacting proteins; glucan water dikinase; phosphoglucan water dikinase; plastidial phosphorylase; potato; *Solanum tuberosum* L.

1. Introduction

Starch is a carbohydrate polymer synthesized by plants and most algae for energy storage. It is naturally produced as a highly ordered and dense packaging of glucan chains, leading to the formation of insoluble granules located mostly in plastids (chloroplasts and amyloplasts) [1]. The mechanism for synthesis of this stunning structure is complex and involves the synchronized activity of several groups of enzymes just for generating the essential glycosidic bonds. From a general perspective, the major groups of enzymes involved in starch synthesis, each with various specific isoforms, can be classified in starch synthases (SSs, EC 2.4.1.21), starch branching enzymes (SBEs, EC 2.4.1.18), and starch debranching enzymes (DBEs, 3.2.1.41) [2–4]. Aside from these core groups of proteins, other enzymes with essential activities in starch biosynthesis take part. ADP-Glc pyrophosphorylase (AGPase, EC 2.7.7.27) is, for instance, the key enzyme on the first steps of starch synthesis, which produces ADP-Glc from ATP and glucose-1-phosphate. ADP-Glc is the substrate used by the SSs to extend the α-1,4 linked glucans. There are five main isoforms of SSs essentially found in all crops, namely SS1, SS2, SS3, SS4, and Granule Bound Starch Synthase (GBSS or WAXY). Two additional isoforms, SS5 and SS6, were recently found to interact with the starch granules of potato tuber [5]. Among these,

SS1, SS2, and SS3 are involved in the synthesis of amylopectin, a polymer made of α-1,4 linked glucan chains connected by α-1,6 linkages [6,7]. On the other side, GBSS is involved in the synthesis of amylose, a linear α-1,4 linked glucose polymer. Recently, a protein named Protein Targeting to Starch 1 (PTST1) was discovered to address GBSS to the starch granules [8]. SBEs, on the other side, are responsible for the trimming of α-1,4 glucan chains transferring segments to an acceptor chain to introduce branching points through α-1,6 linkages, having a major role in the formation of the amylopectin structure [9]. DBEs, in counterpart, hydrolyze the α-1,6 glycosidic bonds releasing glucan chains. In this way, DBE also contributes to the proper amylopectin structure arrangement promoting its crystallization. DBE can be classified in two classes: isoamylase (ISA) and limit-dextrinase (LDA). ISA1 and ISA2 take part in the amylopectin biosynthesis, whereas ISA3 and LDA function in the starch degradation [10–12].

In this regard, to facilitate starch degradation, the external structure of the starch granules must be shifted from a water-insoluble to a soluble state. This is achieved by the action of Glucan, Water Dikinase (GWD) and Phosphoglucan, Water Dikinase (PWD), two plastidial enzymes that mediate these transitions through cycles of phosphorylation of glucans [13,14]. Glucan phosphorylation allows the unfolding of the double helical packaging within the starch, facilitating the access to starch hydrolyzing enzymes [15,16]. GWD catalyzes the phosphorylation of starch at the C-6-OH group of a glucosyl residue, and PWD exerts the same function on C-3-OH. Starch phosphorylation is essential for the mobilization of the transitory starch during the night and is overall regulated by GWD and PWD. In storage starch, in contrast, the role of these two enzymes beyond starch degradation is still unclear. In transgenic potato plants with antisense inhibition of GWD, although a remarkable reduction in the phosphate content was observed, no alterations in the tuber yield or the tuber starch content was detected [17]. Moreover, [18] found that the phosphorylation rate in potato tubers is nearly constant during the whole tuber development, demonstrating at the same time through incubation of tuber discs with Glc and radioisotopes of P^{32} that phosphorylation occurs in de novo synthesis of starch. On the other side, in cassava (*Manihot esculenta*) GWD repression had a significant effect on both leaf and root starch, and the starch excess generated in leaves led to retarded plant and storage root growth [15]. Overall, during starch synthesis phosphorylation also seems to be essential.

Starch dephosphorylation on native starch granules is achieved by the glucan phosphatases SEX4 (Starch Excess 4) and LSF2 (Like Sex Four 2) [19,20]. Arabidopsis plants lacking SEX4 exhibited decreased rates of starch degradation leading to an elevated accumulation of starch in the leaves [21,22].

Other proteins that have been found associated with the starch granules in Arabidopsis and potato tubers are Early Starvation Protein 1 (ESV1) and its homologue Like Early Starvation Protein (LESV) [5,23,24]. Both proteins are involved in the control of starch degradation process by modulating the organization of starch and consequently, affecting the glucan accessibility to catabolic enzymes.

Several proteins, with so far unrevealed catalytic activity, are crucial in the different phases of the starch formation and degradation process. Arabidopsis SS5, for instance, participates in the regulation of the morphology and starch granule number in the chloroplasts, although lacking the glycosyltransferase activity associated with the SSs [25]. Similarly, loss of PTST2 in Arabidopsis causes a phenotype similar to SS4 knockout mutants, in which a reduced number of starch granules is observed [8]. It was proposed that both proteins, PTST2 and SS4, might act together in the starch granule initiation, probably PTST2 providing maltooligosaccharides substrates for SS4 [8].

Plastidial glucan phosphorylase (PHO1), an enzyme with both synthesis and degradation capabilities, has a strong implication on maltooligosaccharide metabolism and, consequently, on the starch metabolic pathway. It catalyzes the addition of glucosyl units to the non-reducing end of α-1,4 linked glucans using glucose-1-phosphate as substrate. In the reverse reaction, PHO1 generates glucose-1-phosphate using α-1,4 linked glucans and orthophosphate as substrates. Transgenic potato tubers with strong repression of PHO1 notably accumulate maltooligosaccharides with a high degree of polymerization, probably due to partial degradation of starch granules during synthesis (e.g., during the trimming of the amylopectin molecules) without being degraded by PHO1 [26]. A distinctive feature of PHO1 is that it has an impact on potato plants during starch biosynthesis when they grow at low temperatures. In potato wild type tubers, increased amounts of short glucans were observed in the chain length distribution of the amylopectin when plants were grown at relatively low conditions (15 °C). Such alteration in the chain length distribution was not observed in transgenic tubers with strong repression of PHO1 [27].

In this work, we investigated the proteins physically interacting with the potato tuber starch under different conditions. First, to analyze if the phosphorylating enzymes GWD and PWD are present during the synthesis of starch, we isolated and characterized the starch proteome in developing tubers, where the synthesis of starch takes place. In addition, as it was demonstrated that PHO1 has a strong implication on the maltodextrin and starch metabolism in potato tubers, the starch proteome of transgenic potato lines repressing the plastidial or cytosolic glucan phosphorylase was characterized and compared with the proteins interacting with the starch of wild type tubers. Moreover, to evaluate the effect of different environmental conditions on the starch proteome, starch-interacting proteins were analyzed in potato tubers stored under different temperatures, revealing that that the expression of specific starch-related proteins are stimulated.

2. Results

2.1. Proteome Profile of Starch-Interacting Proteins

First, we identified all proteins bound to the starch granules isolated from potato tubers. Proteins interacting with the starch granules were isolated from 500 mg of potato tuber starch, concentrated, and separated by SDS-PAGE. To elucidate the nature of the interaction within the starch granules, protein isolation was conducted in two steps. Weakly interacting proteins to starch (WIP), more likely located on the starch surface, were first removed from the starch through successive washes with SDS. Later, strong interacting proteins (SIP) were recovered by mixing the starch samples with protein extraction buffer while disrupting the starch structure by heating at 99 °C. In Figure 1a, SDS-PAGE separation of SIP is presented. As can be observed, several protein bands in the range of 35 to 250 kDa were detected following Coomassie-blue staining. All the observed bands were excised, trypsin digested, and analyzed by mass spectrometry. For protein identification, peptide masses were searched against the Peptide Mass Fingerprint and MS/MS databases included on the Free Mascot server. From the identified proteins, GBSS is clearly the most abundant starch interacting protein, followed by SS2 (Figure 1a,b). Aside from GBSS, additional starch synthases were found to strongly interact with starch: SS1, SS2, and SS3. Interestingly, LESV, a recently discovered protein implicated in the control of starch degradation process, was found to strongly interact with the starch granule in considerable amounts. Similarly, GWD, PWD, and SEX4, also implicated in starch degradation, were not removable from starch by SDS, but only extracted from the SIP fraction. In addition to mass spectrometry analysis, detection of starch-related proteins using specific antibodies was carried out (all detected proteins are given in Tables 1 and 2).

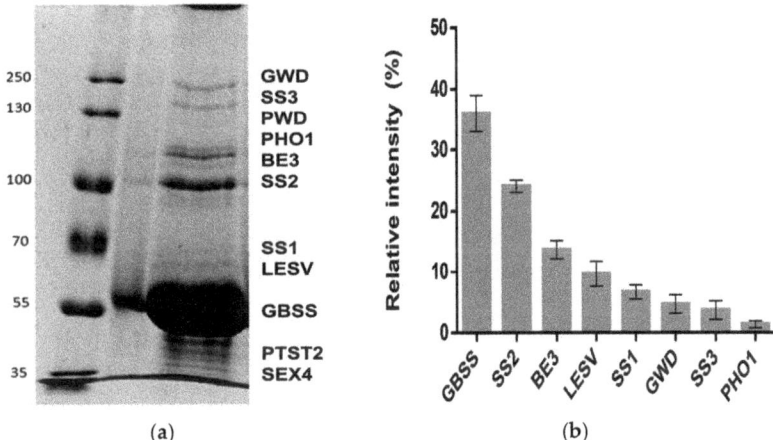

Figure 1. (**a**) A 7.5% SDS-PAGE separation of strong interacting proteins to potato starch. Proteins strongly interacting with starch (SIP) in potato tuber were extracted from 500 mg starch. A volume of 30 µg SIP was separated in a 7.5% SDS-PAGE. For visualization of the protein bands, the separation gel was stained with Roti ®®-Blue (Carl Roth). (**b**) Relative intensity of SIP bands in the separation gel was quantified using the image processor ImageJ. Three different gel images of SIP samples extracted from different starch batches were used for quantitation.

Table 1. Strong interacting proteins to starch (chloro–chloroplast).

Protein Name	Accession Number	Predicted MW (KDa)	Seq. Cov. (%)	Score	Identification Method	Cell Location
GWD	Q9AWA5	163.14	12	105	MALDI-TOF	Chloro
SS3	Q43846	139.02	18	80	MALDI-TOF	Chloro
PWD	D2JRZ6	132.28		N.A.	Immunodetected	Chloro
BE3	P30924	99.02	9	116	MS/MS	Chloro
SS2	Q43847	85.17	27	159	MALDI-TOF	Chloro
SS1	P93568	70.60	13	80	MS/MS	Chloro
GBSS	Q00775	66.58	24	138	MALDI-TOF	Chloro
LESV	Soltu.DM.06G014960.1	63.40	21	77	MS/MS	Chloro
PTST2	XP_006367350	47.26		N.A.	Immunodetected	Chloro
SEX4	Soltu.DM.11G004900.2	41.57		N.A.	Immuno-detected	Chloro

Table 2. Weak interacting proteins to starch.

Protein Name	Accession Number	Predicted MW (KDa)	Seq. Cov. (%)	Score	Identification Method	Cell Location
PHO2	P32811	95.05	23	108	MALDI-TOF	Cytosol
LOX 13 Probable linoleate 9S-11 oxygenate	Q43189	96.98	11		MALDI-TOF	Cytosol
Multicystatin	P37842	86.71	16		MALDI-TOF	N.D.
Probable inactive patatin 3-Kuras PT3	Q3YJS9	41.11	8		MS/MS	Vacuole

Table 2. Cont.

Protein Name	Accession Number	Predicted MW (KDa)	Seq. Cov. (%)	Score	Identification Method	Cell Location
Kunitz type protease inhibitor	AAB32802.1	24.50		51	MS/MS	Vacuole
Cysteine protease inhibitor 10 (fragment)	O24383	20.96	37		MALDI-TOF	Vacuole
Cysteine protease inhibitor 8 (fragment)	O24384	24.69	31		MALDI-TOF	Vacuole

Interestingly, one peptide was identified by MS/MS corresponding to the plastidial glucan phosphorylase PHO1 and it was the least abundant of the SIP.

2.2. Weak Interacting Proteins to Starch

In Table 2, WIP to starch that were removable using SDS are summarized. Few proteins were recovered by SDS washes and mostly included protease inhibitors. Among these, Multicystatin, Kunitz-type, and Cysteine protease inhibitors were identified. Multicystatin, a 86 kDa endopeptidase with capabilities to inhibit cysteine proteases [28], was consistently the predominant protein band in SDS-PAGE gels of WIP fraction. It has been stated that Multicystatin is well distributed throughout tuber tissue [28–30]. Kunitz-type serine protease inhibitor and Cysteine protease inhibitor, both having a mass of nearly 24 kDa, were reported to contain a vacuolar signal peptide [31,32]. In addition, the cytosolic form of glucan phosphorylase, PHO2, was also found to weakly interact with the starch granules. All the WIP removed by SDS had in common that they were not proteins expressed in plastid/amyloplast, and their interaction with starch might have occurred during the starch purification from tubers.

2.3. Starch-Interacting Proteins in Transgenic Lines

To evaluate if changes in the profile of the starch-interacting proteins could be observed, we selected two transgenic potato lines for starch isolation, both sustaining a deficiency in the same catalytic activity but in different cell compartments, and we compared it with the starch-interacting proteins in wild-type tubers. Thus, a transgenic line repressing the glucan phosphorylase PHO1 isozyme, which is expressed in the amyloplast/chloroplast, and a second line repressing the cytosolic phosphorylase PHO2 was used. PHO1 is directly involved in starch and maltodextrin metabolism, whereas PHO2 is involved in starch degradation related maltose metabolism in the cytosol [33]. Repression of the targeted enzymes in the transgenic lines was confirmed through Native-PAGE separation of buffer-soluble proteins and subsequent detection of phosphorylase activity (Figure 2).

In Figure 3, starch-interacting proteins extracted from the three different potato lines and separated in SDS-PAGE gel are shown. The protein profile obtained from the transgenic lines and wild type was comparable, with no substantial differences either on the WIP or SIP. However, on the WIP fraction obtained from wild type starch, a clear band with an apparent mass of 100 kDa was observed, which was not visible in the same fraction of the transgenic lines (Figure 3, See WIP fraction). This band was excised and further identified through mass spectrometry as the cytosolic PHO2 (23% sequence coverage; score: 103) and thus, as a contaminating protein. In potato tuber starch with antisense inhibition of PHO2, the corresponding band was not visible. Similarly, in PHO1-repressed lines this additional band was not observed. Despite not having a precise explanation for the contamination of starch with PHO2, it might be related to the surface properties of the isolated starches. In this regard, it was already shown that various mutants related to starch metabolism showed different starch granule surface properties [34].

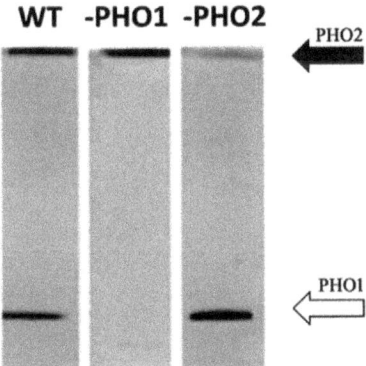

Figure 2. Native-PAGE separation of buffer-soluble proteins extracted from potato (*S. tuberosum*) wild-type tuber, and two transgenic tubers repressing the plastidial (PHO1) and cytosolic (PHO2) phosphorylase enzyme. The procedure used for phosphorylase activity detection was previously described [33]. In each lane, 20 µg buffer-soluble protein from potato tuber was loaded. The separation gel contained 0.2% (*w/v*) glycogen. Black arrow indicates the position of cytosolic PHO2; white arrow, plastidial PHO1. In potato tubers with antisense inhibition of PHO1, phosphorylase activity was strongly reduced in the plastidial isoform. In PHO2-repressed tuber, a reduced phosphorylase activity was detected in the cytosolic isoform compared to wild type.

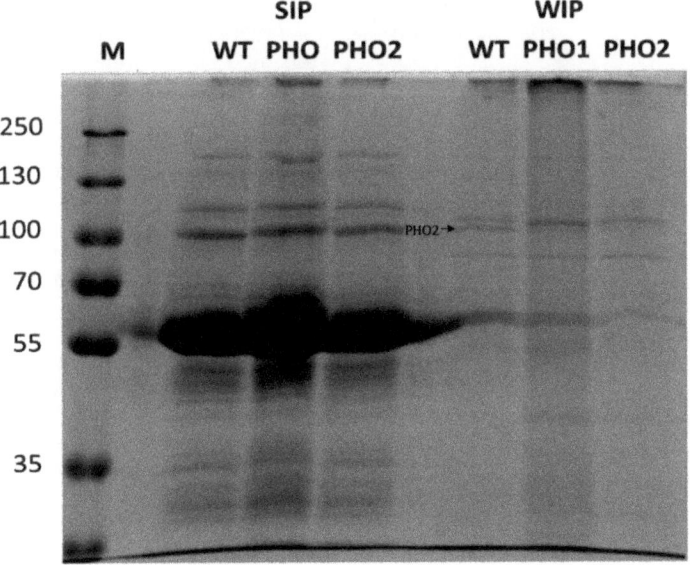

Figure 3. A 9.5% SDS-PAGE of starch-interacting proteins isolated from wild-type, PHO1-repressed and PHO2-repressed transgenic tubers. Starch-interacting proteins isolated from 500 mg starch were separated in a 9.5% SDS-PAGE. Gel was stained with Rotiblue ®® (Carl Roth). A band with an apparent mass of 100 kDa was observed in the protein fraction having a weak interaction with starch (WIP), which corresponded to the cytosolic PHO2.

In addition, we detected starch-bound proteins with antibodies raised against specific enzymes involved in starch degradation (GWD and PWD) or initiation (PTST2) in the three different genotypes. GWD and PWD were not removable from starch by SDS washes and were exclusively found as SIP (Figure 4). PTST2, as was expected for its role on starch initiation, was found in the SIP fraction.

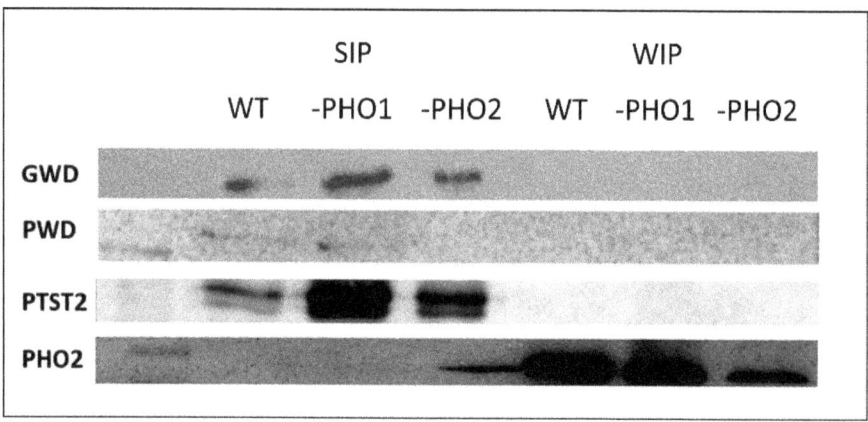

Figure 4. Immunodetection of proteins interacting with starch in potato tuber. Starch-interacting proteins isolated from wild-type, PHO1-repressed, and PHO2-repressed potato tubers were separated in a 9% SDS-PAGE and transferred to a nitrocellulose membrane for western-blot. Specific antibodies raised against *A. thaliana* GWD, PWD, PTST2, and PHO1 were used for immunodetection of starch-interacting proteins in potato. GWD, PWD, and PTST2 were detected as SIP. The cytosolic isoform PHO2 was detected using the AtPHO1 specific antibody.

2.4. Changes in the Protein Band Pattern of WIP to Starch Was Distinguished When Tubers Were Stored under Different Conditions

As shown, PHO1 was detected as SIP, but not as WIP. However, having as antecedent that PHO1 plays an active role on the starch metabolism in potato and rice plants growing at low temperatures, we assayed the isolation of the starch-interacting proteins after storage of wild type and PHO1-repressed tubers under different temperatures. Right after teir harvest, tubers derived from the same plant were stored at room temperature, warm (37 °C), or cold (4 °C) conditions for 15 days in the dark and starch proteins were later extracted. In the protein fraction corresponding to the SIP, no differences in the protein band patterns were observed. However, as it can be observed in the SDS-PAGE gel image of WIP (Figure 5), disparities on the detection of protein bands were evident comparing the wild type samples against PHO1-repressed samples, and even in the same genotype stored under different conditions. For instance, a nearly 55 kDa protein band (Figure 5, lower arrow) was observed in the PHO1-repressed tuber stored at room temperature and faintly at 37 °C, but not at 4 °C, whereas in wild type samples the same protein band was missing in the three cases. More notably, a protein band having a mass near to 130 kDa (Figure 5 upper arrow) was detected only in the WIP of PHO1-repressed tuber stored at 4 °C. Thus, it seems that the WIP protein pattern was altered among the various conditions, which requires further clarifications.

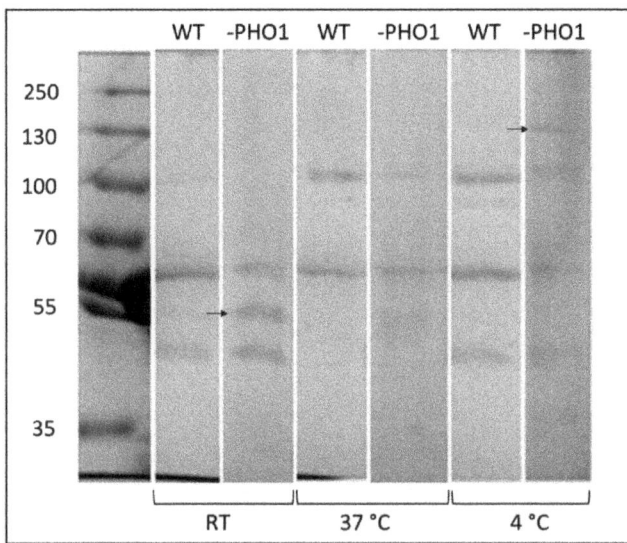

Figure 5. A 7.5% SDS-PAGE separation of weak interacting proteins to starch in wild-type and PHO1-repressed tubers. Immediately after harvesting, wild-type, and PHO1-repressed tubers were stored for 15 days at room temperature (RT), warm (37 °C), or cold (4 °C) conditions. WIP were isolated with 2% (w/v) SDS and 8.0 µg protein recovered and separated by electrophoresis in a 7.5% polyacrylamide gel.

3. Discussion

3.1. Proteins Were Mainly Contained in the SIP Fraction

In this work, we explored the starch-interacting proteins in potato starch under different conditions. We classified the starch proteins in two different fractions based on the nature of the interaction considering the way we isolated them from starch. One fraction contained the proteins that were removable from starch by SDS and thus, more likely found loosely interacting with the starch surface, defined in this work as weak interacting proteins (WIP). Strong interacting protein fraction (SIP) was designated to the proteins which were able to recover only after breaking down the granular structure of the starch by heat-induced gelatinization. All identified proteins from the SIP were described as involved in starch metabolism. The protein profile of the SIP observed in the separation gel mainly resembles that obtained by [5]. GBSS was the most profuse protein, followed by SS2. The least abundant SIPs were PWD and PHO1. Interestingly, in the case of GWD and PWD, which are dikinases mainly involved in starch degradation, it would be assumed that they might act on the starch surface based on the catalytic function they realize. However, no traces of either of the phosphotransferases were recovered through consecutive washes of starch samples with SDS. Thereby, both enzymes remained closely interacting within the starch granules.

Finding these phosphorylating dikinases strongly interacting with potato starch is not surprising, as it was previously demonstrated that phosphorylation occurs along the whole storage starch biosynthesis and accumulation in potato tubers [18]. However, the exact role that phosphorylation might play during the synthesis of starch has not been precisely described but is a focus of interest. Aside from their role in starch degradation, probably for additional processes like, for instance, during the arrangement of the starch structure, the activity of these enzymes might be necessary but not totally indispensable, since the tuber yield and tuber starch content in transgenic plants repressing GWD was not compromised [17]. This statement would be especially applicable in this case, where the accumulation of starch mainly takes place in developing tubers. Thus, finding enzymes strictly and only linked to starch degradation would not be expected.

Regarding the WIP proteins, the protein amount that was possible to isolate from starch was considerably lower than that recovered from the SIP fraction (around 8 µg against ~30 µg of SIP), and less protein bands were detected after gel electrophoresis and staining (Figure 5). Of the seven WIP identified in this study, five corresponded to protease inhibitors. It is believed that protease inhibitors in plants are storage proteins that act as defense mechanisms in case of wounding response and represent around 50% of the total soluble proteins in potato tubers [35,36]. Considering this data, the fact that starch proteases were found interacting with starch is not surprising. On the other hand, since these proteins are not expressed in plastids, where starch is accumulated in potato, we assume that the interaction of these protease inhibitors with the starch granules must have occurred during the starch purification procedure from potato tubers rather than naturally occur under physiological conditions. The same assumption can be followed for the cytosolic phosphorylase PHO2, which we also found in the WIP fraction of potato tuber starch. It is well known that PHO2 possesses a strong affinity towards branched glucans, like starch's amylopectin and glycogen. Therefore, the chance of finding PHO2 interacting with the starch granules after homogenizing of tuber tissue would be very likely.

3.2. PHO1 Is Not a Starch Granule-Bound Protein

As we were able to detect only very small amount of PHO1 through MS/MS analysis in SIP but not in WIP, even though PHO1 was found expressed in significant amounts after extraction of total proteins from potato tuber [37], we assayed the detection and location of the plastidial PHO1 in the two protein fractions using an antibody raised specifically against the homologue protein in *A. thaliana*. Proteins were isolated from potato wild-type starch and transgenic PHO1-repressed and PHO2-repressed potato starch samples. A 100 kDa protein band was detected in the starch surface protein fraction (Figure 6a). However, after PMF analysis and data search against the Mascot protein database, a significant match turned out for the cytosolic PHO2, which had a 95 kDa size. The fact that a reduced immunodetection signal was observed in the starch bound protein sample obtained from potato with strong repression of PHO2 was in accordance with the results (Figure 6a). A possible interaction of the anti-*At*PHO1 with *St*PHO2 epitopes could not be excluded since both the plastidial and cytosolic potato phosphorylase protein sequences share highly conserved regions (see Figure 6b). On the other hand, since PHO2 is expressed in the cytosol and possesses strong affinity towards branched glucans like starch's amylopectin, it can be assumed that its localization within the starch is more related to protein contamination during the starch isolation procedure rather than being a starch-associated protein.

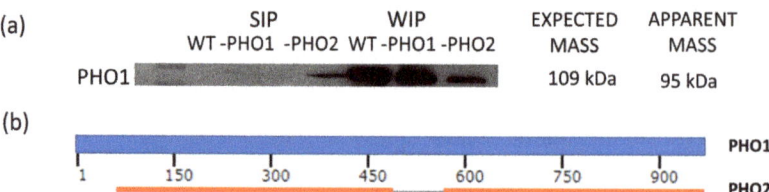

Figure 6. PHO immunodetection in potato starch bound proteins. (**a**) Starch bound proteins separated by SDS-PAGE were transferred to a membrane for immunoblotting and detection using a plastidial phosphorylase (*At*PHO1) specific antibody. A nearly-95-kDa-size protein band was detected in the starch-surface proteins of wild-type and the two transgenic lines with PHO1 and PHO2 antisense inhibition. Through PMF analysis, the corresponding band was identified as the *S. tuberosum* cytosolic phosphorylase isoform PHO2. (**b**) Protein BLAST alignment of *S. tuberosum* PHO1 and PHO2 shows two highly conserved regions among the two sequences. The first region comprises the amino acid 62–487 in PHO1 sequence, and the second region from 566–965.

In case of PHO1, failing to detect it within the WIP indicates that, although this enzyme is fundamental in the carbohydrate and starch metabolism, in potato tuber starch it does not remain attached to the granules. This observation is supported by [2], who stated that the plastidial phosphorylase (SP, PHS1, or PHO1) and Disproportionating Enzyme 1 (DPE1), though localized in the plastid stroma, are in general not granule-associated enzymes. This characteristic would be consistent with the function that both enzymes have in modulating and supplying soluble substrates for other enzymes [2].

3.3. Effects of Genotypes and Storing Conditions on Proteins Bound to Starch Granules

After identification of the proteins which were regularly found interacting with starch, the question arose whether differences could be traceable on the starch-interacting proteins in potato plants containing alterations in enzymes that are closely linked to the carbohydrate metabolism. To investigate this possibility, we used a transgenic potato line with strong repression of the plastidial glucan phosphorylase activity and another repressing the same activity but in the cytosol. However, no obvious differences were observed after SDS-PAGE separation of the SIP and WIP extracted from PHO1-repressed, PHO2-repressed, and wild-type tubers. In wild-type and the two transgenic potato lines, we also assayed the detection of starch-interacting proteins using specific antibodies. GWD and PWD were found in the SIP fraction in the three different potato starch samples. Protein Targeting to Starch 2 (PTST2), a chloroplastic protein involved in starch granule initiation [8], was also immunodetected in the three different genotypes as SIP having a 43.3 kDa mass.

In a second approach seeking for differences on the starch-interacting proteins, we stored PHO1-repressed and wild-type tubers under different temperature conditions. By these means, different patterns of WIP protein bands were visualized between the samples stored at room temperature, 4 °C, or 37 °C, either in wild-type or PHO1-repressed tubers. Differences were basically observed in the loss/detection of four bands having an apparent mass of ~130, ~100, ~50, and ~45 kDa. Unfortunately, due to the low protein concentration of these bands, attempts to identify them were unsuccessful so far. Noteworthy, a nearly 130 kDa protein was detected in the PHO1-repressed protein sample obtained from potato tubers stored at 4 °C, which is close to the molecular weight of PWD. Under such cold storing conditions, the storage starch is more likely under the so called cold-induced sweetening phenomenon, thus the degradation of starch must occur to self-supply reducing sugars. In any case, this band was not detected in wild type tubers stored at 4 °C and a relationship to be attributed to the missing PHO1 activity in the transgenic tuber triggering the expression of PWD is hard to establish.

Considering the findings presented by [23] who reported the presence of ESV1 in both the soluble and insoluble (starch-containing) protein fractions in *A. thaliana* and *N. sylvestris*, and [5], who recently reported the identification of ESV1 (48.9 kDa) in the starch-associated proteins in potato tuber, we consider the probability that one of the WIP bands we observed might correspond to this protein.

Attempts to further investigate the starch proteome under variable environments or simulated stress conditions would be supportive in revealing inducing mechanisms for specific starch-related enzymes and understanding their implication under specific conditions. Like we observed in this assay, expression of different starch-interacting proteins was stimulated by storage of tubers in different temperatures.

4. Conclusions

The proteins identified in this study that were strongly interacting with the potato starch granules corresponded to previously described proteins having a specific role in the starch metabolism. Remarkably, GWD and PWD, despite being enzymes commonly associated with the starch degradation, were found to interact strongly with the starch granules in developing tubers. We assume that the activity of GWD and PWD is not exclusive to the starch breakdown but is also required during the synthesis of storage starch. More likely, these two dikinases play a significant role in the regulation of the starch granule architecture by facilitating the cleavage of glucosyl residues by other enzymes. However, the specific role and relevance of these enzymes during starch synthesis must still be investigated.

Furthermore, we demonstrated that the starch proteome profile can be altered during the storage of potato tubers under different temperature conditions. This assay can serve as a milestone to further investigate the starch proteome under different scenarios. This might be helpful, for instance, in the identification of new proteins or revealing which specific starch-related enzymes are preferentially expressed under particular environmental or stress conditions.

5. Materials and Methods

5.1. Biological Material
Potato Plants (*Solanum tuberosum* L.)

Potato wild-type plants (*Solanum tuberosum* L. cv. Desiree), and transgenic lines containing antisense constructs against the plastidial phosphorylase isozymes (PHO1a + b), and the cytosolic phosphorylase isozyme (PHO2) were grown under controlled conditions (12 h light/12 h dark period, 300 µE m^{-2} s^{-1}, 22 °C) in a grow chamber. Potato tubers were harvested after 3 months and immediately used for starch isolation, except where otherwise stated.

5.2. Antibodies

Polyclonal antibodies used for the detection of AtGWD, AtPWD, and AtPTST2 are described in [8,38,39].

5.3. Methods

5.3.1. Starch Isolation

Potato tubers were peeled and homogenized in 100 mL 4 °C cold distilled water per 10–15 g of tuber material using a blender. The mixture was filtered through a 100 µm net sieve and starch was transferred to 50 mL Falcon tubes. Starch was allowed to settle down for 1 h keeping the sample tubes on ice. The supernatant was discarded, and the starch pellet was resuspended in 40 mL cold distilled water, vortexed, and centrifuged at 1500× g for 3 min to wash the starch. The supernatant was discarded and the washing procedure was repeated three times. Starch was dried by lyophilization.

5.3.2. Detection of Phosphorylase Activity by Native PAGE

Detection of phosphorylase activity in Native PAGE gels was performed through electrophoretic separation of buffer-soluble proteins extracted from wild type and transgenic potato tubers using the method described by [26].

5.3.3. Extraction of Starch-Interacting Proteins

For protein extraction 500 mg starch was used. Proteins weakly interacting to the starch granule (WIP) were extracted by adding 200 µL 2% (w/v) SDS per each 100 mg native starch and recovering the supernatant following vortex and centrifugation (20,000× g for 10 min). The SDS extraction was repeated twice, and the collected supernatants were pooled in a single tube. Later, starch samples were washed twice with 1 mL distilled water to remove the remaining SDS. Then, proteins which interacted strongly with the starch granules (SIP) were extracted by adding 1 mL of protein extraction buffer (0.2 M Tris, 2% (w/v) SDS, 20% (w/v) glycerol, 50 mM DTE, pH 6.8) [5] per sample tube, vortexing, and heating at 99 °C for 20 min. Afterwards, the samples were centrifuged at 20,000× g for 10 min and the supernatant was collected.

5.3.4. Concentration of Protein Samples

The two protein fractions containing the WIP and SIP were separately transferred to 10 kDa filter (Merck Millipore®®, 5 mL volume capacity) for protein concentration following the supplier manual instructions. Following centrifugation, the volume retained in the filter (nearly 200 µL) was recovered and transferred into a 2 mL Eppendorf tube for acetone precipitation. Protein samples were mixed with four volumes of −20 °C cold acetone, vortexed, and kept in −20 °C for at least 1 h. Then, samples were centrifuged at 20,000× g for 5 min. The protein samples were pelleted, and the acetone discarded. Tubes were left uncovered for 10 min for drying.

5.3.5. Determination of Protein Concentration

Starch-interacting protein fractions were resuspended in 100 µL of protein extraction buffer to carry out the Bicinchoninic acid assay (BCA assay; Thermo Fisher Scientific, Waltham, MA, USA) following the supplier protocol. The concentration of protein samples was estimated by comparing with a standard curve generated with known concentrations of Bovine Serum Albumin (BSA). Absorbance was measured at 562 nm as indicated in the supplier manual.

5.3.6. Electrophoretic Separation of Proteins

Starch bound proteins were separated through sodium dodecyl sulfate- polyacrylamide gel electrophoresis (SDS-PAGE) using polyacrylamide concentrations between 7.5 and 10% in the separation gel. The polyacrylamide concentration used in the stacking gel was always 3%. Electrophoresis was performed at 180 V and 40 mA. Following electrophoresis, gels were stained overnight with a commercial Coomassie-blue staining solution (Roti Blue®®-Roth, Karlsruhe, Germany).

5.3.7. Quantitation of Protein Bands

The relative intensity of the protein bands detected by Coomassie staining of SDS-PAGE gels was quantified using the image processing program ImageJ 1.51d.

5.3.8. Tryptic Digestion

SDS-PAGE protein bands were prepared for peptide mass fingerprint (PMF) analysis following the protocol described by [40] with some modifications. Bands were excised from the gel, cut into smaller pieces, and distained in 100 µL of 100 mM NH_4HCO_3/acetonitrile (1:1) for 15 min with a change of distaining solution in between. Later, the gel pieces were dehydrated in 50 µL acetonitrile for 5 min. The acetonitrile was discarded, and gel pieces were dried in vacuum. The gel pieces were rehydrated with 15 µL of pre-cooled trypsin solution (20–30 ng/µL trypsin sequencing grade reconstituted in 50 mM NH_4HCO_3) and incubated on ice for 15 min. Trypsin digestion was performed overnight at 37 °C. For peptide recovering, gel samples were rehydrated with 20 µL distilled water. Later, 20 µL acetonitrile was added and recovered after 5 min. Later, 20 µL of 5% (v/v) formic acid was added to the gel pieces, incubated for 15 min, and recovered. Finally, a second acetonitrile extraction was performed. Peptide samples were completely dried in a vacuum concentrator. The peptide pellet was resuspended in 10 µL of 20% (v/v) acetonitrile dissolved in 0.1% (v/v) trifluoroacetic acid (TFA).

5.3.9. Protein Identification

Peptides were analyzed through mass spectrometry using a MALDI-TOF system (Microflex II RFT, Bruker, Bremen, Germany) and MS/MS MALDI-LTQ XL (Thermo Scientific, Waltham, MA, USA). A 0.35 µL peptide sample was spotted on the target plate and covered with the same volume of α-cyano-4-hydroxycinnamic acid matrix (HCCA) prepared as follows: 3.5 mg HCCA dissolved in 1 mL 84% (v/v) acetonitrile, 13% (v/v) EtOH, and 3% (v/v) trifluoroacetic acid 0.1% (v/v). Peptide mass spectra (m/z 500–4000 Da) were acquired as positive ions using the reflector mode. The peptide mass values were matched against NCBI and Swissprot databases listed in the MASCOT Peptide Mass Fingerprint and MASCOT MS/MS search platforms (http://matrixscience.com; accessed on 1 June 2023). Trypsin was selected as a digesting enzyme and the peptide tolerance set to 0.6 Da. One or no missed cleavage was allowed for the search and no fixed or variable modifications selected. For MS/MS ion search, the peptide charge was set to +1 and the MS/MS tolerance to 0.6 Da. Results with a MASCOT score above the threshold ($p < 0.05$) were considered significant.

5.3.10. Western Blot

The transfer of proteins from SDS-PAGE gels to nitrocellulose membranes was carried out by placing two Whatman filter papers, the membrane, the protein gel, and two Whatman filter papers in a blotting device (Bio-Rad, Hong Kong, China) following the manual instructions. All components were previously wetted with transfer buffer (25 mM Tris-NaOH pH 8.2, 192 mM glycine, 20% (v/v) MeOH). Blotting was performed at 12 V for 12 h at room temperature.

5.3.11. Immunodetection of Proteins

Blots were incubated for one hour in blocking TBST solution (100 mM Tris-HCl pH 7.5, 150 mM NaCl, 0.1% (w/v) Tween 20) containing 3% (w/v) milk powder and then with primary antibodies (against AtGWD, AtPWD and AtSEX4,) diluted 1:1000 (v/v) in blocking solution. After one hour, the primary antibody solution was removed and the membranes were washed for five minutes with TBST buffer (100 mM Tris-HCl pH 7.5, 150 mM NaCl, 0.1% (v/v) Tween 20). This washing step was repeated six times. Later, blot membranes were incubated with secondary antibody (anti-mouse or anti-rabbit conjugated to alkaline phosphatase or horseradish peroxidase (HRP)) for one hour. The membranes were washed again with TBST buffer as mentioned. To visualize alkaline phosphatase activity, BCIP®®/NBT (5-bromo-4-chloro-3-indolyl-phosphate/nitro blue tetrazolium) reagent (Sigma-Aldrich) was used following the manual instructions. For HRP conjugated antibodies, signal was detected using the Supersignal West Pico PLUS Chemiluminescent Substrate kit (Thermo Fisher, Waltham, MA, USA) following the manual instructions.

Author Contributions: Conceptualization, J.F.; methodology, J.F. and J.F.C.; validation, J.F.C.; formal analysis, J.F., J.F.C. and A.K.; investigation, J.F.C. and A.K.; resources, J.F.; writing—original draft preparation, J.F.C. and J.F.; writing—review and editing, J.F.C., A.K. and J.F.; visualization, J.F.C.; supervision, J.F.; project administration, J.F.; funding acquisition, J.F. All authors have read and agreed to the published version of the manuscript.

Funding: Deutscher Akademischer Austauschdienst (DAAD) under the program 'Doctoral Programmes in Germany 2018/19 (57381412) to J.F.C.

Data Availability Statement: Data are available by request from the corresponding author.

Conflicts of Interest: The authors declare no conflict of interest.

Sample Availability: Wild type potato plants and transgenic plants with antisense inhibition of PHO1 and PHO2 are available from the authors.

References

1. Ball, S.; Guan, H.P.; James, M.; Myers, A.; Keeling, P.; Mouille, G.; Buléon, A.; Colonna, P.; Preiss, J. From Glycogen to Amylopectin: A Model for the Biogenesis of the Plant Starch Granule. *Cell* **1996**, *86*, 349–352. [CrossRef]
2. Tetlow, I.J.; Bertoft, E. A Review of Starch Biosynthesis in Relation to the Building Block-Backbone Model. *Int. J. Mol. Sci.* **2020**, *21*, 7011. [CrossRef]
3. Mérida, A.; Fettke, J. Starch Granule Initiation in *Arabidopsis thaliana* Chloroplasts. *Plant J.* **2021**, *107*, 688–697. [CrossRef]
4. Shoaib, N.; Liu, L.; Ali, A.; Mughal, N.; Yu, G.; Huang, Y. Molecular Functions and Pathways of Plastidial Starch Phosphorylase (Pho1) in Starch Metabolism: Current and Future Perspectives. *Int. J. Mol. Sci.* **2021**, *22*, 10450. [CrossRef] [PubMed]
5. Helle, S.; Bray, F.; Verbeke, J.; Devassine, S.; Courseaux, A.; Facon, M.; Tokarski, C.; Rolando, C.; Szydlowski, N. Proteome Analysis of Potato Starch Reveals the Presence of New Starch Metabolic Proteins as Well as Multiple Protease Inhibitors. *Front. Plant Sci.* **2018**, *9*, 746. [CrossRef]
6. Delvallé, D.; Dumez, S.; Wattebled, F.; Roldán, I.; Planchot, V.; Berbezy, P.; Colonna, P.; Vyas, D.; Chatterjee, M.; Ball, S.; et al. Soluble Starch Synthase I: A Major Determinant for the Synthesis of Amylopectin in *Arabidopsis Thaliana* Leaves. *Plant J.* **2005**, *43*, 398–412. [CrossRef] [PubMed]

7. Zhang, X.; Szydlowski, N.; Delvallé, D.; D'Hulst, C.; James, M.G.; Myers, A.M. Overlapping Functions of the Starch Synthases SSII and SSIII in Amylopectin Biosynthesis in *Arabidopsis*. *BMC Plant Biol.* **2008**, *8*, 96. [CrossRef] [PubMed]
8. Seung, D.; Boudet, J.; Monroe, J.; Schreier, T.B.; David, L.C.; Abt, M.; Lu, K.J.; Zanella, M.; Zeeman, S.C. Homologs of PROTEIN TARGETING TO STARCH Control Starch Granule Initiation in Arabidopsis Leaves. *Plant Cell* **2017**, *29*, 1657–1677. [CrossRef] [PubMed]
9. Sawada, T.; Itoh, M.; Nakamura, Y. Contributions of Three Starch Branching Enzyme Isozymes to the Fine Structure of Amylopectin in Rice Endosperm. *Front. Plant Sci.* **2018**, *871*, 1536. [CrossRef]
10. Delatte, T.; Umhang, M.; Trevisan, M.; Eicke, S.; Thorneycroft, D.; Smith, S.M.; Zeeman, S.C. Evidence for Distinct Mechanisms of Starch Granule Breakdown in Plants. *J. Biol. Chem.* **2006**, *281*, 12050–12059. [CrossRef]
11. Lin, Q.; Facon, M.; Putaux, J.L.; Dinges, J.R.; Wattebled, F.; D'Hulst, C.; Hennen-Bierwagen, T.A.; Myers, A.M. Function of Isoamylase-Type Starch Debranching Enzymes ISA1 and ISA2 in the Zea Mays Leaf. *New Phytol.* **2013**, *200*, 1009–1021. [CrossRef]
12. Streb, S.; Zeeman, S.C. Replacement of the Endogenous Starch Debranching Enzymes ISA1 and ISA2 of *Arabidopsis* with the Rice Orthologs Reveals a Degree of Functional Conservation during Starch Synthesis. *PLoS ONE* **2014**, *9*, e92174. [CrossRef] [PubMed]
13. Comparot-Moss, S.; Kötting, O.; Stettler, M.; Edner, C.; Graf, A.; Weise, S.E.; Streb, S.; Lue, W.L.; MacLean, D.; Mahlow, S.; et al. A Putative Phosphatase, LSF1, Is Required for Normal Starch Turnover in *Arabidopsis* Leaves. *Plant Physiol.* **2010**, *152*, 685–697. [CrossRef] [PubMed]
14. Santelia, D.; Kötting, O.; Seung, D.; Schubert, M.; Thalmann, M.; Bischof, S.; Meekins, D.A.; Lutz, A.; Patron, N.; Gentry, M.S.; et al. The Phosphoglucan Phosphatase like Sex Four2 Dephosphorylates Starch at the C3-Position in *Arabidopsis*. *Plant Cell* **2011**, *23*, 4096–4111. [CrossRef]
15. Zhou, W.; He, S.; Naconsie, M.; Ma, Q.; Zeeman, S.C.; Gruissem, W.; Zhang, P. Alpha-Glucan, Water Dikinase 1 Affects Starch Metabolism and Storage Root Growth in Cassava (Manihot Esculenta Crantz). *Sci. Rep.* **2017**, *7*, 9863. [CrossRef] [PubMed]
16. Mak, C.A.; Weis, K.; Henao, T.; Kuchtova, A.; Chen, T.; Sharma, S.; Meekins, D.A.; Thalmann, M.; Vander Kooi, C.W.; Raththagala, M. Cooperative Kinetics of the Glucan Phosphatase Starch Excess4. *Biochemistry* **2021**, *60*, 2425–2435. [CrossRef]
17. Lorberth, R.; Ritte, G.; Willmitzer, L.; Kossmann, J. Inhibition of a Starch-Granule-Bound Protein Leads to Modified Starch and Repression of Cold Sweetening. *Nat. Biotechnol.* **1998**, *16*, 473–477. [CrossRef]
18. Nielsen, T.H.; Wischmann, B.; Enevoldsen, K.; Møller, B.L. Starch Phosphorylation in Potato Tubers Proceeds Concurrently with de Novo Biosynthesis of Starch. *Plant Physiol.* **1994**, *105*, 111–117. [CrossRef]
19. Ritte, G.; Heydenreich, M.; Mahlow, S.; Haebel, S.; Kötting, O.; Steup, M. Phosphorylation of C6- and C3-Positions of Glucosyl Residues in Starch is Catalysed by Distinct Dikinases. *FEBS Lett.* **2006**, *580*, 4872–4876. [CrossRef]
20. Samodien, E.; Jewell, J.F.; Loedolff, B.; Oberlander, K.; George, G.M.; Zeeman, S.C.; Damberger, F.F.; van der Vyver, C.; Kossmann, J.; Lloyd, J.R. Repression of SEX4 and LIKE SEX FOUR2 Orthologs in Potato Increases Tuber Starch Bound Phosphate with Concomitant Alterations in Starch Physical Properties. *Front. Plant Sci.* **2018**, *9*, 1044. [CrossRef]
21. Zeeman, S.C.; Northrop, F.; Smith, A.M.; Ap Rees, T. A Starch-Accumulating Mutant of *Arabidopsis thaliana* Deficient in a Chloroplastic Starch-Hydrolysing Enzyme. *Plant J.* **1998**, *15*, 357–365. [CrossRef] [PubMed]
22. Zeeman, S.C.; Rees, T.A. Changes in Carbohydrate Metabolism and Assimilate Export in Starch-Excess Mutants of *Arabidopsis*. *Plant Cell Environ.* **1999**, *22*, 1445–1453. [CrossRef]
23. Feike, D.; Seung, D.; Graf, A.; Bischof, S.; Ellick, T.; Coiro, M.; Soyk, S.; Eicke, S.; Mettler-Altmann, T.; Lu, K.J.; et al. The Starch Granule-Associated Protein EARLY STARVATION1 Is Required for the Control of Starch Degradation in *Arabidopsis thaliana* Leaves. *Plant Cell* **2016**, *28*, 1472–1489. [CrossRef] [PubMed]
24. Helle, S.; Bray, F.; Putaux, J.L.; Verbeke, J.; Flament, S.; Rolando, C.; D'Hulst, C.; Szydlowski, N. Intra-Sample Heterogeneity of Potato Starch Reveals Fluctuation of Starch-Binding Proteins According to Granule Morphology. *Plants* **2019**, *8*, 324. [CrossRef]
25. Abt, M.R.; Pfister, B.; Sharma, M.; Eicke, S.; Bürgy, L.; Neale, I.; Seung, D.; Zeeman, S.C. STARCH SYNTHASE5, a Noncanonical Starch Synthase-like Protein, Promotes Starch Granule Initiation in *Arabidopsis*. *Plant Cell* **2020**, *32*, 2543–2565. [CrossRef]
26. Flores-Castellanos, J.; Fettke, J. The Plastidial Glucan Phosphorylase Affects the Maltooligosaccharide Metabolism in Parenchyma Cells of Potato (*Solanum Tuberosum*, L.) Tuber Discs. *Plant Cell Physiol.* **2023**, *64*, 422–432. [CrossRef]
27. Orawetz, T.; Malinova, I.; Orzechowski, S.; Fettke, J. Reduction of the Plastidial Phosphorylase in Potato (*Solanum Tuberosum*, L.) Reveals Impact on Storage Starch Structure during Growth at Low Temperature. *Plant Physiol. Biochem. PPB* **2016**, *100*, 141–149. [CrossRef] [PubMed]
28. Green, A.R.; Nissen, M.S.; Mohan Kumar, G.N.; Knowles, N.R.; Kang, C.H. Characterization of Solanum Tuberosum Multicystatin and the Significance of Core Domains. *Plant Cell* **2013**, *25*, 5043–5052. [CrossRef]
29. Rodis, P.; Hoff, J.E. Naturally Occurring Protein Crystals in the Potato. *Plant Physiol.* **1984**, *74*, 907–911. [CrossRef]
30. Nissen, M.S.; Kumar, G.N.M.; Youn, B.; Knowles, D.B.; Lam, K.S.; Ballinger, W.J.; Knowles, N.R.; Kang, C. Characterization of Solanum Tuberosum Multicystatin and Its Structural Comparison with Other Cystatins. *Plant Cell* **2009**, *21*, 861–875. [CrossRef]
31. Stiekema, W.J.; Heidekamp, F.; Dirkse, W.G.; van Beckum, J.; de Haan, P.; ten Bosch, C.; Louwerse, J.D. Molecular Cloning and Analysis of Four Potato Tuber mRNAs. *Plant Mol. Biol.* **1988**, *11*, 255–269. [CrossRef] [PubMed]
32. Ishikawa, A.; Ohta, S.; Matsuoka, K.; Hattori, T.; Nakamura, K. A Family of Potato Genes That Encode Kunitz-Type Proteinase Inhibitors: Structural Comparisons and Differential Expression. *Plant Cell Physiol.* **1994**, *35*, 303–312. [CrossRef]

33. Fettke, J.; Poeste, S.; Eckermann, N.; Tiessen, A.; Pauly, M.; Geigenberger, P.; Steup, M. Analysis of Cytosolic Heteroglycans from Leaves of Transgenic Potato (*Solanum tuberosum*, L.) Plants That under- or Overexpress the Pho 2 Phosphorylase Isozyme. *Plant Cell Physiol.* **2005**, *46*, 1987–2004. [CrossRef]
34. Mahlow, S.; Hejazi, M.; Kuhnert, F.; Garz, A.; Brust, H.; Baumann, O.; Fettke, J. Phosphorylation of Transitory Starch by α-Glucan, Water Dikinase during Starch Turnover Affects the Surface Properties and Morphology of Starch Granules. *New Phytol.* **2014**, *203*, 495–507. [CrossRef] [PubMed]
35. Pouvreau, L.; Gruppen, H.; Piersma, S.R.; Van den Broek, L.A.M.; Van Koningsveld, G.A.; Voragen, A.G.J. Relative Abundance and Inhibitory Distribution of Protease Inhibitors in Potato Juice from Cv. Elkana. *J. Agric. Food Chem.* **2001**, *49*, 2864–2874. [CrossRef] [PubMed]
36. Pouvreau, L.; Gruppen, H.; Van Koningsveld, G.A.; Van Den Broek, L.A.M.; Voragen, A.G.J. The Most Abundant Protease Inhibitor in Potato Tuber (Cv. Elkana) Is a Serine Protease Inhibitor from the Kunitz Family. *J. Agric. Food Chem.* **2003**, *51*, 5001–5005. [CrossRef]
37. Jørgensen, M.; Bauw, G.; Welinder, K.G. Molecular Properties and Activities of Tuber Proteins from Starch Potato Cv. Kuras. *J. Agric. Food Chem.* **2006**, *54*, 9389–9397. [CrossRef]
38. Ritte, G.; Eckermann, N.; Haebel, S.; Lorberth, R.; Steup, M. Compartmentation of the Starch-Related R1 Protein in Higher Plants. *Starch/Staerke* **2000**, *52*, 179–185. [CrossRef]
39. Kötting, O.; Pusch, K.; Tiessen, A.; Geigenberger, P.; Steup, M.; Ritte, G. Identification of a Novel Enzyme Required for Starch Metabolism in *Arabidopsis* Leaves. The Phosphoglucan, Water Dikinase. *Plant Physiol.* **2005**, *137*, 242–252. [CrossRef]
40. Webster, J.; Oxley, D. Protein identification by MALDI-TOF mass spectrometry. *Methods Mol. Biol.* **2012**, *800*, 227–240. [CrossRef]

Disclaimer/Publisher's Note: The statements, opinions and data contained in all publications are solely those of the individual author(s) and contributor(s) and not of MDPI and/or the editor(s). MDPI and/or the editor(s) disclaim responsibility for any injury to people or property resulting from any ideas, methods, instructions or products referred to in the content.

Article

Effects of Different Extraction Methods on the Gelatinization and Retrogradation Properties of Highland Barley Starch

Mengzi Nie [1,†], Chunhong Piao [1,†], Jiaxin Li [2], Yue He [2], Huihan Xi [2], Zhiying Chen [2], Lili Wang [2], Liya Liu [2], Yatao Huang [2], Fengzhong Wang [2,*] and Litao Tong [2,*]

1. College of Food Science and Engineering, Jilin Agricultural University, Changchun 130118, China
2. Key Laboratory of Agro-Products Processing Ministry of Agriculture, Institute of Food Science and Technology, Chinese Academy of Agricultural Sciences, Beijing 100193, China
* Correspondence: wangfengzhong@caas.cn (F.W.); tonglitao@caas.cn (L.T.); Tel./Fax: +86-10-6281-7417 (L.T.)
† These authors contributed equally to this work.

Abstract: The purpose of this study was to compare the gelatinization and retrogradation properties of highland barley starch (HBS) using different extraction methods. We obtained HBS by three methods, including alkali extraction (A-HBS), ultrasound extraction (U-HBS) and enzyme extraction (E-HBS). An investigation was carried out using a rapid viscosity analyzer (RVA), texture profile analysis (TPA), differential scanning calorimetry (DSC), X-ray diffraction (XRD) and Fourier-transform infrared spectrometry (FTIR). It is shown that the different extraction methods did not change the crystalline type of HBS. E-HBS had the lowest damaged starch content and highest relative crystallinity value ($p < 0.05$). Meanwhile, A-HBS had the highest peak viscosity, indicating the best water absorption ($p < 0.05$). Moreover, E-HBS had not only higher G' and G'' values, but also the highest gel hardness value, reflecting its strong gel structure ($p < 0.05$). These results confirmed that E-HBS provided better pasting stability and rheological properties, while U-HBS provides benefits of reducing starch retrogradation.

Keywords: highland barley starch; extraction method; gelatinization; retrogradation

1. Introduction

Highland barley (HB) is a high-quality raw material for the development of functional foods, among which the most abundant component is starch, accounting for around 65% by dry weight of grains [1]. The regular intake of resistant starch (RS) is known to protect against certain diseases and promote colon health [2]. Relevant studies have shown that the content of RS in highland barley starch could amount to 2.27–11.23%, and the content of slow-digestible starch (SDS) amounts to 1.54–40.58% [3–5]. Good processing methods can increase the content of RS and SDS and then reduce the glycemic index, which is beneficial to the preparation of highland barley slow-digestible starch. For the extraction of starch, there are different extraction methods for starch separation from the grain, and the effect of different methods on starch are slightly different. The most commonly used method is alkali extraction (chemical extraction). Pires et al. found that starch obtained by alkali extraction gave higher yields than those from water extraction, but was harmful to the environment [6]. Moreover, alkali extraction has a relatively greater influence on the structure and properties of HBS, which is presumed to be detrimental to the maintenance of starch resistance. Among physical extraction techniques, ultrasound extraction is widely used thanks to its efficiency, chemical-free process and environmental friendliness [7,8]. Ultrasound could act on the amorphous regions of starch granules, changing the relative crystallinity (RC) and weakening the connections between starch molecules by breaking hydrogen bonds and double-helix structures, which in particular also led to a smaller average starch particle size [9]. In addition, some researchers have tried using enzymatic methods to release the starch granules. Buksa found that the yields of rye starch extracted by enzymatic methods (xylanase and protease) were much higher than those extracted by aqueous methods [10]. Ozturk et al. discovered the cellulase, xylanase and protease could cooperatively disrupt and loosen the network around the protein matrix or non-starch polysaccharides by micrographs, resulting in higher purity [11]. However, the implications of the different extraction methods on the physicochemical properties of HBS need to be further clarified.

As is well known, the granular structure of starch is correlated with its physicochemical properties. The pasting behavior of starch is central to many starch-based food matrices and is generally characterized by changes in viscosity during the process of heating, holding and cooling [12]. The degree of gelatinization changes with different extraction methods of starch, which is related to the stability of the starch. Once the initial gelatinization temperature is reached, the granular and crystalline structure begins to break down, and the amylose gradually leaches out. The properties of starch paste were significantly different under different degrees of gelatinizing [13]. For the effect of starch extraction methods on the properties of starch paste, it was found the peak viscosity of Chinese yam starch obtained by aqueous extraction (SBS) was lower than that of those obtained by enzymatic and alkali extraction, which indicated that the water holding capacity of starch extracted by SBS was relatively poor [14]. After gelatinization, the differences in the structural properties of starch make the degree of retrogradation different, and then affect the properties of starch gel after retrogradation. In retrogradation, the starch chains recombine and form a double-helix structure during the cooling phase, which is then packed into crystals [15]. For now, there is little research about the effect of different extraction methods on the gelatinization and retrogradation of starch. The related research might be conductive for the development of grain products, such as bread, noodles and so on, especially for HBS. Therefore, the effects of different extraction methods of starch on the physicochemical properties and paste properties of HBS were explored to further improve the quality of cereal food: 1. to investigate the effect of different extraction methods on the gelatinization characteristics of HBS; 2. to clarify the effect of different extraction methods on the retrogradation characteristics of HBS; and 3. to analyze the influence of HBS structure changes on the gelatinization and retrogradation characteristics.

2. Results and Discussion

2.1. Morphology and Chemical Compositions of HBS

The scanning electron micrographs of HBS obtained from three extraction methods are shown in Figure 1, which shows different granule morphologies. HBS consisted of large (most-part) and small (small-part) granules, similar to a previous report [3]. For all HBS samples, the particle size of the most-part granules were concentrated at 19.89–22.60 μm, and the size of the small-part granules were concentrated at 2.58–3.33 μm, as shown in Figure 2. This is consistent with previous reports that HBS granules' diameter are generally 2–30 μm [16]. The surface of A-HBS (HBS extracted by alkali method) granules and the edge of U-HBS (HBS extracted by ultrasonic method) were relatively smooth, which indicated that the granular morphology was less damaged. In contrast, the E-HBS (HBS extracted by enzymatic method) granules were rough, with some fine fragments in morphology, which might be due to incomplete hydrolysis of partially insoluble dietary fiber (IDF) in the presence of enzymes. The hydrolysis of dietary fiber involved a synergistic attack of multiple enzymes, indicating IDF was hydrolyzed to form numerous cavities or spaces and that a dense network structure also hindered the enzymatic hydrolysis [17,18]. In addition to this, it was found the granular size distribution of A-HBS (D10, D50 and D90) was smaller than other granules (Table 1). The different granular morphology and size distribution of starch granules reflects different physicochemical properties, such as gelatinization and retrogradation behavior, swelling power and so on. [19].

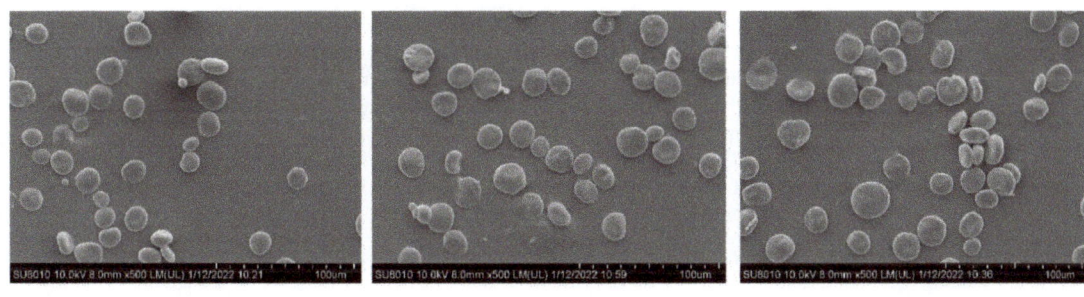

| A-HBS | U-HBS | E-HBS |

Figure 1. Scanning electron micrographs (SEM) (×500) of highland barley starch samples. A-HBS: HBS extracted by alkali method; U-HBS: HBS extracted by ultrasonic method; E-HBS: HBS extracted by enzymatic method.

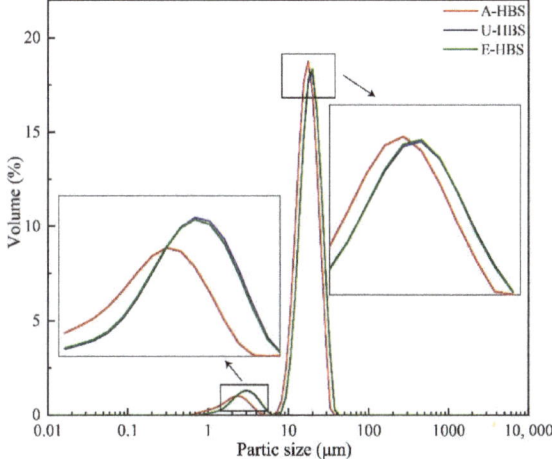

Figure 2. Particle size distribution of highland barley starch samples. A-HBS: HBS extracted by alkali method; U-HBS: HBS extracted by ultrasonic method; E-HBS: HBS extracted by enzymatic method.

Table 1. Chemical composition of highland barley starch samples.

	A-HBS	U-HBS	E-HBS
Total starch (%)	92.33 ± 0.23 [a]	92.15 ± 0.26 [a]	90.68 ± 0.12 [b]
Amylose starch (%)	23.37 ± 0.01 [a]	23.71 ± 0.29 [a]	23.76 ± 0.03 [a]
Damaged starch (%)	1.66 ± 0.02 [b]	2.57 ± 0.06 [a]	1.20 ± 0.03 [c]
Moisture (%)	8.46 ± 0.06 [b]	8.65 ± 0.05 [b]	9.14 ± 0.10 [a]
Crude protein (%)	0.18 ± 0.01 [a]	0.24 ± 0.02 [a]	0.23 ± 0.02 [a]
Crude lipid (%)	0.18 ± 0.04 [a]	0.25 ± 0.00 [a]	0.26 ± 0.02 [a]
Ash (%)	0.35 ± 0.06 [a]	0.25 ± 0.02 [a]	0.20 ± 0.01 [a]
Total dietary fiber (%)	3.04 ± 0.11 [a]	3.14 ± 0.02 [a]	2.96 ± 0.01 [a]
Yield (%)	42.15 ± 0.33 [a]	30.12 ± 0.27 [c]	36.91 ± 0.18 [b]
D10 (μm)	11.35 ± 0.13 [b]	12.02 ± 0.03 [a]	12.20 ± 0.13 [a]
D50 (μm)	19.10 ± 0.12 [b]	20.70 ± 0.00 [a]	20.81 ± 0.10 [a]
D90 (μm)	26.90 ± 0.13 [b]	30.40 ± 0.03 [a]	30.7 ± 0.00 [a]
R1047/1022	1.15 ± 0.00 [a]	1.11 ± 0.07 [a]	1.12 ± 0.02 [a]

A-HBS: HBS extracted by alkali method; U-HBS: HBS extracted by ultrasonic method; E-HBS: HBS extracted by enzymatic method. Mean ± SD is calculated from triplicate measurements. Different lowercase letters represent significant difference ($p < 0.05$).

As shown in Table 1, the purity of HBS obtained from all three extraction methods was larger than 90%, and there was little difference among these three starch samples. It was also found that the extracted method had little effect on amylose content. It is worth noting that the content of damaged starch in U-HBS was significantly higher than that of the other extraction methods, which meant that ultrasonic waves caused some damage to the starch granules. The other experimental values were quite consistent with those of Pina et al., who reported the following content ranges for crude protein, crude lipid, ash and total dietary fiber in HBS: 0.18–0.23%, 0.18–0.26%, 0.20–0.35% and 2.96–3.14%, respectively [20].

2.2. Pasting Properties

Gelatinization is a vital property for starch-based food processing. The pasting curves for HBS were different (Figure 3) and the related parameters are listed in Table 2. The peak viscosity (PV) and final viscosity (FV) of A-HBS were significantly higher than HBS obtained from other extraction methods, which indicated the A-HBS had the best water absorption capacity [6]. According to the particle size distribution, the high water absorption capacity of A-HBS might be related to the larger specific surface area per unit weight of the relatively small granules [19]. Meanwhile, A-HBS had the lowest pasting temperature (PT), implying poor thermal stability [21,22]. The gelatinization curves of U-HBS and E-HBS were relatively gentle with smaller breakdowns (BV), indicating the starch paste had poor shear resistance. The value of BV was the most sensitive index of pasting properties, which represented the degree of starch granule breakage [23]. A higher BV value implied that more starch granules were broken during the heating process, and the internal starch molecules were released [24]. In contrast to A-HBS, U-HBS and E-HBS have a larger average particle diameter (Table 1). This means that water molecules could not easily enter inside the starch granules, leading to a lower swelling force and better stability during heating [25]. Meanwhile, it also suggested that U-HBS and E-HBS had better gelatinization stability and the starch paste, which was presumed to be due to the hydroxyl groups in HBS, combined more closely with the weak water and strengthened the double-helix structure of the starch, and thus the microcrystalline structure of the amylopectin was strengthened [23]. Moreover, U-HBS presented the lowest PV and FV, indicating an application for food products with restricted swelling, such as noodles.

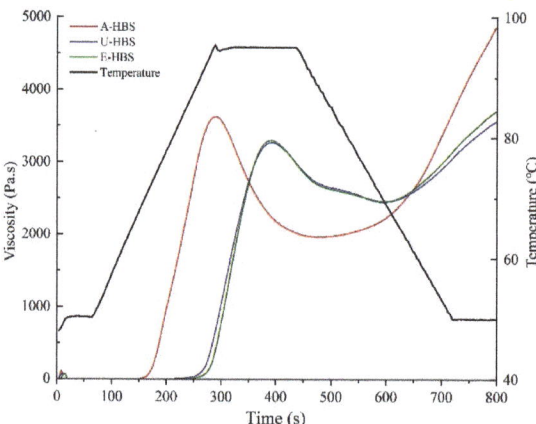

Figure 3. Pasting profiles of highland barley starch samples. A-HBS: HBS extracted by alkali method; U-HBS: HBS extracted by ultrasonic method; E-HBS: HBS extracted by enzymatic method.

Table 2. Pasting and textural properties of highland barley starch samples.

	A-HBS	U-HBS	E-HBS
Pasting properties			
PV (cP)	3597 ± 35 [a]	3258 ± 16 [b]	3311 ± 11 [b]
TV (cP)	1935 ± 34 [b]	2437 ± 43 [a]	2489 ± 36 [a]
FV (cP)	5180 ± 77 [a]	3604 ± 59 [c]	3773 ± 21 [b]
BV (cP)	1662 ± 16 [a]	820 ± 29 [b]	822 ± 7 [b]
SV (cP)	3244 ± 16 [a]	1167 ± 24 [c]	1284 ± 13 [b]
PeT (min)	4.76 ± 0.14 [b]	6.56 ± 0.11 [a]	6.60 ± 0.10 [a]
PeT (°C)	70.76 ± 0.01 [c]	90.50 ± 0.04 [b]	92.34 ± 0.10 [a]
Textural properties			
		1 d	
Hardness (g)	210.28 ± 0.90 [b]	183.47 ± 0.77 [c]	258.58 ± 0.39 [a]
Cohesiveness	0.43 ± 0.05a	0.44 ± 0.00 [a]	0.45 ± 0.01 [a]
Springiness (%)	91.03 ± 1.79 [a]	90.61 ± 0.43 [a]	94.74 ± 0.37 [a]
Gumminess	92.58 ± 1.77 [b]	89.43 ± 0.34 [b]	114.28 ± 1.29 [a]
Chewiness	84.32 ± 1.28 [b]	80.76 ± 0.07 [b]	103.34 ± 0.52 [a]
		7 d	
Hardness (g)	283.77 ± 0.86 [b]	219.75 ± 0.56 [c]	348.30 ± 0.71 [a]
Cohesiveness	0.58 ± 0.02 [a]	0.54 ± 0.04 [a]	0.60 ± 0.01 [a]
Springiness (%)	94.33 ± 1.60 [a]	92.53 ± 0.33 [a]	95.83 ± 0.21 [a]
Gumminess	127.45 ± 1.19 [b]	112.85 ± 1.02 [c]	153.91 ± 1.75 [a]
Chewiness	134.54 ± 0.08 [b]	112.05 ± 1.05 [c]	139.74 ± 0.45 [a]

A-HBS: HBS extracted by alkali method; U-HBS: HBS extracted by ultrasonic method; E-HBS: HBS extracted by enzymatic method; PV: peak viscosity; TV: trough viscosity; FV: final viscosity; BV: breakdown viscosity; SV: setback viscosity; PT (min): pasting time; PeT (°C): pasting temperature. Mean ± SD is calculated from triplicate measurements. Different lowercase letters represent significant difference ($p < 0.05$).

2.3. Gel Properties

As displayed in Table 2, the textural parameters of the HBS gels from different extraction methods were stored at 4 °C for 1 and 7 days. Starch retrogradation occurred during the cooling and storage process, through recrystallization of amylose and amylopectin recrystallization [26]. On the other hand, it can be seen directly through the pictures in Figure 4C that the gel of U-HBS-1d flowed downwards, verifying the results of hardness. The gel hardness of E-HBS was significantly higher than that of A-HBS and U-HBS, since E-HBS had the lowest damaged starch content and larger average particle diameter (Table 1), both implying it had better particle integrity, resulting in incomplete reaction of water with the starch granules [27]. The hardness of all HBS gels increased in varying degrees after storage for 7 days. According to relevant studies, there was a close correspondence between the hardness of starch gels and retrogradation, and the degree of retrogradation during storage was clearly identified by measuring the hardness of HBS gels [28]. Both A-HBS and U-HBS had a lower degree of retrogradation, allowing them to be used as raw material for rice noodles. The cohesiveness was an evaluation indicator of the starch gel maintaining its structural integrity after the first occlusion and reflected the internal gel bonding strength [28]. There was no significant difference in the cohesiveness and springiness of prepared HBS gels.

2.4. Rheological Properties of HBS

To further evaluate the gel systems of different HBS, the rheological properties were characterized. As the angular frequency increased, the storage modulus (G') and the loss modulus (G'') both tended to increase gradually (Figure 4 A,B). Furthermore, the G' values were higher than the G'' values, indicating that all the starch-containing systems had gelled and behaved as viscoelastic solids [29]. U-HBS and E-HBS showed higher G' and G'' values than A-HBS, which also reflected the gel strength of HBS. In addition, it could also be found that the higher the values of PV and SV, the lower the values of G' and G'', which corresponded to the results of pasting properties. The loss tangent (tan δ) values were also an essential rheological parameter, which were evaluated using viscoelastic properties and defined as the ratio of the G″ to the G′ [30]. The tan δ values of HBS samples were far less than 1, which exhibited good elastic behavior. However, the tan δ of starch remarkably exceeded 1, indicating that the elasticity junction zones of the starch gel were destroyed [31]. It is noteworthy that the tan δ value of the A-HBS gels was around 0.15, indicating that the A-HBS pastes had a weaker gel structure. There was also a study reporting that starch pastes with weak gel structures preferred to form more solid doughs and were able to produce starch noodles with excellent dripping properties [31].

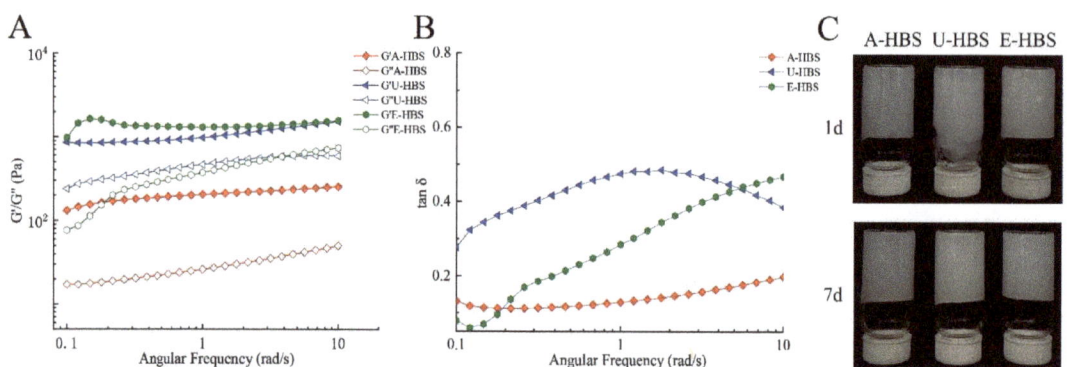

Figure 4. Rheological properties of highland barley starch samples: storage modulus (G′), loss modulus (G″) (**A**); tan δ (**B**) and pictures of HBS gels stored for 1 and 7 days (**C**). A-HBS: HBS extracted by alkali method; U-HBS: HBS extracted by ultrasonic method; E-HBS: HBS extracted by enzymatic method.

2.5. Thermal Properties

To explore the thermal properties of starch, thermograms of all HBS samples treated were revealed in Figure 5. The DSC characteristic values of HBS by different extraction methods are shown in Table 3, and the onset (To), peak (Tp) and conclusion (Tc) temperatures and gelatinization enthalpy (ΔH) of the HBS varied significantly. It was found that the gelatinization temperature (To, Tp and Tc) of E-HBS was highest and that of A-HBS was the lowest, which meant the E-HBS had the best thermal stability and A-HBS the worst. Moreover, the ΔH of E-HBS was highest, followed by A-HBS and U-HBS. This proved the results of the pasting properties: E-HBS had a more ordered molecular or stable crystalline structure. The degree of retrogradation of all HBS samples during the storage process could be expressed by the retrogradation enthalpy ΔH [32]. With longer storage times, the ΔH showed a regular rise, indicating an increased degree of retrogradation. It is worth noting that all U-HBS samples had the lowest ΔH and the difference in enthalpy between day 1 and day 7 decreased (Table 3). This implied that the extent and rate of re-crystallization slowed down towards a certain degree. The differences in ΔH noticed were due to the intrinsic differences in the starch molecular structure once the structure has been completely disrupted by gelatinization [33]. Additionally, regenerated starch had a lower gelatinization temperature compared to natural HBS due to its weaker crystallinity. In the following section, particle structure measurements were carried out to investigate the reasons for the differences in gelatinization and retrogradation properties.

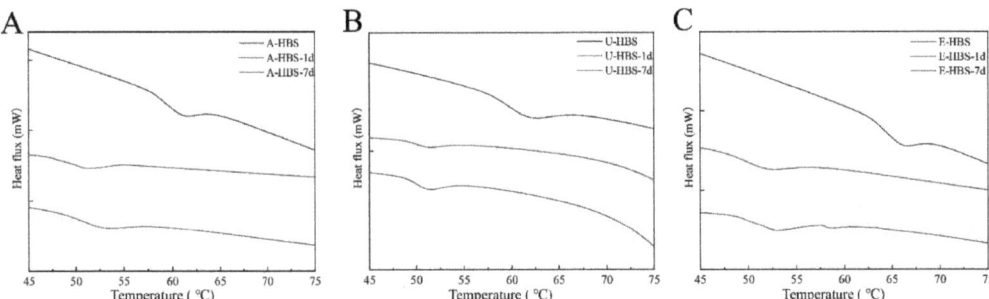

Figure 5. Differential scanning calorimetry (DSC) curves of A-HBS (**A**), U-HBS (**B**), E-HBS (**C**). A-HBS: HBS extracted by alkali method; U-HBS: HBS extracted by ultrasonic method; E-HBS: HBS extracted by enzymatic method.

Table 3. Thermal properties of highland barley starch samples.

	To (°C)	Tp (°C)	Tc (°C)	ΔH (J/g)
A-HBS	57.03 ± 0.53 [b]	61.15 ± 0.61 [b]	65.46 ± 0.77 [b]	5.69 ± 0.11 [b]
U-HBS	59.76 ± 0.16 [b]	61.99 ± 0.42 [b]	65.08 ± 0.45 [b]	5.32 ± 0.10 [b]
E-HBS	62.41 ± 0.08 [a]	65.63 ± 0.18 [a]	69.30 ± 0.96 [a]	6.48 ± 0.37 [a]
A-HBS-1d	47.32 ± 0.74 [b]	50.35 ± 0.53 [a]	52.56 ± 1.93 [a]	1.46 ± 0.18 [b]
U-HBS-1d	48.10 ± 0.34 [a]	50.92 ± 0.39 [a]	51.88 ± 0.29 [a]	1.13 ± 0.04 [b]
E-HBS-1d	47.72 ± 0.64 [b]	51.43 ± 0.78 [a]	53.55 ± 1.63 [a]	2.15 ± 0.12 [a]
A-HBS-7d	47.62 ± 0.17 [b]	52.35 ± 0.89 [a]	54.97 ± 0.08 [a]	2.27 ± 0.16 [b]
U-HBS-7d	48.01 ± 0.28 [a]	51.10 ± 0.56 [b]	52.93 ± 0.32 [b]	1.54 ± 0.18 [c]
E-HBS-7d	48.15 ± 0.16 [a]	53.15 ± 1.13 [a]	55.12 ± 0.35 [a]	3.25 ± 0.09 [a]

A-HBS: HBS extracted by alkali method; U-HBS: HBS extracted by ultrasonic method; E-HBS: HBS extracted by enzymatic method; To: onset temperature; Tp: peak temperature; Tc: conclusion temperature; ΔH: enthalpy of gelatinization. Mean ± SD is calculated from triplicate measurements. Different lowercase letters represent significant difference ($p < 0.05$).

2.6. X-ray Diffraction Pattern

XRD measurement was performed to further investigate the long-range ordered structure of HBS extracted by three methods; the results are shown in Figure 6A. The pattern of the all three HBS showed strong peaks at 2θ of 15.13°, 17.08°, 18.24° and 23.02°, which were characteristic diffraction peaks for the A-type crystalline structure [34]. A-type starch crystals were monoclinic and generally consisted of a double-helix structure formed by amylopectin. XRD patterns were also instrumental in determining the influence of different extraction methods on the crystallinity of starch granules. The relative crystallinity (RC) values were displayed in Figure 6A, which was basically consistent with the values of gelatinization enthalpy (Table 3); the order was as follows: E-HBS (26.37%) > A-HBS (24.78%) > U-HBS (21.04%). It was found that the RC value of U-HBS (21.04%) was lower than that of the two other methods, which was because ultrasonic wave treatment had a high damage intensity to the crystalline structure, and the internal space became loose [8]. The E-HBS had the highest ratio of double helix structure, indicating that the destruction of the crystalline structure of starch by biological extraction of starch is lower than that of physical and chemical methods, and that ultrasonic wave treatment could lead to the disintegration of long-range ordered crystallites, as Wang et al. reported that the RC directly reflected the long-range ordered crystal structure of HBS [4].

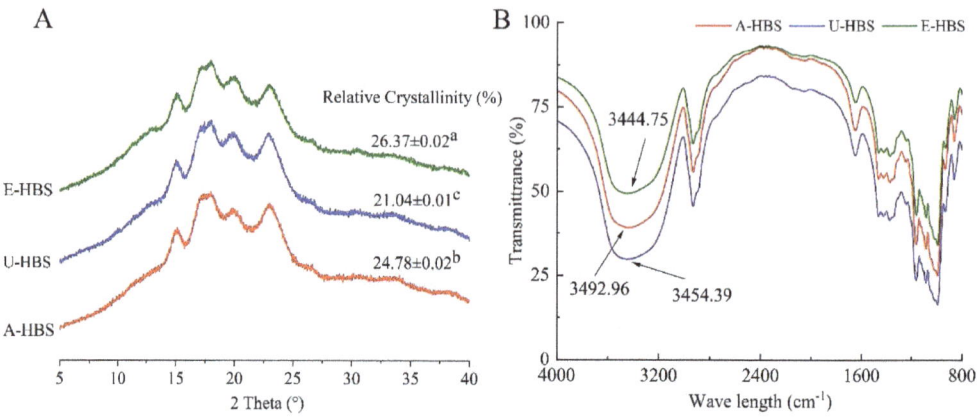

Figure 6. X-ray diffractograms (**A**), FTIR spectra (**B**) of highland barley starch samples. A-HBS: HBS extracted by alkali method; U-HBS: HBS extracted by ultrasonic method; E-HBS: HBS extracted by enzymatic method. Mean ± SD is calculated from triplicate measurements. Different lowercase letters represent significant difference ($p < 0.05$).

2.7. Short-Range Ordered Structure

The FTIR spectra of HBS obtained by different extraction methods are shown in Figure 6B. There was a peak found in the range of 4000 cm^{-1} to 3300 cm^{-1}, which was mainly attributed to the existence of -OH groups occurring normally in carbohydrates [35]. In addition, the -OH absorption peak in A-HBS (3492.96) shifted toward the high-frequency wave number, indicating that the hydrogen bonding of A-HBS was weaker [36], which also explained the phenomenon that A-HBS was more prone to absorb water and swell into a paste. The bands at 1047 cm^{-1} and 1022 cm^{-1} represent the crystalline ordered and amorphous regions of starch, respectively. To reveal short-range order, absorbance ratios were calculated at 1047/1022 cm^{-1} (R1047/1022) [37]. As seen from Table 1, there were marginal differences in HBS obtained by three extraction methods. It was hypothesized that the above findings might be due to differences in the distribution of the amylopectin branch chain length, which in turn affected the composition of the amorphous region [4]. The above observations revealed that the starch granules of E-HBS had a better molecular structure, resulting in a higher gel hardness (Table 2).

2.8. Principal Component Analysis (PCA)

The principal component analysis (PCA) was further applied in this study in order to understand the potential mechanism of pasting and retrogradation. Figure 7A shows the results of PCA on the pasting characteristics of all HBS samples. The two principal components, PC1 and PC2, were able to explain 69.89% and 22.01% of the variable, respectively, for the total score was 91.90. E-HBS was located in the positive quadrant of PC1 and PC2 and had good thermal stability due to its high gelatinization temperature and ΔH, as well as through viscosity (TV). In addition, the results of the loading plot in Figure 7A showed that the short-range ordered (R1042/1022) and RVA parameters were close to each other, indicating that these indicators were positively correlated in the pasting process of HBS. This was further combined with the analysis of different extraction methods with storage time and aging factors in Figure 7B. Overall, the contribution of PC1 and PC2 to the total variation was 90.50%, indicating that the planes of PC1 and PC2 largely reflected the main contribution of the response variables. The factors associated with starch retrogradation (mainly ΔH and hardness) were distributed to the right of PC1, in contrast to the location of A-HBS and U-HBS, suggesting that both had a role in retarding HBS retrogradation in this study.

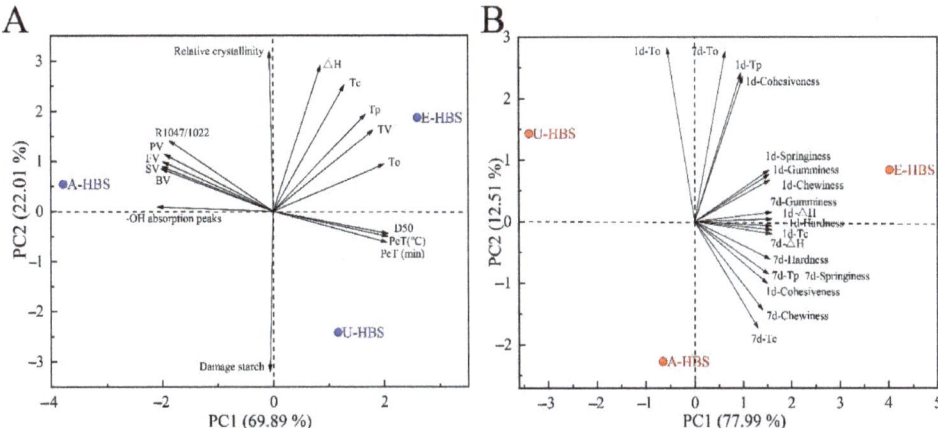

Figure 7. Principal component analysis—Correlation analysis based on different extraction methods with gelatinization (**A**) and retrogradation (**B**) properties. A-HBS: HBS extracted by alkali method; U-HBS: HBS extracted by ultrasonic method; E-HBS: HBS extracted by enzymatic method.

3. Materials and Methods

3.1. Materials

The HB kernels (cultivar: Kunlun 15) were purchased from Xinlvkang food Co., Ltd. (Xining, China). Cellulase (≥400 U/mg protein) was purchased from Yuanye Technology Co., Ltd. (Shanghai, China). Xylanase (≥6500 U/g protein) was purchased from Novozymes (China) Biotechnology Co., Ltd. (Tianjin, China). All other reagents were analytical grade.

3.2. Samples Preparation

HB kernels were milled and sieved through 100 mesh to obtain highland barley flour (HBF). Afterwards, HBS was extracted using three methods based on the conditions of the extraction. The obtained HBS samples were denoted as A-HBS, U-HBS and E-HBS, respectively.

3.2.1. Alkali Extraction

According to Yang et al., with some modifications [5], HBF (100 g) was added to NaOH solution (0.125 M, 600 mL), which was soaked at room temperature for 5 h. After that, the mixture was centrifuged (3000× g, 20 min) to discard the supernatant and the top gray layer was removed. Then, the precipitates were washed with deionized water and ethanol three times to remove soluble impurities, respectively. The precipitate obtained after washing was collected as starch and dried at room temperature, followed by passing through a 100-mesh sieve. The yield was then calculated as the percentage of the HBF weight.

3.2.2. Ultrasound Extraction

HBF (100 g) was added to 600 mL distilled water, which was treated by an ultrasonic device (KQ5200DV, Kunshan Ultrasonic instrument Co., Ltd., Kunshan China) working at a frequency of 20 kHz and input power of 200 W for 20 min. Then, the sample was shaken in a water bath at 50 °C for 5 h. In order to ensure that the purity of HBS was greater than 90%, it was sieved through 300-mesh (0.05 mm) before centrifugation and repeating the above steps.

3.2.3. Enzyme Extraction

According to Ozturk et al., with some modifications [11], HBF (100 g) was added to 600 mL distilled water. Then, 200 U/g cellulase and xylanase were added, and the mixture was shaken in water bath at 50 °C for 8 h, repeating the above steps.

3.3. General Compositions and Particle Size

According to the standard methods described by AACC (2010), the contents of moisture, crude protein, crude fat, ash and total dietary fiber were determined. The contents of total starch, amylose and damaged starch were determined by an assay kit (K-SDAM, Megazyme International Ltd., Wicklow, Ireland). A laser particle size analyzer (MS3000, Malvern Instruments, Worcestershire, UK) was chosen to measure the particle size distribution of HBS samples.

3.4. Scanning Electron Microscopy (SEM)

The morphology of HBS granules was observed by a scanning electron microscope (SEM Hitachi S-570, Hitachi, Co., Ltd., Tokyo, Japan) under the voltage of 10 kV. All samples were sputter-coated with gold and then observed at magnifications of 500×.

3.5. Pasting Property

The pasting properties of HBS samples were analyzed by a rapid viscosity analyzer (RVA-TecMaster, Perten Instruments, Sydney, Australia), following the method of Qin et al. [38]. A 3.0 g HBS sample was dispersed (based on 14% moisture content) in 25 mL of distilled water and then the mixture was measured in an RVA tank. The suspension was stirred at 960 rpm for 10 s and then reduced to 160 rpm for 50 s. The temperature was initially maintained at 50 °C for 1 min, then increased to 95 °C at the rate of 9 °C/min and maintained at 95 °C for 2.7 min. Finally, it was cooled to 50 °C within 3.75 min, and remained there for 2 min.

3.6. Texture Profile Analysis (TPA)

The retrograded starch samples stored at 4 °C for 1 and 7 d in Section 2.5 were collected to determine gel properties by the texture analyzer (TAXT plus, Stable Co., Godalming, UK) coupled with a P/0.5 cylinder probe [39]. The experiment parameters of the texture analyzer were set as follows: the pretest speed, test speed and the latter test speed were set at 1.0 mm/s, the strain was 50%, the trigger type was automatic and the trigger force was 5 g.

3.7. Rheological Characterization

The viscoelastic and flow behaviors were determined by a stress-controlled rheometer (Physica MCR 301, Anton Paar GmbH, Graz, Austria) equipped with parallel plate geometry and a 1 mm gap [40]. The samples were starch pastes prepared in Section 3.5; a fixed amount of sample was transferred to the rheometer plate and excess sample was carefully wiped away. Frequency scan experiments were carried out at 1% strain from 0.1 to 10 Hz in the linear viscoelastic region.

3.8. Thermal Analysis

The thermal properties were measured by differential scanning calorimetry (DSC 8000, PerkinElmer, Norwalk, CT, USA) according to Chen et al., with some modifications [39]. 3 mg HBS was weighed, mixed with 10 μL distilled water in an aluminum pot and equilibrated at 4 °C overnight. The mixture was heated from 30 °C to 130 °C at a rate of 10 °C/min. A sealed empty pan was used as a reference. After testing, the samples were stored at 4 °C for 1 and 7 d, repeating the above assay measurements.

3.9. X-ray Diffraction (XRD) Analysis

The crystalline structure of HBS samples extracted by different methods was measured by an X-ray diffractometer (XRD, Bruker D8 Advance, Salbuluken, Germany) equipped with a copper tube operating at 40 kV and 200 mA and producing Cu-Kα radiation at a wavelength of 0.1542 nm. The diffractograms were obtained by scanning from 5° to 45° (2θ) at room temperature at a rate of 10°/min and in steps of 0.02°. The crystalline peak and total area of the diffractogram were analyzed using MDI Jade 5.0 software (Materials Data, Inc., Livermore, CA, USA). The relative crystallinity was calculated as the percentage of the area of the crystalline region to the area of total diffraction, as represented in Equation (1):

$$\mathrm{RC}\ (\%) = \frac{T_a - A_{f_a}}{T_a} \times 100 \tag{1}$$

where RC is the relative crystallinity, T_a is the total area and A_{fa} is the area of the amorphous fraction.

3.10. Fourier-Transform Infrared Spectrometry (FTIR) Analysis

The HBS samples (2 mg) and the dried KBr powders (100 mg) were mixed and ground thoroughly. The mixture was then pressed into tablets and observed by Fourier-transform infrared spectroscopy (TENSOR 27, Borken, Germany). The blank background was completed by the KBr powder alone. The scanning conditions were 400–4000 cm^{-1} of wavelength, 4 cm^{-1} of resolution and 64 scans.

3.11. Statistical Analysis

Analysis of variance (ANOVA) was used to measure statistical differences by SPSS Version 16.0 software (IBM software, Chicago, IL, USA). Significant difference ($p < 0.05$) was determined using the Duncan procedure. PCA was performed to visualize gelatinization and retrogradation properties of three kinds of HBS samples, respectively. PCA was achieved using Origin 2018b software (Origin-Lab, Inc., Northampton, USA). All experiments were carried out at least in triplicate.

4. Conclusions

In conclusion, the gelatinization and retrogradation characteristics of HBS obtained by the different extraction methods were significantly different. It was attributed to the granular structure of the HBS. E-HBS had the highest relative crystallinity value, thus it exhibited better thermal stability, which increased the crispness of foods as cookies. On the other hand, A-HBS showed the weakest hydrogen bonding, which resulted in the highest viscosity and could be used as a food thickener. Moreover, U-HBS had the highest damaged starch content, as well as the lowest gelatinization enthalpy after 1 and 7 days of storage, respectively, providing anti-aging properties, which made it suitable for foods with a short shelf life such as bread or noodles. The findings above provide a theoretical basis for the development of HBS and highland barley food with different expected properties.

Author Contributions: M.N.: Conceptualization, Data curation, Writing—original draft. C.P.: Data curation, Formal analysis, Validation, Writing—review and editing; J.L.: Methodology, Investigation; Y.H. (Yue He): Validation, Methodology; H.X.: Resources, Methodology; Z.C.: Software, Resources; L.W.: Software, Methodology; L.L.: Resources; Y.H. (Yatao Huang): Conceptualization; F.W.: Supervision, Resources, Funding acquisition; L.T.: Conceptualization, Funding acquisition, Project administration, Supervision. All authors have read and agreed to the published version of the manuscript.

Funding: This study was funded by Special National Key Research and Development Plan (2021YFD 1600101), and Key Science and technology Project of Qinghai Province (2021-0101NCC-0001).

Institutional Review Board Statement: Not applicable.

Informed Consent Statement: Not applicable.

Data Availability Statement: Not applicable.

Conflicts of Interest: The authors declare no conflict of interest.

Sample Availability: Samples of the compounds are available from the authors.

References

1. Obadi, M.; Sun, J.; Xu, B. Highland barley: Chemical composition, bioactive compounds, health effects, and applications. *Food Res. Int.* **2021**, *140*, 110065. [CrossRef] [PubMed]
2. Zaman, S.A.; Sarbini, S.R. The potential of resistant starch as a prebiotic. *Crit. Rev. Biotechnol.* **2016**, *36*, 578–584. [CrossRef] [PubMed]
3. Liu, K.; Zhang, B.; Chen, L.; Li, X.; Zheng, B. Hierarchical structure and physicochemical properties of highland barley starch following heat moisture treatment. *Food Chem.* **2019**, *271*, 102–108. [CrossRef]
4. Wang, H.; Li, Y.; Wang, L.; Wang, L.; Li, Z.; Qiu, J. Multi-scale structure, rheological and digestive properties of starch isolated from highland barley kernels subjected to different thermal treatments. *Food Hydrocoll.* **2022**, *129*, 107630. [CrossRef]
5. Yang, Y.; Jiao, A.; Zhao, S.; Liu, Q.; Fu, X.; Jin, Z. Effect of removal of endogenous non-starch components on the structural, physicochemical properties, and in vitro digestibility of highland barley starch. *Food Hydrocoll.* **2021**, *117*, 106698. [CrossRef]
6. Pires, M.B.; Amante, E.R.; Oliveira Petkowicz, C.L.; Esmerino, E.A.; Cruz Rodrigues, A.M.; Silva, L.H. Impact of extraction methods and genotypes on the properties of starch from peach palm (*Bactris gasipaes* Kunth) fruits. *LWT-Food Sci. Technol.* **2021**, *150*, 111983. [CrossRef]
7. Umego, E.C.; He, R.; Ren, W.; Xu, H.; Ma, H. Ultrasonic-assisted enzymolysis: Principle and applications. *Process Biochem.* **2021**, *100*, 59–68. [CrossRef]
8. Wang, L.Y.; Wang, M.; Zhou, Y.; Wu, Y.; Ouyang, J. Influence of ultrasound and microwave treatments on the structural and thermal properties of normal maize starch and potato starch: A comparative study. *Food Chem.* **2022**, *377*, 131990. [CrossRef]
9. Lin, X.; Sun, S.; Wang, B.; Zheng, B.; Guo, Z. Structural and physicochemical properties of lotus seed starch nanoparticles prepared using ultrasonic-assisted enzymatic hydrolysis. *Ultrason. Sonochemistry* **2020**, *68*, 105199.
10. Buksa, K. Extraction and characterization of rye grain starch and its susceptibility to resistant starch formation. *Carbohydr. Polym.* **2018**, *194*, 184–192. [CrossRef]
11. Ozturk, O.K.; Kaasgaard, S.G.; Palmén, L.G.; Vidal, B.C.; Hamaker, B.R. Enzyme treatments on corn fiber from wet-milling process for increased starch and protein extraction. *Ind. Crops Prod.* **2021**, *168*, 113622. [CrossRef]
12. Li, Z.; Wang, L.; Chen, Z.; Yu, Q.; Feng, W. Impact of binding interaction characteristics on physicochemical, structural, and rheological properties of waxy rice flour. *Food Chem.* **2018**, *266*, 551–556. [CrossRef]
13. Chang, Q.; Zheng, B.; Zhang, Y.; Zeng, H. A comprehensive review of the factors influencing the formation of retrograded starch. *Int. J. Biol. Macromol.* **2021**, *186*, 163–173. [CrossRef] [PubMed]
14. Zhang, P.; Wang, L.; Qian, Y.; Wang, X.; Zhang, S.; Chang, J.; Ruan, Y.; Ma, B. Influences of Extraction Methods on Physicochemical and Functional Characteristics of Three New Bulbil Starches from *Dioscorea opposita* Thunb. cv. Tiegun. *Molecules* **2019**, *24*, 2232. [CrossRef] [PubMed]
15. Wang, S.; Li, C.; Copeland, L.; Niu, Q.; Wang, S. Starch Retrogradation: A Comprehensive Review. *Compr. Rev. Food Sci. Food Saf.* **2015**, *14*, 568–585. [CrossRef]
16. Obadi, M.; Qi, Y.; Xu, B. Highland barley starch (Qingke): Structures, properties, modifications, and applications. *Int. J. Biol. Macromol.* **2021**, *185*, 725–738. [CrossRef]
17. Liu, Y.; Zhang, H.; Yi, C.; Quan, K.; Lin, B. Chemical composition, structure, physicochemical and functional properties of rice bran dietary fiber modified by cellulase treatment. *Food Chem.* **2021**, *342*, 128352. [CrossRef]
18. Bernardes, A.; Pellegrini, V.O.A.; Curtolo, F.; Camilo, C.M.; Mello, B.L.; Johns, M.A.; Scott, J.L.; Guimaraes, F.E.C.; Polikarpov, I. Carbohydrate binding modules enhance cellulose enzymatic hydrolysis by increasing access of cellulases to the substrate. *Carbohydr. Polym.* **2019**, *211*, 57–68. [CrossRef]
19. Li, M.; Daygon, V.D.; Solah, V.; Dhital, S. Starch granule size: Does it matter? *Crit. Rev. Food Sci. Nutr.* **2021**, 1–21. [CrossRef]
20. Punia, S. Barley starch: Structure, properties and in vitro digestibility—A review. *Int. J. Biol. Macromol.* **2020**, *155*, 868–875. [CrossRef]
21. Liu, S.; Xiao, Y.; Shen, M.; Zhang, X.; Wang, W.; Xie, J. Effect of sodium carbonate on the gelation, rheology, texture and structural properties of maize starch-Mesona chinensis polysaccharide gel. *Food Hydrocoll.* **2019**, *87*, 943–951. [CrossRef]
22. Ma, Y.-S.; Pan, Y.; Xie, Q.-T.; Li, X.-M.; Zhang, B.; Chen, H.-Q. Evaluation studies on effects of pectin with different concentrations on the pasting, rheological and digestibility properties of corn starch. *Food Chem.* **2019**, *274*, 319–323. [CrossRef] [PubMed]
23. Cao, H.; Sun, R.; Liu, Y.; Wang, X.; Guan, X.; Huang, K.; Zhang, Y. Appropriate microwave improved the texture properties of quinoa due to starch gelatinization from the destructed cyptomere structure. *Food Chem.-X* **2022**, *14*, 100347. [CrossRef] [PubMed]
24. Devraj, L.; Natarajan, V.; Ramachandran, S.V.; Manickam, L.; Sarvanan, S. Influence of microwave heating as accelerated aging on physicochemical, texture, pasting properties, and microstructure in brown rice of selected Indian rice varieties. *J. Texture Stud.* **2020**, *51*, 663–679. [CrossRef] [PubMed]
25. Hug-Iten, S.; Escher, F.; Conde-Petit, B. Structural properties of starch in bread and bread model systems: Influence of an antistaling alpha-amylase. *Cereal Chem.* **2001**, *78*, 421–428. [CrossRef]

26. Yu, Z.; Wang, Y.-S.; Chen, H.-H.; Li, Q.-Q.; Wang, Q. The gelatinization and retrogradation properties of wheat starch with the addition of stearic acid and sodium alginate. *Food Hydrocoll.* **2018**, *81*, 77–86. [CrossRef]
27. Xie, H.; Ying, R.; Huang, M. Effect of arabinoxylans with different molecular weights on the gelling properties of wheat starch. *Int. J. Biol. Macromol.* **2022**, *209*, 1676–1684. [CrossRef]
28. Tian, Y.; Li, Y.; Manthey, F.A.; Xu, X.; Jin, Z.; Deng, L. Influence of beta-cyclodextrin on the short-term retrogradation of rice starch. *Food Chem.* **2009**, *116*, 54–58. [CrossRef]
29. Dudu, O.E.; Oyedeji, A.B.; Oyeyinka, S.A.; Ma, Y. Impact of steam-heat-moisture treatment on structural and functional properties of cassava flour and starch. *Int. J. Biol. Macromol.* **2019**, *126*, 1056–1064. [CrossRef]
30. Wang, Y.; Ye, F.; Liu, J.; Zhou, Y.; Lei, L.; Zhao, G. Rheological nature and dropping performance of sweet potato starch dough as influenced by the binder pastes. *Food Hydrocoll.* **2018**, *85*, 39–50. [CrossRef]
31. Wang, N.; Wu, L.; Zhang, F.; Kan, J.; Zheng, J. Modifying the rheological properties, in vitro digestion, and structure of rice starch by extrusion assisted addition with bamboo shoot dietary fiber. *Food Chem.* **2022**, *375*, 131900. [CrossRef] [PubMed]
32. Lin, Y.S.; Yeh, A.I.; Li, C.Y. Correlation between starch retrogradation and water mobility as determined by differential scanning calorimetry (DSC) and nuclear magnetic resonance (NMR). *Cereal Chem.* **2001**, *78*, 647–653. [CrossRef]
33. Gong, B.; Cheng, L.; Gilbert, R.G.; Li, C. Distribution of short to medium amylose chains are major controllers of in vitro digestion of retrograded rice starch. *Food Hydrocoll.* **2019**, *96*, 634–643. [CrossRef]
34. Qin, W.; Lin, Z.; Wang, A.; Xiao, T.; He, Y.; Chen, Z.; Wang, L.; Liu, L.; Wang, F.; Tong, L.-T. Influence of damaged starch on the properties of rice flour and quality attributes of gluten-free rice bread. *J. Cereal Sci.* **2021**, *101*, 103296. [CrossRef]
35. Qin, W.; Xi, H.; Wang, A.; Gong, X.; Chen, Z.; He, Y.; Wang, L.; Liu, L.; Wang, F.; Tong, L. Ultrasound Treatment Enhanced Semidry-Milled Rice Flour Properties and Gluten-Free Rice Bread Quality. *Molecules* **2022**, *27*, 5403. [CrossRef]
36. Zhang, J.; Luo, D.; Xiang, J.; Xu, W.; Xu, B.; Li, P.; Huang, J. Structural Variations of Wheat Proteins under ultrasound treatment. *J. Cereal Sci.* **2021**, *99*, 103219. [CrossRef]
37. Zhou, D.; Ma, Z.; Yin, X.; Hu, X.; Boye, J.I. Structural characteristics and physicochemical properties of field pea starch modified by physical, enzymatic, and acid treatments. *Food Hydrocoll.* **2019**, *93*, 386–394. [CrossRef]
38. Qin, W.; Lin, Z.; Wang, A.; Chen, Z.; He, Y.; Wang, L.; Liu, L.; Wang, F.; Tong, L.-T. Influence of particle size on the properties of rice flour and quality of gluten-free rice bread. *LWT-Food Sci. Technol.* **2021**, *151*, 112236. [CrossRef]
39. Cheng, W.; Sun, Y.; Xia, X.; Yang, L.; Fan, M.; Li, Y.; Wang, L.; Qian, H. Effects of β-amylase treatment conditions on the gelatinization and retrogradation characteristics of wheat starch. *Food Hydrocoll.* **2022**, *124*, 107286. [CrossRef]
40. Iqbal, S.; Wu, P.; Kirk, T.V.; Chen, X.D. Amylose content modulates maize starch hydrolysis, rheology, and microstructure during simulated gastrointestinal digestion. *Food Hydrocoll.* **2021**, *110*, 106171. [CrossRef]

Article

Physicochemical Properties and In Vitro Digestibility of Starch from a Trace-Rutinosidase Variety of Tartary Buckwheat 'Manten-Kirari'

Takahiro Noda [1,*], Koji Ishiguro [1], Tatsuro Suzuki [2] and Toshikazu Morishita [1]

1 Hokkaido Agricultural Research Center, NARO, Shinsei, Memuro, Kasai-gun, Hokkaido 082-0081, Japan
2 Kyushu-Okinawa Agricultural Research Center, NARO, Suya, Koshi, Kumamoto 861-1192, Japan
* Correspondence: noda@affrc.go.jp

Abstract: We recently developed a novel Tartary buckwheat variety, 'Manten-Kirari', with trace-rutinosidase activity. The use of 'Manten-Kirari' enabled us to make rutin-rich food products with low bitterness. This study was intended to evaluate the physicochemical properties and in vitro digestibility of starch isolated from 'Manten-Kirari'. For comparison, the representative common buckwheat variety 'Kitawasesoba' and Tartary buckwheat variety 'Hokkai T8' in Japan were also used. The lowest content of amylose was found in 'Manten-Kirari' starch (18.1%) while the highest was in 'Kitawasesoba' starch (22.6%). 'Manten-Kirari' starch exhibited a larger median granule size (11.41 μm) and higher values of peak viscosity (286.8 RVU) and breakdown (115.2 RVU) than the others. The values of onset temperature for gelatinization were 60.5 °C for 'Kitawasesoba', 61.3 °C for 'Manten-Kirari', and 64.7 °C for 'Hokkai T8'. 'Manten-Kirari' and 'Hokkai T8' starches were digested more slowly than 'Kitawasesoba' starch. Our results will provide fundamental information concerning the expanded use of 'Manten-Kirari' in functional foods.

Keywords: Tartary buckwheat; rutinosidase; starch; physicochemical properties; digestibility

Citation: Noda, T.; Ishiguro, K.; Suzuki, T.; Morishita, T. Physicochemical Properties and In Vitro Digestibility of Starch from a Trace-Rutinosidase Variety of Tartary Buckwheat 'Manten-Kirari'. *Molecules* 2022, 27, 6172. https://doi.org/10.3390/molecules27196172

Academic Editor: Litao Tong

Received: 23 August 2022
Accepted: 15 September 2022
Published: 20 September 2022

Publisher's Note: MDPI stays neutral with regard to jurisdictional claims in published maps and institutional affiliations.

Copyright: © 2022 by the authors. Licensee MDPI, Basel, Switzerland. This article is an open access article distributed under the terms and conditions of the Creative Commons Attribution (CC BY) license (https://creativecommons.org/licenses/by/4.0/).

1. Introduction

Buckwheat belongs to the Polygonaceae family and is grown as a traditional crop around the world. It has starch-rich grains that are processed, utilized, and consumed like cereals, for example, rice, wheat, and maize. Two popular buckwheat species, common buckwheat (*Fagopyrum esculentum* Moench) and Tartary buckwheat (*Fagopyrum tataricum* Gaertn.), are used for food and feed purposes. In Japan, buckwheat food products are mainly buckwheat noodles, which are made from a mixture of buckwheat wheat and flours. Buckwheat has high nutritional values of protein, dietary fiber, phenolic compounds, and minerals in its grains. Rutin, the major phenolic compound of buckwheat, is well known for its health-promoting properties, such as antioxidative [1,2], anti-hypertensive [3], and α-glucosidase inhibitory activities [4]. Tartary buckwheat has received extensive attention for its remarkably high rutin content as compared to that of common buckwheat [2,5]. However, Tartary buckwheat seeds exhibit markedly high rutinosidase activity, which easily hydrolyzes rutin to quercetin and rutinose within just a few minutes of adding water to Tartary buckwheat flour [6,7] (Figure 1). Products made from Tartary buckwheat contain several bitter compounds, leading to the limited use of Tartary buckwheat. The enzymatic hydrolysis of rutin plays a large role in the bitterness because quercetin, a rutin hydrolysate, is a bitter compound [8]. Recently, our research group released a new Tartary buckwheat variety, 'Manten-Kirari', that has trace-rutinosidase activity [9,10]. Our previous studies demonstrated that rutin-rich food products with low bitterness could be produced using 'Manten-Kirari' flour [11–13]. Additionally, Nishimura et al. [14] reported that the intake of rutin-rich noodles containing 'Manten-Kirari' lowered body weight, BMI (body mass index), and TBARS (2-thiobarbituric acid reactive substances) levels.

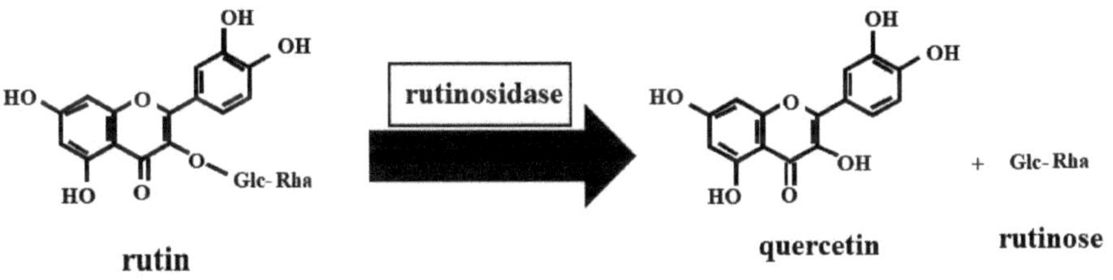

Figure 1. Rutin hydrolysis in Tartary buckwheat flour.

As starch is the major component of buckwheat grains, the starch properties have a great impact on the quality of buckwheat products. Comprehending the starch structure and functionality of buckwheat grains is critical for their suitable application in foods. The molecular structures and physicochemical properties of purified buckwheat starch have been studied extensively [15–21]. Review articles that deal with the present knowledge of buckwheat starch quality have been published in recent years [22,23]. However, no detailed studies of characteristics of starch from Tartary buckwheat grains with trace-rutinosidase activity have been carried out to date. Thus, in this study, the physicochemical properties and in vitro digestibility of starch from a trace-rutinosidase variety of Tartary buckwheat, 'Manten-Kirari', were evaluated and compared with those from a representative common buckwheat variety, 'Kitawasesoba', and a Tartary buckwheat variety, 'Hokkai T8', in Japan.

2. Results and Discussion

Table 1 presents data on the amylose content, median granule size, and color components of starches from two Tartary buckwheat varieties, 'Manten-Kirari' and 'Hokkai T8', and one common buckwheat variety, 'Kitawasesoba'. The amylose content of 'Manten-Kirari' was manifestly lower (18.1%) than that of 'Kitawasesoba' (22.6%) and was slightly lower than that of 'Hokkai T8' (18.9%). These values were lower than those of previous research on buckwheat starch [15,17,19–21]. This could be due to differences in variety, cultivation conditions, and/or methods for measuring amylose content. Generally, the amylose content affects the physicochemical parameters of a starch because amylose possesses the ability to form a firm gel. The median granule size differed significantly among buckwheat varieties. 'Manten-Kirari' starch had the largest median granule size (11.41 μm), whereas 'Kitawasesoba' starch had the smallest (8.14 μm). In our present study, the median granule sizes of buckwheat starches were somewhat larger than those reported by Zheng et al. [18] and Qian et al. [19], who revealed that the average granule size of common buckwheat starch is around 6 μm. Values of L* (lightness), a* (greenness–redness), and b* (blueness–yellowness) are given as evaluated by a color meter. All buckwheat starches examined exhibited higher L* (89.87–91.48) as well as low a* (0.70–1.18) and b* (1.78–3.79). Higher L* values (close to 100) represented strong white color, while a* and b* values were close to zero, which indicated a neutral color. Thus, it was suggested that all of the starches had satisfactorily high purity. 'Manten-Kirari' starch showed a lower L* value (89.87) than that of 'Hokkai T8' (91.32) and 'Kitawasesoba' (91.48) starches. Higher b* values were observed in 'Manten-Kirari' (3.79) and 'Hokkai T8' (3.29) starches than in 'Kitawasesoba' (1.78), indicating that two Tartary buckwheat starches had a slightly stronger yellow cast. Li et al. [15] studied the color parameters of common and Tartary buckwheat starches, and they observed that Tartary buckwheat starches were more yellow than common buckwheat starches, as shown by our data. This suggested that the yellowness could not be removed completely during the isolation of starch from Tartary buckwheat flour.

Table 1. Amylose content, median granule size, and color components (L*, a*, and b*) of buckwheat starches.

Varieties	Amylose Content (%)	Median Granule Size (μm)	Color Components		
			L*	a*	b*
Manten-Kirari	18.1 ± 0.2 b	11.41 ± 0.02 a	89.87 ± 0.09 b	1.18 ± 0.04 a	3.79 ± 0.11 a
Hokkai T8	18.9 ± 0.7 b	9.93 ± 0.01 b	91.32 ± 0.07 a	1.09 ± 0.02 a	3.29 ± 0.05 b
Kitawasesoba	22.6 ± 1.4 a	8.14 ± 0.00 c	91.48 ± 0.05 a	0.70 ± 0.00 b	1.78 ± 0.04 c

Values are means ± SD (n = 3). Means with similar letters in a column do not differ significantly ($p > 0.05$).

Rapid visco-analyzer (RVA), in which viscosity is monitored over time during a standard heating–cooling cycle under a constant rate of shear, is becoming popular for analyzing the pasting properties of starches. The pasting properties of buckwheat starches assessed by RVA with 8% starch suspension (dry weight basis, w/w) are given in Table 2. The peak viscosity, breakdown, setback, and pasting temperature differed significantly among the buckwheat starches examined. 'Manten-Kirari' starch displayed the highest values of peak viscosity (286.8 RVU) and breakdown (115.2 RVU), while 'Kitawasesoba' starch had the lowest (peak viscosity, 251.1 RVU; breakdown, 80.4 RVU). This implied that 'Manten-Kirari' starch had greater swelling during heating and lower resistance to shearing at a high temperature. The setback, the ability of starch granules to re-associate after heating and cooling, varied in the following order: 'Kitawasesoba' (128.4 RVU) > 'Manten-Kirari' (119.0 RVU) > 'Hokkai T8' (109.6 RVU). The lowest pasting temperature was observed in 'Manten-Kirari' (71.9 °C) starch, while the highest was in 'Hokkai T8' starch (73.8 °C). Starch-pasting properties have been shown to be influenced by the amylose content, as starch swelling is inhibited by amylose. Generally, higher values of peak viscosity and breakdown and reduced values of setback and pasting temperature are associated with lower amylose content [24,25]. Our present data on peak viscosity, breakdown, and pasting temperature supported this concept. However, the setback of 'Manten-Kirari' starch with the lowest amylose content was not the highest.

Table 2. RVA pasting properties of buckwheat starches.

Varieties	Peak Viscosity (RVU)	Breakdown (RVU)	Setback (RVU)	Pasting Temperature (°C)
Manten-Kirari	286.8 ± 1.1 a	115.2 ± 4.1 a	119.0 ± 0.4 ab	71.9 ± 0.0 b
Hokkai T8	274.9 ± 1.4 b	104.3 ± 1.2 b	128.4 ± 4.7 a	73.8 ± 0.5 a
Kitawasesoba	251.1 ± 1.4 c	80.4 ± 1.9 c	109.6 ± 2.3 b	72.6 ± 0.1 b

Values are means ± SD (n = 3). Means with similar letters in a column do not differ significantly ($p > 0.05$).

Differential scanning calorimetry (DSC), which determines thermal properties, has been used to analyze the gelatinization properties of starches. The gelatinization properties of buckwheat starches monitored by DSC with 30% starch suspension (dry weight basis, w/w) are presented in Table 3. 'Manten-Kirari' starch showed lower onset temperature (To) (61.3 °C) and peak temperature (Tp) (67.1 °C) for gelatinization than 'Hokkai T8' starch (To, 64.7 °C; Tp, 69.3 °C), whereas 'Kitawasesoba' starch exhibited lower To (60.5 °C) and Tp (66.4 °C) than 'Manten-Kirari' starch. No significant difference in enthalpy (ΔH) (10.6–11.1 J/g) for gelatinization was recognized among buckwheat starch samples examined. Our present results of To and Tp were comparable to the values reported previously [15,18–21]. Li et al. [15] and Qian et al. [19] obtained the results that buckwheat starches had ΔH of around 10 J/g, which were in good agreement with our present data. Zhang et al. [18] and Yoshimoto et al. [20] observed higher ΔH of 12.7 J/g and 14.5–15.0 J/g in buckwheat starches, respectively, while lower ΔH of 7.5–8.3 J/g was reported by Gao et al. [20]. Contrary to these results, we previously found great diversities in To (51.5–62.3 °C), Tp (57.2–66.7 °C), and ΔH (9.4–13.9 J/g) among starches from 17 samples of common buckwheat and 10 samples of Tartary buckwheat [17]. The differences in DSC gelatinization parameters between the present and previous data on buckwheat starches might be attributed to differences in variety, cultivation conditions, and/or experimental conditions for DSC analysis. The decrease in the amylose content, which indicates a high relative crystal concentration, usually causes higher ΔH [25]. In contrast, 'Manten-Kirari' starch with the lowest amylose content did not show higher ΔH.

Table 3. DSC gelatinization properties of buckwheat starches.

Varieties	To (°C)	Tp (°C)	ΔH (J/g)
Manten-Kirari	61.3 ± 0.2 b	67.1 ± 0.4 b	10.6 ± 0.3 a
Hokkai T8	64.7 ± 0.1 a	69.3 ± 0.1 a	11.0 ± 0.3 a
Kitawasesoba	60.5 ± 0.3 c	66.4 ± 0.2 b	11.1 ± 0.2 a

Values are means ± SD (n = 3). Means with similar letters in a column do not differ significantly ($p > 0.05$).

Starch resistant to amylolytic enzymes has been associated with a decrease in postprandial glycemic responses, resulting in reductions in obesity, diabetes, and cardiovascular disease. Previous studies found that buckwheat products had lower rates of starch hydrolysis as compared with those of wheat products [26,27]. Therefore, buckwheat shows promise as a material of functional food based on its low starch digestibility. However, little information is available on the in vitro digestibility of the purified buckwheat starch granules except for the report of Acquistucci and Fornal [16]. Thus, we performed in vitro enzymatic digestion of the purified buckwheat starch granules by porcine pancreatic amylase and amyloglucosidase at different time points (20, 60, 120, and 240 min), and the results are shown in Figure 2. A rapid enhancement of the hydrolysis percentage of three buckwheat starches was found during the first hydrolysis time of 20 min. The hydrolysis percentage of three buckwheat starches increased gradually with hydrolysis times of 20–240 min. Namely, hydrolysis percentages at 20, 60, 120, and 240 min were observed to be in the ranges of 25.3–27.2%, 49.1–55.7%, 76.5–81.4%, and 95.5–98.3%, respectively. 'Manten-Kirari' and 'Hokkai T8' starches exhibited lower hydrolysis patterns than that of 'Kitawasesoba' starch. Until a hydrolysis time of 120 min, 'Manten-Kirari' starch showed a hydrolysis pattern similar to that of 'Hokkai T8' starch. However, for the hydrolysis time of 240 min, the hydrolysis percentage of 'Manten-Kirari' starch was somewhat lower (95.5%) than that of 'Hokkai T8' starch (97.1%). Assumedly, 'Manten-Kirari' containing starch slightly resistant to amylolytic enzymes at a late hydrolysis stage could be beneficial for health. It is well known that granule size and amylose content can affect the enzymatic digestibility of native starch granules. Smaller starch granules have been reported to digest faster than large granules [28–31]. Starch granules with lower amylose content have been shown to have higher enzymatic digestibility [16,25,32–34]. Thus, the higher digestibility of 'Kitawasesoba' starch could be related to its smaller granule size. However, contradictory data—that 'Manten-Kirari' starch with a lower amylose content did not display higher digestibility—were obtained in this study. Our present results of in vitro enzymatic digestion suggested that granule size had a stronger effect on starch digestibility than did the amylose content.

Most domestic buckwheat in Japan is marketed as flour and is used to manufacture noodles called soba. Previously, we have developed rutin-rich noodles from a novel Tartary buckwheat variety, 'Manten-Kirari', with trace-rutinosidase activity [12,13]. Starch properties appear to be the most critical determinant of the quality of buckwheat noodles, as starch is the main component. The present study has focused on the characteristics of starch prepared from 'Manten-Kirari' for its expanded use in rutin-rich foods, especially noodles. It is confirmed that 'Manten-Kirari' starch has lower amylose content, larger granule size, higher peak viscosity, and relatively lower enzymatic digestibility as compared with other varieties. Reduced amylose content [35–37] and increased peak viscosity [36,38] of starch are beneficial to the quality of white salted noodles made from wheat flour. 'Manten-Kirari', containing starch with low amylose content and high peak viscosity, would be desirable for making rutin-rich buckwheat noodles with good texture. As the relationship of starch properties with the quality of buckwheat noodles remains unknown, further investigation would be needed.

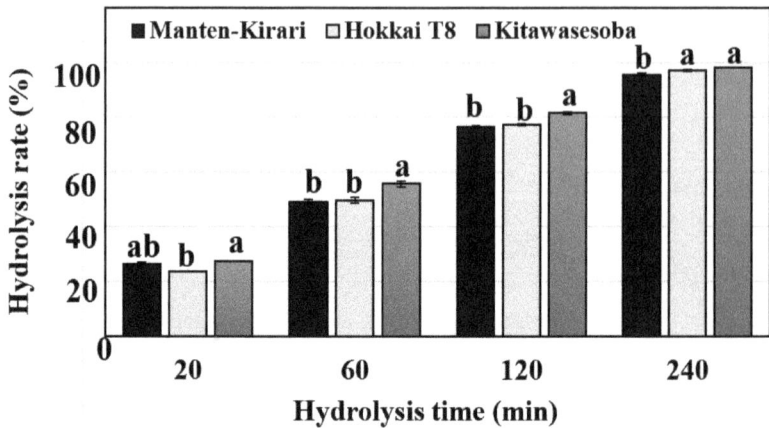

Figure 2. Hydrolysis rate of buckwheat starches by amylases. Values are means (n = 3). Error bars indicate SD. For each hydrolysis time, values with similar letters do not differ significantly ($p > 0.05$).

3. Materials and Methods

3.1. Materials

Tartary buckwheat flours from the varieties 'Manten-Kirari' and 'Hokkai T8' were purchased from Kobayashi Shokuhin Co., Ltd., Okoppe, Hokkaido, Japan. The common buckwheat variety 'Kitawasesoba' was grown in an experimental field at the Hokkaido Agricultural Research Center, NARO, Memuro, Hokkaido, Japan. Common buckwheat flour from 'Kitawasesoba' was obtained by milling the grains. Starch was isolated from each buckwheat flour using 0.2% sodium hydroxide as described previously [17].

3.2. Physicochemical Properties of Starch

The amylose content was estimated based on the blue value at 680 nm in accordance with the method previously described [17]. The median granule size was determined using a Sympatec HELOS Particle Size Analyzer by a previously reported method [39]. Color value analysis was carried out using a color meter as described previously [40]. Pasting properties and gelatinization properties were measured using RVA and DSC, respectively, as described earlier [41], except that RVA measurement was performed with 8% starch suspension (dry weight basis, w/w), not with 4%.

3.3. In Vitro Digestibility of Starch

In vitro digestibility of starch was determined by the Megazyme Resistant Starch Assay Kit, 05/2008 (Megazyme International Ireland Ltd., Co. County Wicklow, Ireland) in accordance with AOAC method 2002.02 [42]. After digestion at given time points (20, 60, 120, and 240 min), the percentage of starch digestion was calculated by measuring the content of glucose in the digestion solution.

3.4. Statistical Analysis

The determinations of all starch quality parameters were conducted in triplicate. Duncan t-tests were computed to measure variations in the average of all starch quality parameters of each variety.

4. Conclusions

The physicochemical properties and in vitro digestibility of starches isolated from a novel Tartary buckwheat variety, 'Manten-Kirari', with trace-rutinosidase activity, as well as the common buckwheat variety, 'Kitawasesoba', and a Tartary buckwheat variety, 'Hokkai T8', were determined. 'Manten-Kirari' starch exhibited the lowest amylose content, largest median granule size, and highest peak viscosity. The hydrolysis rates of 'Manten-Kirari' and 'Hokkai T8' starches were lower than that of 'Kitawasesoba' starch. Some properties, such as lower amylose content and higher peak viscosity, of 'Manten-Kirari' starch may be advantageous for making buckwheat noodles with good texture. Moreover, the beneficial health effects of 'Manten-Kirari'-containing foods derived from the content of starch fraction resistant to amylolytic enzymes as well as the rutin content can be expected.

Author Contributions: Conceptualization, T.N., T.S. and T.M.; investigation, T.N.; writing-original draft preparation, T.N.; writing—review and editing, K.I., T.S. and T.M.; funding acquisition, T.N., K.I., T.S. and T.M. All authors have read and agreed to the published version of the manuscript.

Funding: This work was partially supported by a grant from the Research Project on Development of Agricultural Products and Foods with Health-promoting benefits (NARO), Japan.

Institutional Review Board Statement: Not applicable.

Informed Consent Statement: Not applicable.

Data Availability Statement: Not applicable.

Acknowledgments: We thank K. Asano and Y. Terasawa, Hokkaido Agricultural Research Center, NARO, for the use of Sympatec HELOS Particle Size Analyzer. We also thank M. Saito for their technical assistance.

Conflicts of Interest: All author has no conflict of interest.

References

1. Jiang, P.; Burczynski, F.; Campbell, C.; Pierce, G.; Austria, J.A.; Briggs, C.J. Rutin and flavonoid contents in three buckwheat species *Fagopyrum esculentum*, *F. tataricum*, and *F. homotropicum* and their protective effects against lipid peroxidation. *Food Res. Int.* **2007**, *40*, 356–364. [CrossRef]
2. Morishita, T.; Yamaguchi, H.; Degi, K. The contribution of polyphenols to antioxidative activity in common and Tartary buckwheat grain. *Plant Prod. Sci.* **2007**, *10*, 99–104. [CrossRef]
3. Matsubara, Y.; Kumamoto, H.; Iizuka, Y.; Murakami, T.; Okamoto, K.; Miyake, H.; Yokoi, K. Structure and hypotensive effect of flavonoid glycosides in *Citrus unshiu* peelings. *Agric. Biol. Chem.* **1985**, *49*, 909–914.
4. Li, Y.Q.; Zhou, F.C.; Gao, F.; Bian, J.S.; Shan, F. Comparative evaluation of quercetin, isoquercetin and rutin as inhibitors of α-glucosidase. *J. Agric. Food Chem.* **2009**, *57*, 11463–11468. [CrossRef]
5. Fabjan, T.; Rode, J.; Kosir, I.J.; Wang, Z.; Zhang, Z.; Kreft, I. Tartary buckwheat (*Fagopyrum tataricum* Gaertn.) as a source of dietary rutin and quercetin. *J. Agric. Food Chem.* **2003**, *51*, 6452–6455. [CrossRef]
6. Yasuda, T.; Nakagawa, H. Purification and characterization of rutin-degrading enzymes in Tartary buckwheat seeds. *Phytochem* **1994**, *37*, 133–136. [CrossRef]
7. Suzuki, T.; Honda, Y.; Funatsuki, W.; Nakatsuka, K. Purification and characterization of flavonol 3-glucosidase, and its activity during ripening in Tartary buckwheat seeds. *Plant Sci.* **2002**, *163*, 417–423. [CrossRef]
8. Kawakami, A.; Kayahara, H.; Ujihara, A. Properties and elimination of bitter components derived from Tartary buckwheat (*Fagopyrum tataricum*) flour. *J. Jpn. Soc. Food Sci.* **1995**, *11*, 892–898. [CrossRef]
9. Suzuki, T.; Morishita, T.; Mukasa, Y.; Takigawa, S.; Yokota, S.; Ishiguro, K.; Noda, T. Discovery and genetic analysis of non-bitter Tartary buckwheat (*Fagopyrum tataricum* Gaertn.) with trace-rutinosidase activity. *Breed. Sci.* **2014**, *64*, 339–343. [CrossRef]
10. Suzuki, T.; Morishita, T.; Mukasa, Y.; Takigawa, S.; Yokota, S.; Ishiguro, K.; Noda, T. Breeding of 'Mantenkirari', a non-bitter and trace-rutinosidase variety of Tartary buckwheat (*Fagopyrum tataricum* Gaertn.). *Breed. Sci.* **2014**, *64*, 344–350. [CrossRef]
11. Suzuki, T.; Morishita, T.; Takigawa, S.; Noda, T.; Ishiguro, K. Characterization of rutin-rich bread made with 'Mantenkirari', a trace-rutinosidase variety of Tartary buckwheat (*Fagopyrum tataricum* Gaertn.). *Food Sci. Technol. Res.* **2015**, *21*, 733–738. [CrossRef]
12. Ishiguro, K.; Morishita, T.; Ashizawa, J.; Suzuki, T.; Noda, T. Antioxidative activities in rutin-rich noodles and cookies made with a trace-rutinosidase variety of Tartary buckwheat (*Fagopyrum tataricum* Gaertn.), 'Mantenkirari'. *Food Sci. Technol. Res.* **2016**, *22*, 557–562. [CrossRef]
13. Suzuki, T.; Morishita, T.; Takigawa, S.; Noda, T.; Ishiguro, K. Development of rutin-rich noodles using trace-rutinosidase variety of Tartary buckwheat (*Fagopyrum tataricum* Gaertn.) 'Mantenkirari'. *Food Sci. Technol. Res.* **2020**, *25*, 915–920. [CrossRef]

14. Nishimura, M.; Ohkawara, T.; Sato, Y.; Satoh, H.; Suzuki, T.; Ishiguro, K.; Noda, T.; Morishita, T.; Nishihira, J. Effectiveness of rutin-rich Tartary buckwheat (*Fagopyrum tataricum* Gaertn.) 'Manten-Kirari' in body weight reduction related to its antioxidant properties: A randomised, double-blind, placebo-controlled study. *J. Funct. Foods* **2016**, *26*, 460–469. [CrossRef]
15. Li, W.; Lin, F.; Corke, H. Physicochemical properties of common and tartary buckwheat starch. *Cereal Chem.* **1997**, *74*, 79–82. [CrossRef]
16. Acquistucci, R.; Fornal, J. Italian buckwheat (*Fagopyrum esculentum*) starch: Physicochemical and functional characterization and in vitro digestibility. *Nahrung* **1997**, *5*, 281–284. [CrossRef]
17. Noda, T.; Takahata, Y.; Sato, T.; Suda, I.; Morishita, T.; Ishiguro, K.; Yamakawa, O. Relationships between chain length distribution of amylopectin and gelatinization properties within the same botanical origin for sweet potato and buckwheat. *Carbohydr. Polym.* **1998**, *37*, 153–158. [CrossRef]
18. Zheng, G.H.; Sosulski, F.W.; Tyler, R.T. Wet-milling, composition and functional properties of starch and protein isolated from buckwheat groats. *Food Res. Int.* **1998**, *30*, 493–502. [CrossRef]
19. Qian, J.; Rayas-Duarte, P.; Grant, L. Partial characterization of buckwheat (*Fagopyrum esculentum*) starch. *Cereal Chem.* **1998**, *75*, 365–373. [CrossRef]
20. Yoshimoto, Y.; Egashira, T.; Hanashiro, I.; Ohinata, H.; Takase, Y.; Takeda, Y. Molecular structure and some physicochemical properties of buckwheat starches. *Cereal Chem.* **2004**, *81*, 515–520. [CrossRef]
21. Gao, J.; Kreft, I.; Chao, G.; Wang, Y.; Liu, X.; Wang, L.; Wang, P.; Gao, X.; Feng, B. Tartary buckwheat (*Fagopyrum tataricum* Gaertn.) starch, a side product in functional food production, as a potential source of retrograded starch. *Food Chem.* **2016**, *190*, 552–558. [CrossRef]
22. Zhu, F. Buckwheat starch: Structures, properties, and applications. *Trends Food Sci. Technol.* **2016**, *49*, 121–135. [CrossRef]
23. Suzuki, T.; Noda, T.; Morishita, T.; Ishiguro, K.; Otsuka, S.; Brunori, A. Present status and future perspectives of breeding for buckwheat quality. *Breed. Sci.* **2020**, *70*, 48–66. [CrossRef]
24. Zeng, M.; Morris, C.F.; Batey, I.L.; Wrigley, C.W. Sources of variation for starch gelatinization, pasting, and gelation properties in wheat. *Cereal Chem.* **1997**, *74*, 63–71. [CrossRef]
25. Mangalika, W.H.A.; Miura, H.; Yamauchi, H.; Noda, T. Properties of starches from near-isogenic wheat lines with different Wx protein deficiencies. *Cereal Chem.* **2003**, *80*, 662–666. [CrossRef]
26. Kreft, I.; Skrabanja, V. Nutritional properties of starch in buckwheat noodles. *J. Nutr. Sci. Vitaminol.* **2002**, *48*, 47–50. [CrossRef]
27. Wolter, A.; Hager, A.S.; Zannini, E.; Arendt, E.K. In vitro starch digestibility and predicted glycaemic indexes of buckwheat, oat, quinoa, sorghum, teff and commercial gluten-free bread. *J. Cereal Sci.* **2013**, *58*, 431–436. [CrossRef]
28. MacGregor, A.W.; Balance, D.L. Hydrolysis of large and small starch granules from normal and waxy barley cultivars by alpha-amylases from barley malt. *Cereal Chem.* **1980**, *57*, 397–402.
29. Kang, M.Y.; Sugimoto, Y.; Kato, I.; Sakamoto, S.; Fuwa, H. Some properties of large and small starch granules of barley (*Hordeum vulgare* L.) endosperm. *Agric. Biol. Chem.* **1985**, *49*, 1291–1297. [CrossRef]
30. Noda, T.; Takigawa, S.; Matsuura-Endo, C.; Kim, S.J.; Hashimoto, N.; Yamauchi, H.; Hanashiro, I.; Takeda, Y. Physicochemical properties and amylopectin structures of large, small and extremely small potato starch granules. *Carbohydr. Polym.* **2005**, *60*, 245–251. [CrossRef]
31. Noda, T.; Takigawa, S.; Matsuura-Endo, C.; Suzuki, T.; Hashimoto, N.; Kottearachchi, N.S.; Yamauchi, H.; Zaidul, I.S.M. Factors affecting the digestibility of raw and gelatinized potato starches. *Food Chem.* **2008**, *68*, 793–796. [CrossRef]
32. Evers, A.D.; Juliano, B.O. Varietal differences in surface ultrastructure of endosperm cells and starch granules of rice. *Starch/Staerke* **1976**, *28*, 160–166. [CrossRef]
33. Noda, T.; Kimura, T.; Otani, M.; Ideta, O.; Shimada, T.; Saito, A.; Suda, I. Physicochemical properties of amylose-free starch from transgenic sweet potato. *Carbohydr. Polym.* **2002**, *49*, 253–260. [CrossRef]
34. Noda, T.; Nishiba, Y.; Sato, T.; Suda, I. Properties of starches from several low-amylose rice cultivars. *Cereal Chem.* **2003**, *80*, 193–197. [CrossRef]
35. Oda, M.; Yasuda, Y.; Okazaki, S.; Yamaguchi, Y.; Yokoyama, Y. A method of flour quality assessment for Japanese noodles. *Cereal Chem.* **1980**, *57*, 253–254.
36. Noda, T.; Tohnooka, T.; Taya, S.; Suda, I. Relationships between physicochemical properties of starches and white salted noodle quality in Japanese wheat flours. *Cereal Chem.* **2001**, *78*, 395–399. [CrossRef]
37. Baik, B.K.; Lee, M.R. Effects of starch amylose content of wheat on textural properties of white salted noodles. *Cereal Chem.* **2003**, *80*, 304–309. [CrossRef]
38. Crosbie, G.B. The relationship between starch swelling properties, paste viscosity and boiled noodle quality in wheat flours. *J. Cereal Sci.* **1991**, *13*, 145–150. [CrossRef]
39. Noda, T.; Tsuda, S.; Mori, M.; Takigawa, S.; Matsuura-Endo, C.; Kim, S.J.; Hashimoto, N.; Yamauchi, H. Determination of the phosphorus content in potato starch using an energy-dispersive X-ray fluorescence method. *Food Chem.* **2006**, *95*, 632–637. [CrossRef]
40. Noda, T.; Matsuura-Endo, C.; Ishiguro, K. Preparation of iron-fortified potato starch and its properties. *J. Food Sci. Technol.* **2018**, *55*, 1360–1365. [CrossRef]

41. Noda, T.; Tsuda, S.; Mori, M.; Takigawa, S.; Matsuura-Endo, C.; Saito, K.; Mangalika, W.H.A.; Hanaoka, A.; Suzuki, Y.; Yamauchi, H. The effect of harvest date on starch properties in various potato cultivars. *Food Chem.* **2004**, *86*, 119–125. [CrossRef]
42. McCleary, B.V.; Monaghan, D.A. Measurement of resistant starch. *J. Assoc. Off. Anal. Chem.* **2002**, *86*, 665–675. [CrossRef]

Article

Reactive Extrusion as a Pretreatment in Cassava (*Manihot esculenta* Crantz) and Pea (*Pisum sativum* L.) Starches to Improve Spinnability Properties for Obtaining Fibers

David Tochihuitl-Vázquez *, Rafael Ramírez-Bon, José Martín Yáñez-Limón and Fernando Martínez-Bustos *

Centro de Investigación y de Estudios Avanzados del Instituto Politécnico Nacional (CINVESTAV-Unidad Querétaro), Libramiento Norponiente 2000, Fraccionamiento Real de Juriquilla, Querétaro 76230, Mexico
* Correspondence: david.tochi@cinvestav.mx (D.T.-V.); fmartinez@cinvestav.mx (F.M.-B.)

Abstract: Starch is a biocompatible and economical biopolymer in which interest has been shown in obtaining electrospun fibers. This research reports that cassava (CEX) and pea (PEX) starches pretreated by means of reactive extrusion (REX) improved the starches rheological properties and the availability of amylose to obtain fibers. Solutions of CEX and PEX (30–36% w/v) in 38% v/v formic acid were prepared and the rheological properties and electrospinability were studied. The rheological values indicated that to obtain continuous fibers without beads, the entanglement concentration (Ce) must be 1.20 and 1.25 times the concentration of CEX and PEX, respectively. In CEX, a higher amylose content and lower viscosity were obtained than in PEX, which resulted in a greater range of concentrations (32–36% w/v) to obtain continuous fibers without beads with average diameters ranging from 316 ± 65 nm to 394 ± 102 nm. In PEX, continuous fibers without beads were obtained only at 34% w/v with an average diameter of 170 ± 49 nm. This study showed that starches (20–35% amylose) pretreated through REX exhibited electrospinning properties to obtain fibers, opening the opportunity to expand their use in food, environmental, biosensor, and biomedical applications, as vehicles for the administration of bioactive compounds.

Keywords: electrospinning; fibers; pea starch; cassava starch; reactive extrusion; entanglement concentration

Citation: Tochihuitl-Vázquez, D.; Ramírez-Bon, R.; Yáñez-Limón, J.M.; Martínez-Bustos, F. Reactive Extrusion as a Pretreatment in Cassava (*Manihot esculenta* Crantz) and Pea (*Pisum sativum* L.) Starches to Improve Spinnability Properties for Obtaining Fibers. *Molecules* **2022**, *27*, 5944. https://doi.org/10.3390/molecules27185944

Academic Editors: Litao Tong and Lili Wang

Received: 2 August 2022
Accepted: 27 August 2022
Published: 13 September 2022

Publisher's Note: MDPI stays neutral with regard to jurisdictional claims in published maps and institutional affiliations.

Copyright: © 2022 by the authors. Licensee MDPI, Basel, Switzerland. This article is an open access article distributed under the terms and conditions of the Creative Commons Attribution (CC BY) license (https://creativecommons.org/licenses/by/4.0/).

1. Introduction

Electrospinning is a versatile and cost-effective electrodynamic technique, capable of producing micro- to nanoscale fibers with structural and functional properties such as high porosity, high surface area/volume ratio, and encapsulation efficiency that can be obtained from biopolymers, synthetic polymers, or a combination of both [1]. Due to their renewability, biodegradability, biocompatibility, and low cost, electrospun fibers from biopolymers have aroused interest in applications that include the food [2], environment [3], biosensors [4], and biomedical [5] sectors as supply-systems bioactive compounds.

Starch is one of the most abundant biopolymers in nature. It consists of glucose molecules linked by glycosidic bonds, whose classification is based on their content of amylose and amylopectin, in waxy starches (>90% amylopectin), high in amylose (>40% amylose), and in normal starches (20–35% amylose) [6]. It can be obtained from various vegetable sources such as corn, wheat, potato, rice, pea, and cassava [7,8].

Compared with potato, wheat, and corn starch, cassava and pea starches are in high demand due to their low cost and unique functional characteristics. The production of cassava (*Manihot esculenta* Crantz) for starch is increasing rapidly (by over 3% per year) and accounts for about 7% of the starch produced globally. Cassava roots are rich in starch (approximately 84.5%) and have an amylose content that ranges between 15.2% and 26.5%. The increasing demand is due to its special characteristics, such as its being allergy-free, its having a blunt taste, clarity, and freeze–thaw stability, which are very attractive to the food and non-food industries [9].

Pea (*Pisum sativum* L.) is one of the most important food crops in agricultural grain production, and around 25 million hectares of peas are planted each year worldwide. Pea starch, a by-product of pea protein extraction, accounts for about 40% of the dry weight of the seeds and is considered an inexpensive source of starch with an amylose content that ranges between 29.4% and 65.0%. In addition, it is often used in food formulations due to its slow digestion and high content of resistant starch [10,11].

Several researchers have reported that obtaining fibers from starch via electrospinning depends on the amylose content and the rheological properties of the solution, which facilitate the interactions among the chains to achieve the required level of molecular entanglement.

Consequently, the production of electrospun fibers from starch has focused mainly on high amylose starches and the use of high concentrations of strong polar solvents or mixed solvents (dimethyl sulfoxide or formic acid) to dissolve the starch. For example, Fonseca et al. [12] obtained electrospun fibers from corn starch with an amylose content of 70% in formic acid (75% v/v) during long aging times (24, 48, and 72 h) to dissolve the starch. Lancuški et al. [13], reported that solutions of Hylon starch (70% amylose) in 100% and 90% v/v formic acid formed uniform fibers. Similarly, Kong and Ziegler [14] electrospun high amylose corn starch (80% w/w) in dimethyl sulfoxide (95% and 100% v/v).

Unfortunately, the high cost of high amylose starches and the use of high concentrations of strong polar solvents limit the production of fibers by electrospinning. To obtain electrospun starch fibers with economic viability on a commercial scale, approaches are needed that allow normal starch to be processed and to overcome the limitations that it entertains in its native form. For example, we find it has low solubility in water and high viscosity in aqueous solvents, due to the strong hydrogen bonds between the starch chains, a high degree of polymerization, and complex semi-crystalline structures that generate an effective volume fraction much higher than its actual volume fraction [15].

One strategy to overcome these drawbacks and improve the physicochemical properties of starch is through physical, chemical, and enzymatic treatments. Physical treatments are more widely accepted as an ecological, profitable, and highly efficient alternative [16], and among the physical treatments REX is being increasingly adopted due to its being a process with a low operating cost. In addition, its adoption may encompass the possibility of a continuous process that implies high shear and rapid heat transfer and efficient mixing, with significant changes in the properties of starch, such as partial gelatinization, fusion, fragmentation, loss of crystallinity, and a decrease in molecular order [17].

Researchers [18,19] have reported that starch under REX conditions exhibits a decrease in molecular order, allowing the improvement of the starch's end-use properties by breaking the hydrogen bonds between straight-chain amylose and branched-chain amylopectin, with amylopectin more susceptible to changes in size distribution. For example, Lai and Kokini [20] found that starch granules after extrusion were fragmented by high shear and that they exhibited a decrease in viscosity and an increase in solubility in solution.

Fasheun et al. [21] applied REX to a mixture of cassava starch with sugarcane bagasse and reported that the physicochemical properties, such as the solubility index and swelling power, improved.

Therefore, in this study, our objective was to apply REX in cassava (NCS) and pea (NPS) starches to improve the physicochemical and rheological properties for obtaining fibers by the electrospinning technique. Electrospun fibers were successfully manufactured from CEX and PEX starches and electrospun fibers were characterized.

Additionally, the spinnability and rheological properties of starch solutions were studied. Among the rheological properties, the semi-dilute unentangled rate, the semi-dilute entangled rate, and the C_e were identified for obtaining electrospun fibers. These results provide the opportunity to expand the use of normal starches in obtaining electrospun fibers as a vehicle for bioactive molecules. To our knowledge, there are no reports in the literature that use REX as a pretreatment in starches for the manufacture of fibers by means of the electrospinning technique.

2. Results and Discussion

2.1. Morphological Characterization and Particle-Size Distribution of Starches

To observe the effect of REX on starches, the microstructure of CEX and PEX was analyzed by scanning electron microscopy (SEM) and was compared with NCS and NPS. Figure 1a shows that the NCS granule is intact and has round ellipsoidal-like morphology, with a truncated outer surface with a mean size of 12.7 µm. Similar results were reported for tapioca starch [22]. Figure 1b reveals that the NPS granule presented elliptical and spherical morphology with a smooth external surface and no cracks, with an average size of 20.3 µm. These results are in agreement with the information reported by Zhang et al. [23].

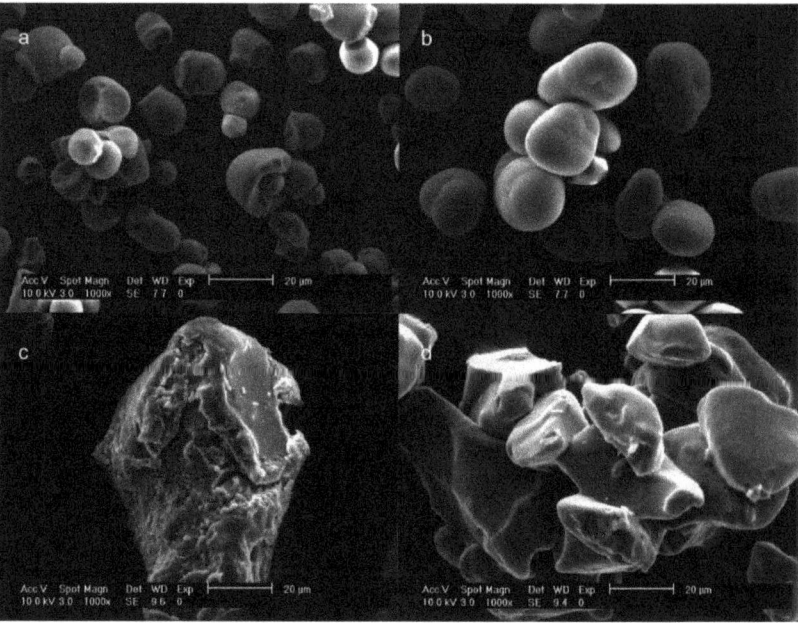

Figure 1. SEM micrographs of native and extruded starches: (**a**) native cassava starch (NCS); (**b**) native pea starch (NPS); (**c**) extruded cassava starch (CEX); and (**d**) extruded pea starch (PEX).

The CEX and PEX starches (Figure 1c,d) presented damaged granular structures, with irregular fractures associated with the fusion of the crystallites due to the high temperature and the shear force applied during the REX process; these could favor rapid diffusion of the solvent into starches during the preparation of solutions.

Figure 2 reveals that CEX presented a higher degradation rate than PEX. This is attributed to the reduced particle-size distribution (PSD) at the maximal peak and bimodal-size distribution with values of 183.2 nm and 3438 nm, due to the granular fractionation of the starch attributed to its size and to the damage of amylopectin molecules due in turn to its high molecular weight; meanwhile, PEX exhibited broad distribution in the maximal peak and in three size-distribution peaks (4036 nm, 1124 nm, and 123.1 nm), indicating minor degradation and a change in size from larger to smaller. However, PEX demonstrated a PSD with a maximal peak of 123.1 nm, smaller than that of CEX (183.1 nm). This can be attributed to the cleavage of the low-molecular-weight amylopectin backbone of PEX comprising a small number of long branches because of starch biosynthesis in the amylose/amylopectin ratio [24].

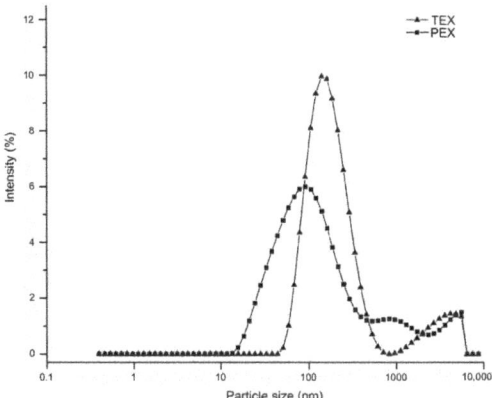

Figure 2. Particle-size distribution (by intensity %) of the extruded starches: extruded cassava starch, (CEX) and extruded pea starch (PEX).

2.2. Physicochemical Properties of Starches

The values of the apparent amylose content (AAC), solubility (S), and swelling power (SP) of the NCS, NPS, CEX, and PEX starches are presented in Table 1. The AAC values obtained in the NCS and NPS starches were 21.41 ± 0.06% and 32.24 ± 0.12%, respectively, indicating that they belong to the classification of normal starches. Similar values of amylose content have been reported in cassava [25] and pea starches [10]. The CEX and PEX starches demonstrated significantly higher values ($p < 0.05$) of ACC, S, and SP than the NPS and NCS starches. This is due to the partial gelatinization of the starch that led to the release of amylose and changes in the structure of the starch's amylopectin during the REX process [26,27]. A similar effect was found in rice and waxy rice flours after the extrusion process [28].

Table 1. Apparent amylose content (AAC) and properties of native and extruded starches [1].

Starches [1]	AAC (%)	S (%)	SP (g gel/g Starch)
NPS	32.24 ± 0.12 d	15.02 ± 0.10 d	16.90 ± 0.13 d
NCS	21.41 ± 0.06 c	22.56 ± 0.21 b	22.95 ± 0.16 b
PEX	43.36 ± 0.13 b	21.30 ± 0.13 c	21.77 ± 0.07 c
CEX	49.02 ± 0.13 a	48.12 ± 0.04 a	37.41 ± 0.29 a

[1] S, solubility; SP, swelling power; NPS, native pea starch; NCS, native cassava starch; PEX, pea extruded starch, and CEX, cassava extruded starch. Assays were performed in triplicate, and results are expressed as mean ± standard deviation (SD). Different lowercase letters in the same column are significantly different ($p < 0.05$) using Tukey test.

In this study, the AAC values obtained in CEX (49.02 ± 0.13) and PEX (43.36 ± 0.13) permitted sufficient molecular entanglement in electrospinning solutions to obtain fibers with continuous morphology and without beads. Other investigators confirmed that high amylose content in starches is essential for fiber formation due to their ease of molecular entanglement in the electrospinning solution [29]. For example, Fonseca et al. [12] obtained fibers with different morphologies from Hylon starches with amylose contents of 70% and 55%, using aqueous formic acid (75% v/v) as a solvent to induce chemical gelatinization during an aging time of 24, 48, and 72 h.

The starches CEX and PEX revealed that when they were pretreated by REX, they improved their S and SP due to the breaking of hydrogen bonds and covalent bonds between molecules [16], which facilitated the diffusion of the aqueous solvent toward the starches in electrospinning solutions and the use of a higher proportion of water in the solvent system (water/formic acid).

The values of AAC, S, and SP were significantly higher ($p < 0.05$) in CEX than in PEX.

This can be explained as due to the molecular size of the amylopectin present in cassava, which is more susceptible to REX conditions [18,19], therefore allowing for the improvement of the rheological properties in the electrospinning solution and the amylose content necessary for the chains to interact with each other.

On the other hand, the values of S and SP were lower in NPS than in NCS. This is attributed to the characteristic of low solubility and to the higher content of amylose present in pea starches, which restricts the swelling of starch granules during gelatinization [30].

2.3. Rheological Properties and Electrospinability of Starches

Rheological properties such as viscosity and molecular entanglements in polymer solutions exert a great influence on electrospinning [14]. For example, low viscosity leads to the formation of beads or droplets during electrospinning; otherwise, the flow will cause a blockage in the capillary [27], while sufficient molecular entanglement is necessary to stabilize the long-range network and facilitate fiber formation from the ejected polymer solutions [31]. The flow curves of the CEX and PEX solutions at concentrations ranging from 0.1–38% (w/v) are depicted in Figure 3a,b. It was observed that the viscosity of the solutions increased as the concentration of CEX and PEX increased; this was more notable in PEX due to its granulometric heterogeneity, generated by the lesser amount of damage undergone during REX that may contribute to a greater hydrodynamic volume and therefore high resistance to flow. When the concentrations of CEX and PEX were low they exhibited a trend toward Newtonian behavior, where the concentration of the starches increased by more than 24% and shear thinning was shown, resulting in pseudoplastic behavior that was more evident in PEX.

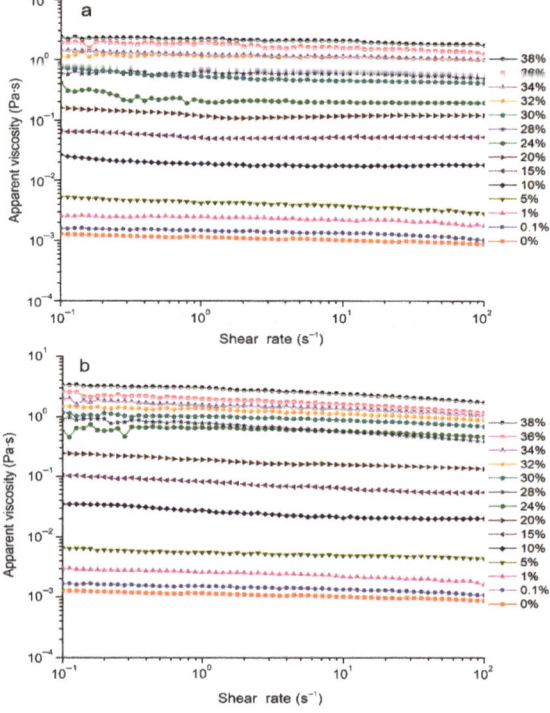

Figure 3. Flow curves of extruded starch in aqueous formic acid (38% v/v) as a function of concentration (w/v) at 25 °C: (**a**) extruded cassava starch (CEX) and (**b**) extruded pea starch (PEX).

Figure 4a,b show the adjusted slopes of the semi-dilute unentangled rate, the semi-dilute entangled rate, and the Ce for the CEX and PEX solutions. Theoretical predictions suggest that in the dilute regime n is 1.0, while in the semi-dilute unentangled regime n is 1.25 when the C of the polymer is greater than C* and less than Ce. For the semi-dilute entangled regime n is 4.8, that is, the C* is less than the C of the polymer for linear and neutral polymer solutions.

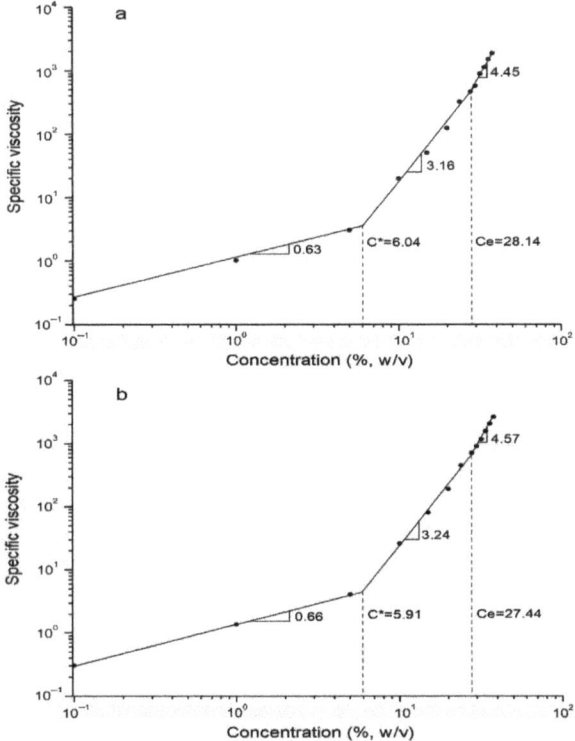

Figure 4. Plot of specific viscosity vs. starch concentration (0.1–38% w/v) in aqueous formic acid 38% (v/v): (**a**) extruded cassava starch (CEX) and (**b**) extruded pea starch (PEX). The overlap concentration (C*), the entanglement concentration (Ce), and the slopes of the fitted lines in three regimes are illustrated.

The semi-dilute unentangled rate, the semi-dilute entangled rate, and the Ce for electrospun fibers were identified [31]. It was observed that the value of the diluted regime of CEX and PEX was 0.63 and 0.66, respectively, indicating that there was no overlap of polymeric chains according to the theoretical prediction [32].

The semi-dilute entangled value of the CEX and PEX starches was 3.16 and 3.24, respectively, which was higher than the theoretically predicted $n = 1.25$ for linear and neutral polymers; this demonstrated a stronger concentration dependence than the theoretical prediction. The latter could be attributed to a reduced viscosity due to the decrease in the molecular weight of amylopectin by means of the REX process and redistribution in portions of the fractions with different degrees of polymerization.

In the semi-dilute entangled regime, concentration dependencies were 4.45 and 4.57 for CEX and PEX, respectively. These values are near the theoretical prediction reported ($n = 4.8$) for linear polymers, indicating the presence of molecular entanglements in the concentration of starches with a strong interaction, possibly attributed to a greater presence of linear amylose molecules that allow the overlap and entanglement of the chains [32,33].

Kong and Ziegler [14] reported similar values for Gelose 80 in 75% and 70% dimethyl sulfoxide aqueous solutions. However, in these solutions the starch could not be completely dissolved due to the significant increase in viscosity.

Likewise, a lower concentration-dependence value was observed in CEX that, despite its having a greater number of amylose chains that strongly contribute to the extended coils, it exerted an influence on the existence of chains with lower molecular weight that translates into a lower hydrodynamic volume in the polymer solution. Therefore, Ce was defined as the point at which the molecular chains significantly overlap to form interlocking couplings. Ce values were 28.14% and 27.44% for the CEX and PEX solutions, respectively. It was observed that the Ce value decreased in PEX, and this could be due to a higher molecular weight of the chains responsible for molecular entanglement; therefore, these require a lower concentration to become entangled. Furthermore, these results were much higher than the values reported for solutions of amylose and amylopectin in formic acid [34], suggesting that CEX and PEX starches must be more concentrated to establish significant molecular entanglement in solutions with higher ratios of water. If we consider the critical entanglement concentration (C**) at which fibers without defects begin to form, C**/Ce values can be obtained for starches in 38% (v/v) formic acid. These values are 1.14 and 1.25 for CEX and PEX, respectively. Our findings are consistent with reported C**/Ce values for without beads electrospun fibers for Gelosa 80 [14].

In this study, the solutions of CEX and PEX at 38% could not be electrospun, because in PEX, the electrospinning jet did not form, and in CEX the jet was unstable. Therefore, electrospinning solutions were considered within a concentration range of 30–36%.

These solutions exhibited non-Newtonian behavior where, as the shear rate increased, the apparent viscosity (η_{app}) decreased (Figure 3a,b) with values of $n < 1$ from 0.95 ± 0.00 Pa·s to 0.85 ± 0.00 Pa·s (Table 2) described by the Herschel-Bulkley model. Similar values of $n < 1$ were reported in solutions of pea flour, hydroxypropyl methyl-cellulose, and polyethylene oxide (when the pH was increased in electrospinning solutions) to obtain nanofibers by electrospinning [35]. Other researchers [12] studied the behavior of soluble potato-starch solutions in formic acid employing the Herschel–Bulkley model and reported similar values of $n < 1$.

Table 2. Rheological parameters of solutions at different concentrations of extruded starches.

Rheological Parameters	Concentration (% w/v)				R^2
	30	32	34	36	
Yield stress τ_0 [Pa]					
CEX	0 ± 0.10 aA	0 ± 0.07 aA	0 ± 0.08 aA	0 ± 0.2 aA	0.99
PEX	0 ± 0.04 aA	0 ± 0.05 aA	0 ± 0.06 aA	0 ± 0.08 aA	0.99
Consistency coefficient (K) [Pa·s^n]					
CEX	0.50 ± 0.00 cB	1.29 ± 0.02 bB	1.33 ± 0.02 bB	2.18 ± 0.08 aA	0.99
PEX	1.14 ± 0.01 dA	1.46 ± 0.02 cA	1.75 ± 0.02 bA	2.37 ± 0.03 aB	0.99
Flow behavior index (n) [-]					
CEX	0.95 ± 0.00 aA	0.94 ± 0.00 bA	0.94 ± 0.00 bA	0.88 ± 0.01 cA	0.99
PEX	0.89 ± 0.00 aB	0.89 ± 0.00 aB	0.88 ± 0.00 bB	0.85 ± 0.00 cB	0.99

CEX, extruded cassava starch; PEX, extruded pea starch; CF, cassava fibers, and PF, pea fibers. Assays were performed in triplicate, and results are expressed as mean ± standard deviation (SD). Different capital letters in the columns and different lowercase letters in the rows indicate significant differences ($p < 0.05$) using Tukey test.

On the other hand, it is shown that by increasing the concentration from 30 to 36% in the electrospinning solutions, the η_{app} increased from 0.72–1.90 Pa·s for CEX and from 1.15–2.62 Pa·s for PEX (Figure 3a,b). The 38% CEX solutions demonstrated better stability of the electrospun jet and the production of continuous fibers without beads (Figure 5a). In addition, the suitable viscosity for electrospun continuous fibers without beads for each polymer was of ~1.11–1.90 Pa·s for CEX and ~2.01 Pa·s for PEX.

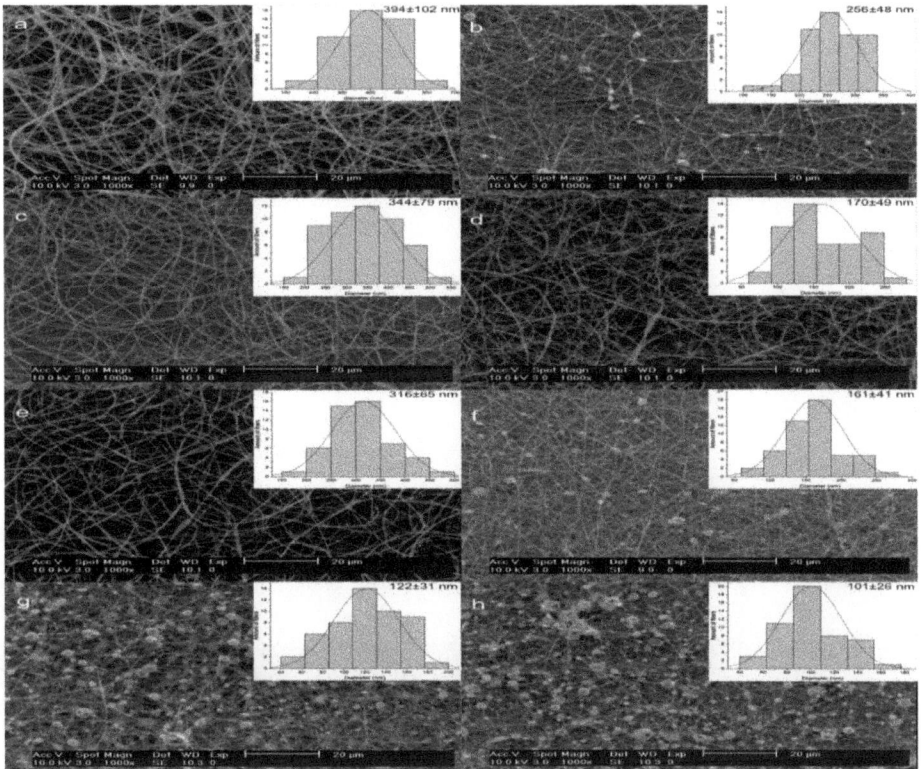

Figure 5. SEM micrographs and histograms with the average diameter of electrospun fibers from extruded starches at different concentrations. Extruded cassava starch (CEX): (**a**) 36%; (**c**) 34% (**e**) 32%, and (**g**) 30%, and PEX extruded pea starch: (**b**) 36%; (**d**) 34%; (**f**) 32%, and (**h**) 30%.

Obtaining fibers without beads in PEX was limited to a specific viscosity, because a lower amylose content is sufficient for molecular entanglement at a given concentration. A similar trend was reported for a solution of Hylon VII starch in 95% dimethyl sulfoxide when it was electrospun at concentrations ranging from 8–20% (w/v), and this range became smaller as the amylose content of the starch decreased [14]. It was also observed that the η_{app} in PEX and CEX solutions is close to that reported in obtaining starch fibers from solutions of Hylon VII (15% w/v) aged from 24–48 h in formic acid with a viscosity range from 1.7–0.9 Pa·s [12].

Unlike what was reported in Hylon VII starch solutions at 17% w/v in formic acid at 100% and 90% v/v aged with viscosity values of 0.79 and 0.74 Pa·s, respectively, which showed homogeneous fibers without beads [13].

In this study, we observed that the viscosity of the CEX and PEX solutions (Figure 3a,b) exerted a greater influence on the diameter of the fibers (Table 3) than the electrical conductivity. The values of the consistency coefficient (K) of CEX and PEX (Table 2) entertained a significant difference ($p < 0.05$). As the concentration of PEX in the solutions increased, a higher viscosity was observed in a range of 81.14 ± 0.01 to 2.37 ± 0.03 Pa·s^n; which could be attributed to the particle size; that is, to a certain concentration of starch where the smaller the particle size, the more particles there are in the solution, and consequently more interaction between the particles. In a previous study [35], pea-flour solutions for obtaining nanofibers showed values of K similar to those obtained in PEX and CEX solutions, while the elastic limit in these solutions did not reveal any significant difference ($p > 0.05$), implying low resistance to the initial electrospinning flow.

Table 3. Electrical conductivity of solutions of extruded starches at different concentrations and diameter distribution of electrospun starch fibers.

Concentration (% w/v)	Values of Electrical Conductivity (mS/cm)		Diameter Distribution of Electrospun Fibers (nm)	
	CEX	PEX	CF	PF
30	3.41 ± 0.09 aA	3.46 ± 0.33 aA	122 ± 31 aB	101 ± 26 aB
32	2.97 ± 0.27 aAB	3.23 ± 0.32 aAB	316 ± 65 aA	161 ± 41 bAB
34	2.97 ± 0.09 aAB	3.03 ± 0.07 aAB	344 ± 79 Aa	170 ± 49 bAB
36	2.77 ± 0.20 aB	2.58 ± 0.24 aB	394 ± 102 aA	256 ± 48 aA

CEX, extruded cassava starch; PEX, extruded pea starch; CF, cassava fibers, and PF, pea fibers. Assays were performed in triplicate, and results are expressed as mean ± standard deviation (SD). Different capital letters in the columns and different lowercase letters in the rows indicate significant differences ($p < 0.05$) using Tukey's test.

2.4. Electrical Conductivity of Starch Solution

Electrical conductivity determines the ability of a material to allow the passage of electrical current and to move towards the surface of the hanging droplet, in order to generate an electrostatic repulsive force that is critical to initiate the jet during electrospinning [36]. Therefore, electrical conductivity is essential in the formation of electrospun fibers. Increasing electrical conductivity in electrospinning solutions could improve fiber uniformity and decrease bead generation [37]. Table 3 shows that the solutions of CEX and PEX in concentrations of 30% and 36% presented a significant difference ($p < 0.05$), with an increase in electrical conductivity in the 30% concentration with values of 3.41 ± 0.09 mS/cm and 3.46 ± 0.33 mS/cm for CEX and PEX, respectively. However, at this concentration, the electrical conductivity did not influence the uniformity of the fibers. Thus, it was correlated with poor consistency in the morphology of the fibers (Figure 5g,h), while in the concentration at 36% of CEX and PEX low electrical-conductivity values of 2.77 ± 0.20 mS/cm and 2.58 ± 0.24 mS/cm, respectively, were obtained, with uniform morphology in the CEX and PEX fibers; this is possibly attributable to the increase in free hydroxyl groups.

In addition, in this study it was observed that the range of electrical conductivity was correlated with the uniformity of the fibers without beads: for CEX this was 2.77 ± 0.20 to 2.97 ± 0.27 mS/cm, while for PEX, it was only 3.03 ± 0.07 mS/cm. On the other hand, there was no significant difference ($p > 0.05$) in the electrical conductivity of the CEX and PEX solutions at 30%. Fonseca et al. [12] reported similar values of electrical conductivity in electrospinning solutions of corn starches with different amylose contents aged for 24 and 48 h. In Hylon V starch, these authors reported values of 2.83 mS/cm and 3.03 mS/cm, and for Hylon VII starch the values were 2.82 mS/cm and 3.01 mS/cm.

2.5. Morphological Characterization and Size Distribution of Electrospun Fibers

Table 3 and Figure 5 show the average diameter and morphology of the electrospun fibers at different concentrations of PEX and CEX. In Figure 5g,h, mixtures of fibers and beads (at a similar proportion) are observed, without significant difference ($p > 0.05$). The average diameter of the cassava fibers (CF) was 122 ± 31 nm, while the average diameter for the pea fibers (PF) was 101 ± 26 nm (Table 3). The presence of beads may be due to weak molecular-chain entanglement, generated by low polymer concentration and the insufficient viscosity (Figure 3a,b) of ~0.72 Pa·s and ~1.15 Pa·s in solutions of electrospinning for CEX and PEX. However, when the concentration of CEX and PEX was increased to 34% in the solutions, an increase in fiber-diameter distribution and a viscosity of ~1.42 Pa·s and ~2.01 Pa·s, respectively, was observed; this permitted the production of homogeneous and continuous fibers with random orientation and without beads (Figure 5c,d), with a significant difference ($p < 0.05$) in the average diameter of 344 ± 79 nm and 170 ± 49 nm, respectively. The latter may possibly be attributed to a higher number of chains and a higher molecular weight, responsible for molecular entanglement in CEX.

Likewise, in Table 3, it was observed that the average diameters of PF and CF obtained from the concentrations of 30% and 36% of the PEX and CEX starches presented a significant difference ($p < 0.05$), indicating that the concentration exerted an influence on the average diameter of the fibers and on the reduction of the beads. Furthermore, it was observed that the fiber-diameter distribution broadened with an increasing concentration, regardless of the starch source. Jia et al. [38] observed a similar behavior: as the concentration of poly/(vinyl alcohol) in chitosan solutions increased, nanofibers with the presence of beads changed to a uniform structure without beads.

In another study [13], the authors reported fiber diameters of 300 nm and 150 nm for solutions of Hylon VII starch in formic acid at 100% and 90% *v/v*, values very similar to those obtained in our research. Continuous fibers without beads were obtained in the CEX solutions at 32%, 34%, and 36% *w/v* (Figure 5a,c,e), with an average diameter of 394 ± 102 nm, 344 ± 79 nm, and 316 ± 65 nm, respectively. This result was due to the high content of amylose present in the CEX concentrations and a lower viscosity compared to PEX, allowing for a higher molecular entanglement within a range of viscosities for fiber formation, and increasing the fiber diameter as the concentration increased. The same trend was found in zein solutions: as the zein concentration increased, the fiber diameter increased [39].

In Table 3, it was also observed that concentrations at 32%, 34%, and 36% *w/v* for CEX and PEX did not present a significant difference ($p > 0.05$) in the average diameter of the electrospun fibers. However, in PEX it was observed that defects were generated in the fibers on increasing the concentration to 36%, due to an increase in viscosity that prevented the electrospinning jet from being maintained stable.

2.6. Fourier Transform Infrared Spectroscopy (FT-IR)

The FT-IR spectra of NCS, NPS, CEX, and PEX starches and CF and PF electrospun fibers are presented in Figure 6. CEX, PEX, CF, and PF materials exhibited the characteristic spectra of starches (NCS and NPS), suggesting that they exhibited the same type of molecular vibrations without altering the functional groups of the native starch. Assuming that the OH groups of the water molecule absorb energy (3000–3600 cm^{-1}), the bands observed in the region of 3294 cm^{-1} correspond to the stretching modes of the OH groups [40,41]. The bands that appear at around 2927 cm^{-1} indicate the existence of methyl groups in the starch chains [42,43]. Starches demonstrate that C-H stretching modes in the region of 2800–3000 cm^{-1}; the different intensities located between 2800 and 3000 cm^{-1} can be attributed to the variation in the amounts of amylose and amylopectin contained in the starch [43]. Therefore, the intensity in the bands suggests a higher amount of amylose in CEX and PEX starches than in NCS and NPS, as a result of extrusion [44].

The vibrational bands of flexion and deformation, related to carbon and hydrogen atoms, can be observed in the region of 1500–1300 cm^{-1}. The 1363 cm^{-1} band is attributed to the C=O and O-H deformations. The band observed at 1638 cm^{-1} could be assigned to water absorbed in the amorphous regions of the starches. The small differences in the location and intensity of these bands can be attributed to REX causing a loss in molecular weight, especially in amylopectin, as well as a loss of crystallinity and consequently, an increase in the quantity of amylose molecules. The 1712 cm^{-1} band suggests weaker inter- and intramolecular hydrogen bonds, probably due to strong C=O stretching on the carboxylic groups of the starch formed during electrospinning [45]. The infrared bands observed in the 900–1000 cm^{-1} region, with the appearance of vibrational peaks at 996 and 928 cm^{-1}, are related to the vibrations that originate in the C-O-C of the glycosidic bond. The subtle changes in peak location and intensity of the glycosidic-bond band can be attributed to the presence of an α-1,6 bond in amylopectin that shifts the band at higher wavenumbers [40].

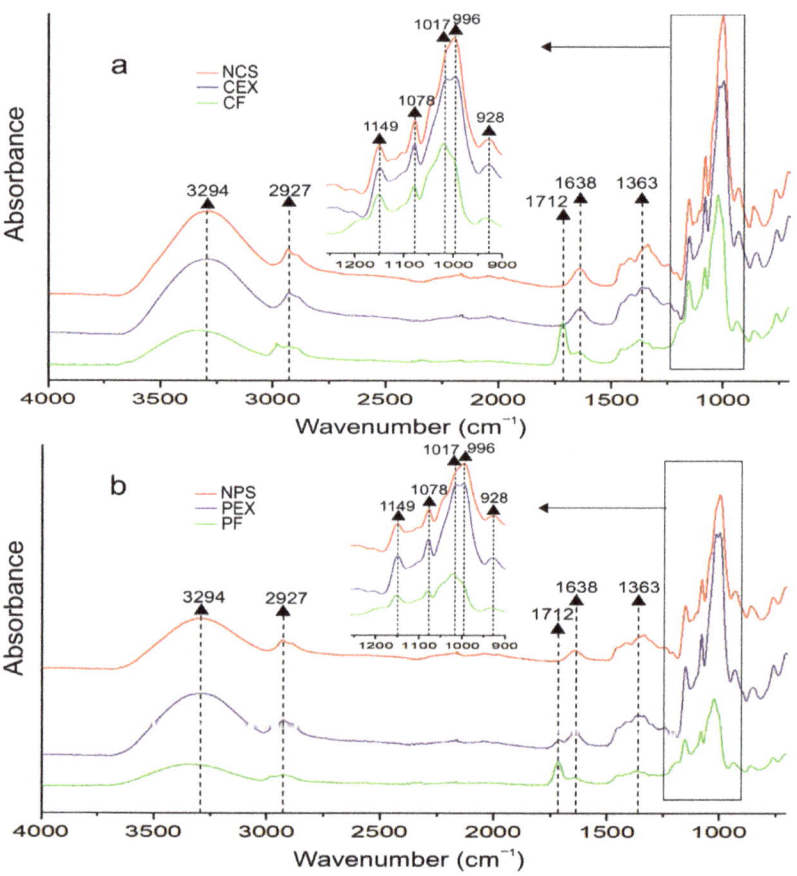

Figure 6. FT-IR spectra of starches and electrospun starch fibers: (**a**) native cassava starch (NCS), extruded cassava starch (CEX), and cassava fibers (CF) and (**b**) native pea starch (NPS) extruded pea starch (PEX) and pea fibers (PF).

The bands at 1149, 1078, and 1017 cm^{-1} are attributed to the contribution of two main vibrational modes C-O and C-O-H, stretching in the glucose monomer [40,46].

2.7. X-ray Diffraction (XRD)

XRD diffraction patterns have permitted the quantification of the amorphous/crystalline relationship in starches and the characterization of the crystalline phases type A, B, and C, the latter being more complex due to the combination of phases A and B [47,48]. Figure 7a shows that NCS presented XRD type A patterns; this indicates a degree of polymerization in ranges from 23–29 of the average length of the amylopectin branched chain. Its branched amylopectin chains are in the form of double helices organized in an orthorhombic structure [49], with a crystallinity value of 18.10% and with predominant peaks at diffraction angles (2θ) of 15.08°, 17.25°, 18.00°, 23.02° and weak reflection at 2θ of 26.66°. These results are in agreement with what has been reported [47], while the NPS (Figure 7b) exhibited type C XRD patterns, indicating a degree of polymerization of 15–17 of the average length of the branched chain [48]. This is characteristic of legumes, with a proportion (crystalline phases A and B) closest to phase A, with a crystallinity value of 18.02%, and with predominant peaks at diffraction angles (2θ) 15.14°, 17.25°, 23.02°, and weak reflections in 2θ of

18.20° and 26.66° [50]. In this respect, the modification in type C starches by REX is more complicated than that in type A starches [48].

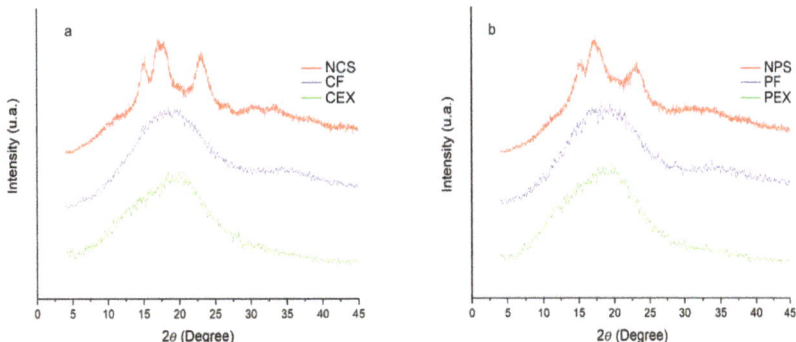

Figure 7. XRD patterns of starches and electrospun starch fibers: (**a**) native cassava starch (NCS), extruded cassava starch (CEX), and cassava fibers (CF), and (**b**) native pea starch (NPS), extruded pea starch (PEX), and pea fibers (PF).

On the other hand, we know that native starches possess poor solubility in aqueous solvents, due to the strong hydrogen bonds between starch chains and to their semi-crystalline nature. The REX process promoted the creation of amorphous structures in CEX and PEX (Figure 7a,b), favoring the properties of solubility and viscosity necessary for electrospinning. Furthermore, CF and PF (Figure 7a,b) did not demonstrate recrystallization, which resulted in a network of fibers with an amorphous tendency.

3. Materials and Methods

3.1. Materials

Formic acid (purity ≥ 95%) was purchased from Sigma-Aldrich Chemical Co. (Toluca, Mexico City, Mexico), NCS (CAS 9005-25-8) was obtained from Ingredion México, S.A. de C.V., and NPS was isolated from dehydrated pea seeds (13% moisture) acquired at a local market (Querétaro, México) according to the methodology reported by Simsek et al. [51]. To obtain the CEX and PEX starches, distilled water was sprayed on the NCS and the NPS starch, respectively, to achieve a moisture level of 23%. The samples were mixed for 5 min to allow for moisture distribution, and the starches were stored in polyethylene bags and allowed to equilibrate for 24 h before their use in an extruder designed and manufactured in the Organic Materials Processing Laboratory of Cinvestav, Querétaro, Mexico. The extruder had a barrel (20-mm internal diameter), a screw (428 mm in length and a 19-mm diameter), three heating zones, and a 4-mm diameter round die. The spindle ratio utilized was 3:1 and the speed was constant at 40 rpm. The temperatures in the feeding, transition, and extrusion zones of the die were maintained constant at 65 °C, 130 °C, and 170 °C, respectively. Subsequently, the CEX and PEX starches were brought to constant weight at 40 °C in a conventional oven (Felisa, model FE-291, Zapopan, Jalisco, Mexico) and ground in a hammer mill (Pulvex, model 200, Mexico City, Mexico) with integrated mesh (150 μm).

3.2. Physicochemical Properties of Starches

The AAC, the S, and the SP were determined for the starches NCS, NPS, CEX, and PEX. The AAC was calculated by the iodine colorimetric method [52]. The S and SP were obtained according to Equations (1) and (2) [53].

$$S\ (\%) = (A/W) * 100 \tag{1}$$

$$SP = (P * 100)/W(100 - S) \tag{2}$$

where A refers to the constant weight of the supernatant (g), W represents the weight of the sample (g), and P indicates the weight of the precipitate after centrifugation (g).

3.3. Particle Size Distribution

PSD was determined in CEX and PEX by the dynamic light scattering method, using a Zetasizer (Nano-ZS, Malvern Instruments, Ltd., Malvern, UK). The CEX and PEX starches were suspended in distilled water and mixed with a vortex. Particle-refractive and -absorption indices of 1.3 and 0.01, respectively, were employed. The dispersion index of water was 1.33.

3.4. Preparation of Starch Solutions

Solutions of the CEX and PEX starches were prepared at concentrations ranging from 0.1–38% w/v in aqueous formic acid at 38% v/v with stirring at 1000 rpm on a hot plate (Super-Nuova Multi-place, Thermo Scientific, Waltham, MA, USA) at 85 °C for 15 min until homogeneous solutions were obtained.

3.5. Rheological Characterization and Electrical Conductivity of Starch Solutions

Starch solutions were measured in a stress-controlled rheometer (ARES, TA Instruments, New Castle, DE, USA) with 16-mm diameter couette geometry and a 10-mm gap. The solutions were poured into the measurement system and allowed to stand for 5 min to equilibrate the temperature at 25 °C. Then an increasing shear rate ranging from 0.1–100 s^{-1} was applied to obtain the flow curves; that is, apparent viscosity (η_{app}) vs. shear rate.

Moreover, the specific viscosity (η_{sp}) was calculated with Equation (3), considering the shear viscosity of 0.1 s^{-1} as near the shear rate of the electrospinning process. The resulting values were plotted against the concentration of each starch, and the Ce was determined from the intersection of the fitted slopes in the semi-dilute unentangled and semi-dilute entangled regions. The rheological parameters were analyzed according to the Herschel–Bulkley model (Equation (4)), for the semi-dilute entangled regime, considered as the electrospinning region of CEX and PEX starch solutions due to its significant high degree of entanglement and its influence in determining the Ce in the electrospinning process.

$$\eta_{sp} = (\eta_0 - \eta_s)/\eta_s \qquad (3)$$

where η_{sp} is specific viscosity, η_0 is the shear viscosity (Pa·s) of the polymer solutions at 0.1 s^{-1}, and η_s is the viscosity of the solvent (Pa·s).

$$\tau = \tau_0 + K(\dot{\gamma})^n \qquad (4)$$

where τ is the shear stress (Pa), τ_0 is the yield stress (Pa), K is the consistency coefficient (Pa·s^n), $\dot{\gamma}$ is the shear rate (s^{-1}), and n is the flow behavior index (-). The electrical conductivity of the electrospinning operating-region solutions was determined according to Ledezma-Oblea et al. [54].

3.6. Electrospinning

A horizontal-configuration electrospinning apparatus consisting of a voltage source (Bertan 230 series, Hauppauge, NY, USA), a syringe feed pump (KD Scientific, Holliston, MA, USA), and an aluminum plate as a grounded collector was utilized. CEX and PEX starch solutions that were within the entanglement semi-dilute entangled regime were chosen as electrospinning solutions. Each solution was individually loaded into a 5 mL syringe (Plastipak BD, Mexico City, Mexico) with a 21-gauge blunt needle as a spinneret. The feed rate was set at 0.2 mL/h, the distance between the needle tip and the collector was 15 cm, and the voltage was 19 kV. The electrospinning process was carried out at room temperature (23 ± 2 °C). CF and PF fibers were stored at room temperature until their characterization.

3.7. Characterization of Electrospun Fibers and Starches

Fibers (CF and PF) and starches (NCS, NPS CEX, and PEX) were analyzed by SEM, FT-IR, and XRD.

The morphological characterization of the samples was analyzed by SEM (ESEM Philips, XL30, Tokyo, Japan) at an accelerating voltage of 10 kV. Previously, the powders of the starches and fibers were placed and fixed directly on a metallic support with double-sided carbon tape and covered with gold/palladium in a vacuum chamber (Denton Vacuum in Sputtering DESK V, Moorestown, NJ, USA). Fiber-diameter distribution and starch-granule sizes were determined for each micrograph using ImageJ 1.53k (National Institutes of Health, Bethesda, MD, USA).

The samples were analyzed by FT-IR with the attenuated total reflection (ATR) technique employing a spectrometer (Perkin Elmer, Spectrum GX, Inc., Waltham, MA USA). Each sample was held in a diamond crystal of the instrument and 24 scans were recorded for each spectrum, with a wave-number resolution of 4 cm^{-1}. Spectra were measured in the wave-number range of 600–4000 cm^{-1}.

For the XRD analysis, an X-ray diffractometer (Rigaku, DMAX-2100, Tokyo, Japan) was used, each sample was placed in a glass sample holder, and the analysis was performed with a Bragg scanning angle of 0–45° by scanning 2θ; with CuKα radiation, the irradiation conditions were at a voltage of 40 kV with scanning-speed intervals of 0.05 2θ/min.

3.8. Statistic Analysis

Measurements of ACC, SI, SP, rheological parameters, and electrical conductivity were performed in triplicate, and the average diameters of the electrospun fibers were evaluated using 50 randomly selected fibers. All data were statistically analyzed by analysis of variance (ANOVA) with Tukey test. Significant differences were determined with a 95% confidence interval (95% CI) ($p < 0.05$), expressed as mean ± standard deviation (SD).

4. Conclusions

In this study, the application of REX in cassava and pea starches improved the properties of solubility, viscosity, and amylose content in electrospinning solutions; as a result, there was a successful manufacture of continuous fibers without beads with an average diameter of 316 ± 65 at 394 ± 102 nm for CF and at 170 ± 49 nm for PF with amorphous structure and characteristic functional groups of starches. We observed that the content of amylose in the electrospinning solutions comprises an important parameter that defines the diameter of the fibers and the viscosity of the solutions. Rheological analysis of the electrospinning solutions indicated that to obtain well-formed fibers, the concentration of CEX and PEX must be 1.14 and 1.25 times the Ce, respectively. Therefore, the experimental results revealed that REX as a pretreatment can potentially be used to improve the attributes of cassava and pea starch for obtaining fibers by electrospinning, thus expanding its possible use in food and non-food applications as vehicles of administration of bioactive compounds.

Author Contributions: D.T.-V.: formal analysis, investigation, methodology, writing-original draft; R.R.-B.: resources, validation; J.M.Y.-L.: conceptualization, data curation; F.M.-B.: resources, supervision. All authors have read and agreed to the published version of the manuscript.

Funding: This research was funded by the Secretaría de Educación Pública and the Centro de Investigación y Estudios Avanzados del Instituto Politécnico Nacional, (SEP–CINVESTAV Querétaro), grant number 2018/267.

Institutional Review Board Statement: Not applicable.

Informed Consent Statement: Not applicable.

Data Availability Statement: Not applicable.

Acknowledgments: David Tochihuitl-Vázquez thanks CONACYT-México for his PhD scholarship. The authors would like to thank Martín Adelaido Hernández-Landaverde for his support in the XRD analysis, Reina Araceli Mauricio-Sánchez for her support in the FT-IR analysis, José Eleazar Urbina-Álvarez for his technician support in SEM, José Juan Velés-Medina and Verónica Flores-Casamayor for their technicial support in laboratory. We also thank the Laboratorio Nacional de Investigación y Desarrollo Tecnológico de Recubrimientos Avanzados, LIDTRA, CINVESTAV Querétaro, México, for technicial support.

Conflicts of Interest: The authors declare no conflict of interest.

Sample Availability: Samples of the compounds are available from the authors.

References

1. Anu Bhushani, J.; Anandharamakrishnan, C. Electrospinning and electrospraying techniques: Potential food based applications. *Trends Food Sci. Technol.* **2014**, *38*, 21–33. [CrossRef]
2. Wen, P.; Wen, Y.; Zong, M.H.; Linhardt, R.J.; Wu, H. Encapsulation of bioactive compound in electrospun fibers and its potential application. *J. Agric. Food Chem.* **2017**, *65*, 9161–9179. [CrossRef] [PubMed]
3. Balusamy, B.; Sarioglu, O.F.; Senthamizhan, A.; Uyar, T. Rational design and development of electrospun nanofibrous biohybrid composites. *ACS Appl. Bio Mater.* **2019**, *2*, 3128–3143. [CrossRef] [PubMed]
4. Zhang, M.; Zhao, X.; Zhang, G.; Wei, G.; Su, Z. Electrospinning design of functional nanostructures for biosensor Applications. *J. Mater. Chem. B* **2017**, *5*, 1699–1711. [CrossRef]
5. Iravani, S.; Varma, R.S. Plants and plant-based polymers as scaffolds for tissue engineering. *Green Chem.* **2019**, *21*, 4839–4867. [CrossRef]
6. Angel, N.; Li, S.; Yan, F.; Kong, L. Recent advances in electrospinning of nanofibers from bio-based carbohydrate polymers and their applications. *Trends Food Sci. Technol.* **2022**, *120*, 308–324. [CrossRef]
7. Mitchell, J.R.; Hill, S.E.S. *Handbook of Hydrocolloids*; Phillips, G.O., Williams, P.A., Eds.; Woodhead Publishing Series in Food Science, Technology and Nutrition; Elsevier: London, UK, 2021; pp. 239–271. ISBN 9780128201046.
8. Bertoft, E. Understanding starch structure: Recent progress. *Agronomy* **2017**, *7*, 56. [CrossRef]
9. Wang, Z.; Mhaske, P.; Farahnaky, A.; Kasapis, S.; Majzoobi, M. Cassava starch: Chemical modification and its impact on functional properties and digestibility, a review. *Food Hydrocoll.* **2022**, *129*, 107542. [CrossRef]
10. Cui, W.; Ma, Z.; Li, X.; Hu, X. Structural rearrangement of native and processed pea starches following simulated digestion in vitro and fermentation characteristics of their resistant starch residues using human fecal inoculum. *Int. J. Biol. Macromol.* **2021**, *172*, 490–502. [CrossRef]
11. Gao, L.; Wu, Y.; Wan, C.; Wang, P.; Yang, P.; Gao, X.; Eeckhout, M.; Gao, J. Structural and physicochemical properties of pea starch affected by germination treatment. *Food Hydrocoll.* **2022**, *124*, 107303. [CrossRef]
12. Fonseca, L.M.; de Oliveira, J.P.; de Oliveira, P.D.; da Rosa Zavareze, E.; Dias, A.R.G.; Lim, L.-T. Electrospinning of native and anionic corn starch fibers with different amylose contents. *Food Res. Int.* **2019**, *116*, 1318–1326. [CrossRef] [PubMed]
13. Lancuški, A.; Vasilyev, G.; Putaux, J.-L.; Zussman, E. Rheological properties and electrospinnability of high-amylose starch in formic acid. *Biomacromolecules* **2015**, *16*, 2529–2536. [CrossRef] [PubMed]
14. Kong, L.; Ziegler, G.R. Role of molecular entanglements in starch fiber formation by electrospinning. *Biomacromolecules* **2012**, *13*, 2247–2253. [CrossRef]
15. Rostamabadi, H.; Falsafi, S.R.; Jafari, S.M. Starch-based nanocarriers as cutting-edge natural cargos for nutraceutical delivery. *Trends Food Sci. Technol.* **2019**, *88*, 397–415. [CrossRef]
16. Sun, X.; Sun, Z.; Guo, Y.; Zhao, J.; Zhao, J.; Ge, X.; Shen, H.; Zhang, Q.; Yan, W. Effect of twin-xuscrew extrusion combined with cold plasma on multi-scale structure, physicochemical properties, and digestibility of potato starches. *Innov. Food Sci. Emerg. Technol.* **2021**, *74*, 102855. [CrossRef]
17. Arora, B.; Yoon, A.; Sriram, M.; Singha, P.; Rizvi, S.S.H. Reactive extrusion: A review of the physicochemical changes in food systems. *Innov. Food Sci. Emerg. Technol.* **2020**, *64*, 102429. [CrossRef]
18. Maliger, R.B.; Halley, P.J. Reactive Extrusion for Thermoplastic Starch-Polymer Blends. In *Starch Polymers: From Genetic Engineering to Green Applications*; Halley, P.J., Avérous, L., Eds.; Elsevier: Alpharetta, GA, USA, 2014; pp. 291–317. ISBN 9780444537300.
19. Song, D.; Thio, Y.S.; Deng, Y. Starch nanoparticle formation via reactive extrusion and related mechanism study. *Carbohydr. Polym.* **2011**, *85*, 208–214. [CrossRef]
20. Lai, L.S.; Kokini, J.L. Physicochemical changes and rheological properties of starch during extrusion (a review). *Biotechnol. Prog.* **1991**, *7*, 251–266. [CrossRef]
21. Fasheun, D.O.; de Oliveira, R.A.; Bon, E.P.S.; da Silva, A.S.A.; Teixeira, R.S.S.; Ferreira-Leitão, V.S. Dry extrusion pretreatment of cassava starch aided by sugarcane bagasse for improved starch saccharification. *Carbohydr. Polym.* **2022**, *285*, 119256. [CrossRef]
22. Mishra, S.; Rai, T. Morphology and functional properties of corn, potato and tapioca starches. *Food Hydrocoll.* **2006**, *20*, 557–566. [CrossRef]
23. Zhang, H.; Hou, H.; Liu, P.; Wang, W.; Dong, H. Effects of acid hydrolysis on the physicochemical properties of pea starch and its film forming capacity. *Food Hydrocoll.* **2019**, *87*, 173–179. [CrossRef]

24. Yoo, S.-H.; Jane, J. Molecular weights and gyration radii of amylopectins determined by high-performance size-exclusion chromatography equipped with multi-angle laser-light scattering and refractive index detectors. *Carbohydr. Polym.* **2002**, *49*, 307–314. [CrossRef]
25. Sánchez, T.; Salcedo, E.; Ceballos, H.; Dufour, D.; Mafla, G.; Morante, N.; Calle, F.; Pérez, J.C.; Debouck, D.; Jaramillo, G.; et al. Screening of starch quality traits in cassava (Manihot esculenta Crantz). *Starch-Stärke* **2009**, *61*, 12–19. [CrossRef]
26. Liu, W.-C.-C.; Halley, P.J.; Gilbert, R.G. Mechanism of degradation of starch, a highly branched polymer, during extrusion. *Macromolecules* **2010**, *43*, 2855–2864. [CrossRef]
27. Liu, Y.; Chen, J.; Luo, S.; Li, C.; Ye, J.; Liu, C.; Gilbert, R.G. Physicochemical and structural properties of pregelatinized starch prepared by improved extrusion cooking technology. *Carbohydr. Polym.* **2017**, *175*, 265–272. [CrossRef]
28. Jongsutjarittam, O.; Charoenrein, S. The Effect of moisture content on physicochemical properties of extruded waxy and non-waxy rice flour. *Carbohydr. Polym.* **2014**, *114*, 133–140. [CrossRef] [PubMed]
29. Hemamalini, T.; Giri Dev, V.R. Comprehensive review on electrospinning of starch polymer for biomedical applications. *Int. J. Biol. Macromol.* **2018**, *106*, 712–718. [CrossRef]
30. Sun, Q.; Xiong, C.S.L. Functional and pasting properties of pea starch and peanut protein isolate blends. *Carbohydr. Polym.* **2014**, *101*, 1134–1139. [CrossRef]
31. Chang, K.H.; Lin, H.L. Electrospin of polysulfone in N,N'-dimethyl acetamide solutions. *J. Polym. Res.* **2009**, *16*, 611–622. [CrossRef]
32. McKee, M.G.; Wilkes, G.L.; Colby, R.H.; Long, T.E. Correlations of solution rheology with electrospun fiber formation of linear and branched polyesters. *Macromolecules* **2004**, *37*, 1760–1767. [CrossRef]
33. Li, S.; Kong, L.; Ziegler, G.R. Electrospinning of octenylsuccinylated starch-pullulan nanofibers from aqueous dispersions. *Carbohydr. Polym.* **2021**, *258*, 116933. [CrossRef]
34. Vasilyev, G.; Vilensky, R.; Zussman, E. The ternary system amylose-amylopectin-formic acid as precursor for electrospun fibers with tunable mechanical properties. *Carbohydr. Polym.* **2019**, *214*, 186–194. [CrossRef]
35. Oguz, S.; Tam, N.; Aydogdu, A.; Sumnu, G.; Sahin, S. development of novel pea flour-based nanofibres by electrospinning method. *Int. J. Food Sci. Technol.* **2017**, *53*, 1269–1277. [CrossRef]
36. Vega-Lugo, A.-C.; Lim, L.-T. Effects of poly(ethylene oxide) and pH on the electrospinning of whey protein isolate. *J. Polym. Sci. Part B Polym. Phys.* **2012**, *50*, 1188–1197. [CrossRef]
37. Thenmozhi, S.; Dharmaraj, N.; Kadirvelu, K.; Kim, H.Y. Electrospun nanofibers: New generation materials for advanced applications. *Mater. Sci. Eng. B Solid-State Mater. Adv. Technol.* **2017**, *217*, 36–48. [CrossRef]
38. Jia, Y.-T.; Gong, J.; Gu, X.-H.; Kim, H.-Y.Y.; Dong, J.; Shen, X.-Y. Fabrication and characterization of poly (vinyl alcohol)/chitosan blend nanofibers produced by electrospinning method. *Carbohydr. Polym.* **2007**, *67*, 403–409. [CrossRef]
39. Neo, Y.P.; Ray, S.; Easteal, A.J.; Nikolaidis, M.G.; Quek, S.Y. Influence of solution and processing parameters towards the fabrication of electrospun zein fibers with sub-micron diameter. *J. Food Eng.* **2012**, *109*, 645–651. [CrossRef]
40. Kizil, R.; Irudayaraj, J.; Seetharaman, K. Characterization of irradiated starches by using FT-raman and FTIR spectroscopy. *J. Agric. Food Chem.* **2002**, *50*, 3912–3918. [CrossRef]
41. Pelissari, F.M.; Andrade-Mahecha, M.M.; Sobral, P.J.D.A.; Menegalli, F.C. Comparative study on the properties of flour and starch films of plantain bananas (Musa paradisiaca). *Food Hydrocoll.* **2013**, *30*, 681–690. [CrossRef]
42. Amini, M.; Arash Haddadi, S.; Ghaderi, S.; Ahmad Ramazani, S.A.; Ansarizadeh, M. Preparation and characterization of PVDF/starch nanocomposite nanofibers using electrospinning method. *Mater. Today Proc.* **2018**, *5*, 15613–15619. [CrossRef]
43. Movahedi, M.; Asefnejad, A.; Rafienia, M.; Khorasani, M.T. Potential of novel electrospun core-shell structured polyurethane/starch (hyaluronic acid) nanofibers for skin tissue engineering: In vitro and in vivo evaluation. *Int. J. Biol. Macromol.* **2020**, *146*, 627–637. [CrossRef]
44. Mościcki, L.; Mitrus, M.; Wójtowicz, A.; Oniszczuk, T.; Rejak, A.; Janssen, L. Application of extrusion-cooking for processing of thermoplastic starch (TPS). *Food Res. Int.* **2012**, *47*, 291–299. [CrossRef]
45. Lancuški, A.; Ammar, A.A.; Avrahami, R.; Vilensky, R.; Vasilyev, G. Design of starch-formate compound fibers as encapsulation platform for biotherapeutics. *Carbohydr. Polym.* **2017**, *158*, 68–76. [CrossRef]
46. Estevez-Areco, S.; Guz, L.; Famá, L.; Candal, R.; Goyanes, S. Bioactive starch nanocomposite films with antioxidant activity and enhanced mechanical properties obtained by extrusion followed by thermo-compression. *Food Hydrocoll.* **2019**, *96*, 518–528. [CrossRef]
47. Palavecino, P.M.; Penci, M.C.; Ribotta, P.D. Impact of chemical modifications in pilot-scale isolated sorghum starch and commercial cassava starch. *Int. J. Biol. Macromol.* **2019**, *135*, 521–529. [CrossRef]
48. He, W.; Wei, C. Progress in C-type starches from different plant sources. *Food Hydrocoll.* **2017**, *73*, 162–175. [CrossRef]
49. Rodriguez-Garcia, M.E.; Hernandez-Landaverde, M.A.; Delgado, J.M.; Ramirez-Gutierrez, C.F.; Ramirez-Cardona, M.; Millan-Malo, B.M.; Londoño-Restrepo, S.M. Crystalline structures of the main components of starch. *Curr. Opin. Food Sci.* **2021**, *37*, 107–111. [CrossRef]
50. Zhou, D.; Ma, Z.; Yin, X.; Hu, X.; Boye, J.I. Structural Characteristics and physicochemical properties of field pea starch modified by physical, enzymatic, and acid treatments. *Food Hydrocoll.* **2019**, *93*, 386–394. [CrossRef]
51. Simsek, S.; Tulbek, M.C.; Yao, Y.; Schatz, B. Starch characteristics of dry peas (Pisum sativum L.) grown in the USA. *Food Chem.* **2009**, *115*, 832–838. [CrossRef]

52. Hoover, R.; Ratnayake, W.S. Starch characteristics of black bean, chick pea, lentil, navy bean and pinto bean cultivars grown in Canada. *Food Chem.* **2002**, *78*, 489–498. [CrossRef]
53. Yang, Z.; Hao, H.; Wu, Y.; Liu, Y.; Ouyang, J. Influence of moisture and amylose on the physicochemical properties of rice starch during heat treatment. *Int. J. Biol. Macromol.* **2021**, *168*, 656–662. [CrossRef] [PubMed]
54. Ledezma-Oblea, J.G.; Morales-Sánchez, E.; Gaytán-Martúnez, M.; Figueroa-Cárdenas, J.D.; Gaona-Sánchez, V.A. Corn starch nanofilaments obtained by electrospinning. *Rev. Mex. Ing. Quím.* **2015**, *14*, 497–502.

Article

Ultrasound Treatment Enhanced Semidry-Milled Rice Flour Properties and Gluten-Free Rice Bread Quality

Wanyu Qin [†], Huihan Xi [†], Aixia Wang, Xue Gong, Zhiying Chen, Yue He, Lili Wang, Liya Liu, Fengzhong Wang * and Litao Tong *

Key Laboratory of Agro-Products Processing Ministry of Agriculture, Institute of Food Science and Technology, Chinese Academy of Agricultural Sciences, Beijing 100193, China
* Correspondence: wangfengzhong@caas.cn (F.W.); tonglitao@caas.cn (L.T.); Tel./Fax: +86-10-6281-7417 (L.T.)
† These authors contributed equally to this work.

Abstract: The structural and functional properties of physical modified rice flour, including ultrasound treated rice flour (US), microwave treated rice flour (MW) and hydrothermal treated rice flour (HT) were investigated with wet-milled rice flour (WF) used as a positive control. The results showed the presence of small dents and pores on the rice flour granules of US and MW while more fragments and cracks were showed in HT. XRD and FTIR revealed that moderate ultrasonic treatment promoted the orderly arrangement of starch while hydrothermal treatment destroyed the crystalline structure of rice flour. In addition, the significant decrease of gelatinization enthalpy and the narrowing gelatinization temperature were observed in US. Compared to that of SF, adding physical modified rice flour led to a batter with higher viscoelasticity and lower tan δ. However, the batter added HT exhibited highest G' and G'' values and lowest tan δ, which led to a harder texture of bread. Texture analysis demonstrated that physical modified rice flour (except HT) reduced the hardness, cohesion, and gumminess of rice bread. Especially, the specific volume of bread with US increased by 15.6% and the hardness decreased by 17.6%. This study suggested that ultrasound treatment of rice flour could improve texture properties and appearance of rice bread.

Keywords: ultrasound; semidry-milled rice flour; rice bread; properties

1. Introduction

Due to the increase in the prevalence of gluten-related diseases, such as celiac disease, gluten-free bread has a growing demand and consumption [1]. Rice flour is the major ingredient frequently used in the manufacture of commercially available gluten-free bread since it is colorless, bland taste, easy digestibility and hypoallergenic [2]. However, rice flour lacks gluten to impart viscoelasticity to the dough, which limited the industrial range of native rice flour. Therefore, some modification methods, including mechanical, physical, chemical and enzymatic methods, are usually used to improve the physicochemical properties of rice flour [3–5]. Physical modification is the preferred modification method for environmental and economic considerations, which can change the properties of water absorption, gelatinization and thermal properties by affecting the molecular structure of starch or protein, so as to achieve satisfactory processing performance.

Studies have shown that hydrothermal treatment, microwave and ultrasound are commonly used as physical modifications. Hydrothermal treatment is based on treating materials with moisture content less than 35% (w/w) within a certain temperature range, which would destroy the crystallization structure of starch and change the solubility, viscosity and expansion ability of rice flour. Ruiz, Srikaeo and Revilla [6] found that the hydrothermal treatment reduced the solubility, expansion capacity and peak viscosity of rice starch, and increased its gelatinization temperature. Microwave technology is based on the dielectric heating effect, leading to the rapid dipole reorientation of polar molecules in the starch system, which changes the structure and properties of starch. Li et al. [7] observed that pores and cracks appeared in the starch particles with 30% water content under microwave treatment. Villanueva et al. [8] adopted microwave-assisted hydrothermal treatment to modify rice flour, which found that the pasting temperature was delayed and the pasting viscosity was reduced, and the deformation resistance and recovery ability of the dough were improved. Moreover, ultrasound wave could cause structural changes to the starch by producing cavitation effect and generating free radicals [9,10]. Cui and Zhu [11] discovered that ultrasound treatment changed the morphology of starch particles and decreased its crystallinity, and the gelation enthalpy and pasting viscosity of sweet potato powder were significantly reduced with increasing treatment time. To our knowledge, level 6 structure of starch refers to the whole grain structure of starch particles after interaction with other biomolecules [12]. Although most studies have focused on starch particles after modified treatment, the change of the complex whole grain structure during processing is the main determinant of food quality. Thus, how these three physical modification methods may change the physicochemical properties of rice flour and further affect the rice bread quality remains to be better studied.

In this research, three physical modified methods, including hydrothermal treatment, microwave and ultrasound, were applied on semidry-milled rice flour. The effects of physical modification on the structural characteristics, including morphology, particle size, crystal structure and short-range ordered structure, were analyzed. Changes in physicochemical properties of modified rice flour were investigated for hydration properties, thermal and pasting properties, and rheological properties. The quality of gluten-free bread added physically modified rice flour were evaluated. As one of the common commercial rice flour raw materials, wet-milled rice flour was employed as a reference for comparison. It is expected that this research may provide new perspectives for the application of physically modified flour in gluten-free rice bread.

2. Results and Discussion

2.1. Particle Size, Damaged Starch (DS)

According to Qin et al. [13], rice flour with particle size <100 μm (D50) showed more suitable properties for bread making. The D50 of all the samples was lower than 100 μm in Table 1. The particle size distribution of US was more uniform in Figure 1, and its average particle size was almost 2.47 times lower than that of SF, which may be conducive to the uniform distribution of starch granules. Cui et al. [11] found that with the increase of ultrasonic power, the particle size of sweet potato and wheat flours gradually decreased, which was consistent with our study.

Table 1. Particle size, damaged starch, short-range ordered degree (R1047/1022) and protein secondary structure of rice flour with different physical modifications.

Sample	WF	SF	HT	MW	US
D50 (μm)	50.7 ± 0.8c	59.4 ± 0.2a	37.7 ± 1.1d	57.0 ± 1.2b	17.1 ± 0.9e
Damage starch content (g/100 g)	2.06 ± 0.12d	3.96 ± 0.04c	24.41 ± 1.60a	3.33 ± 0.07cd	9.91 ± 0.20b
$R_{1047/1022}$	1.34 ± 0.15ab	1.18 ± 0.05cd	1.28 ± 0.10bc	1.14 ± 0.03d	1.44 ± 0.23a
β-Sheet (%)	26.8 ± 0.7c	30.8 ± 0.2b	30.0 ± 0.5b	35.0 ± 0.3a	30.1 ± 0.3b
α-Helix (%)	25.6 ± 0.5c	27.7 ± 0.4b	15.2 ± 0.5d	28.1 ± 0.3b	31.7 ± 0.7a
β-Turn (%)	27.1 ± 0.4a	21.0 ± 0.6b	27.4 ± 0.5a	18.6 ± 0.2c	18.3 ± 0.7c
Random coil (%)	20.5 ± 0.4b	20.4 ± 0.6b	27.4 ± 0.4a	18.3 ± 0.2c	19.9 ± 0.7bc

Different letters in the same row indicate significant differences ($p < 0.05$). WF: wet-milled rice flour, SF: semidry-milled rice flour, HT: semidry-milled rice flour treated by hydrothermal, US: semidry-milled rice flour treated by ultrasound, MW: semidry-milled rice flour treated by microwave. D50 reflect the particle size by volume at 50% of all particles.

The DS content determined in rice flour was also shown in Table 1. The relatively lower content of DS in WF (2.06 g/100 g) and SF (3.96 g/100 g) was due to the soaking and tempering process that could soften rice grains, reduce the input of mechanical force and obtain more complete starch granules [14]. Meanwhile, MW had similar contents of DS (3.33 g/100 g), which was associated with the low output power of the microwave process. However, ultrasonication and hydrothermal treatment create significant damage to the integrity of flour particles, the corresponding DS content was 9.91 g/100 g and 24.41 g/100 g, respectively. The observed results are consistent with the smaller particle size of US and HT. During ultrasonic treatment, the granules of cereal starch easily disrupted by the generation and rapid disintegration of gas bubbles in the cavitation effect, and eventually lead to damaged particle [15]. As for HT, the treatment temperature was higher than the gelatinization temperature, the starch underwent an irreversible transition and lost crystallinity. Freeze-drying followed by grinding further led to an increase in DS content.

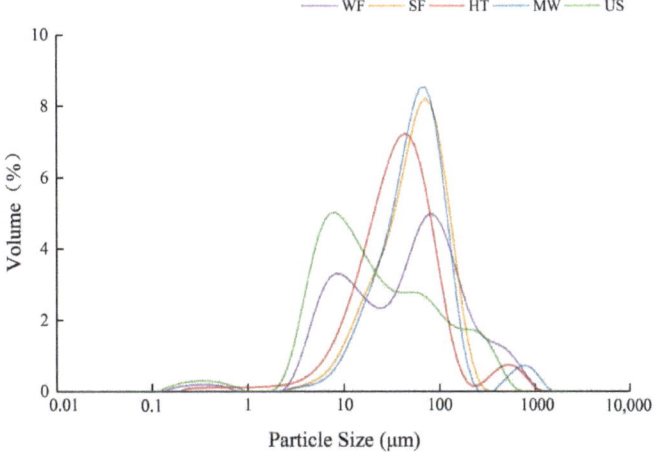

Figure 1. Particle size distribution of rice flour with different physical modifications.

2.2. Granule Morphology

The micrographs of SEM analysis are presented in Figure 2. WF and SF granules were polyhedral in form and regular in shape with a smooth surface. Rice starch were tightly wrapped in rice granule cells and closely linked to protein bodies and lipids [16]. Additionally, the microwave treated rice flour maintained the same particle integrity as the control flour (WF, SF). The subsequent magnification revealed that there were some dents and roughness on the surface of MW (white arrow in Figure 2B-MW). After ultrasonication treatment, the surface depression, pores and channels of rice flour became more obvious, which go in line with the results of DS (Table 1). The cavitation effect and mechanical effect of ultrasonication resulted in slight disruption on the surface of rice starch, which was in agreement with the observation of Kaur and Gill [15].

The modified granular structure using hydrothermal as compared to the other rice flour exhibited significant variations in shape and degree of agglomeration. These granules appeared to have a honeycomb-like structure (Figure 2A-HT). More fragments and cracks appeared on the surface shape, and the starch particles were swollen and gelatinized, resulting in particle aggregation (Figure 2B-HT). The gelatinization of starch was due to the higher temperature of the hydrothermal process. The microstructure changes of rice flour may affect its physicochemical properties, such as water absorption capacity and so on.

2.3. Crystal Structure

The X-ray diffraction patterns of native and physical modified rice flour are presented in Figure 3. Physically modified rice flour showed similar X-ray diffraction pattern to the control groups except HT, and they displayed the typical A-type crystal characteristics with strong diffraction peaks around 15°, 17°, 18° and 23° at 2θ. However, the crystalline structure of HT was completely damaged, showing a non-crystalline state [17]. The relative crystallinity calculated from the X-ray diffractogram pattern slightly decreased in US and MW, whereas it significantly decreased in HT. Yang et al. [18] reported that the non-crystalline regions of starch granules were more easily damaged than the crystalline regions. Therefore, we speculated that ultrasound mainly destroyed the non-crystalline regions of US in this study. In addition, the high-density energy input makes the water molecules in the starch particles boil and evaporate quickly without moving out, and high internal vapor pressure was generated during microwave treatment. The existence of pressure and the rapid migration of water may cause the decrease in MW crystallinity [19].

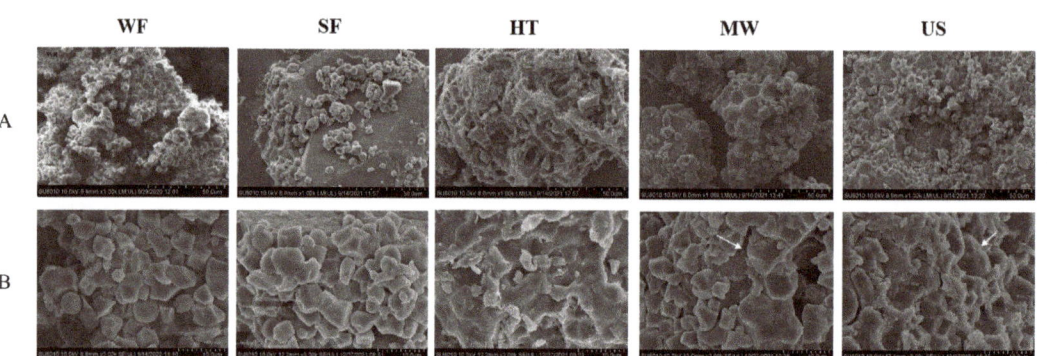

Figure 2. SEM images of rice flour with different physical modifications at magnification of (**A**) 1000×, (**B**) 3000×, respectively.

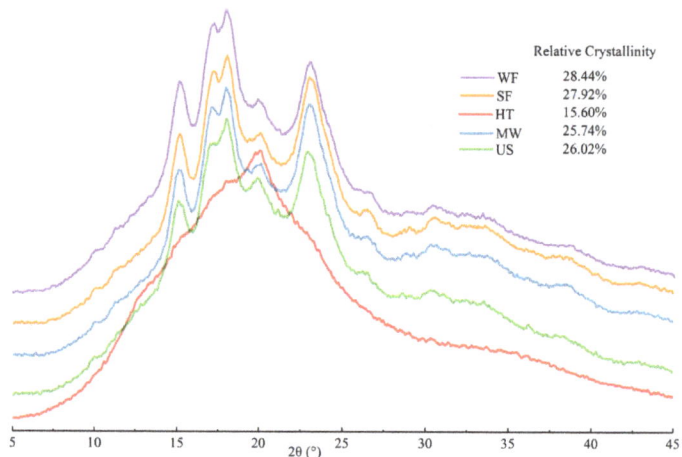

Figure 3. X-ray diffraction spectrum of rice flour with different physical modifications.

2.4. Short-Range Ordered Structure of Starch and Secondary Structure of Protein

Besides investigating the long-range order of the double helices in rice starch by XRD, the short-range order structure was also determined by FT-IR. The absorbance ratio of 1047/1022 cm^{-1} has been widely applied to describe the degree of short-range order in starch. The bands at 1047 and 1022 cm^{-1} represented the crystalline order and the amorphous region of starch, respectively. As shown in Table 1, compared to SF, the R1047/1022 value of MW remained unchanged, but increased in HT and US. This indicated that appropriate ultrasonic irradiation enhanced the recrystallization of starch and improved the ordered crystalline region. However, the study of Vela, Villanueva and Ronda [20] reported that the R1047/1022 value of rice starch was reduced after ultrasound treatment, owing to the breaking of hydrogen bonds. The inconsistent result may be due to the differences in varieties and experimental conditions, such as flour concentration, treatment time and treatment temperature. Besides, the R1047/1022 value of MW was slightly lower than that of SF, which might be attributed to the rapid boiling and evaporation of water caused by microwave heating. This was similar to the result of Han et al. [21].

The absorption peak of -OH (3200–3500 cm^{-1}) represents the strengths of the intermolecular hydrogen bonds in rice flour, and is presented in Figure 4. The shift of the peak to the low frequency wave number (red shift) implies an increase in the strength of intermolecular hydrogen bonds, while the shift to the high frequency wave number (blue shift) indicates a decrease in the strength of hydrogen bonds [22]. Figure 4 showed that the -OH absorption peaks in HT and MW blue shifted to the high frequency wave numbers, 3418.43 and 3406.44, respectively. This result may be due to the fast vibration of polar molecules caused by microwave heating and the disintegration of the double helix structure after gelatinization caused by hydrothermal treatment. While the -OH absorption peaks in US red shifted to the high frequency wave number compared to that of SF, indicating that ultrasound strengthened the intermolecular hydrogen bond in rice flour. With the increase in ultrasonic intensity and time, the loosened proteins reassembled again to form new hydrogen bonds, and the starch short chains were further linked by hydrogen bonds to form a new ordered structure [23].

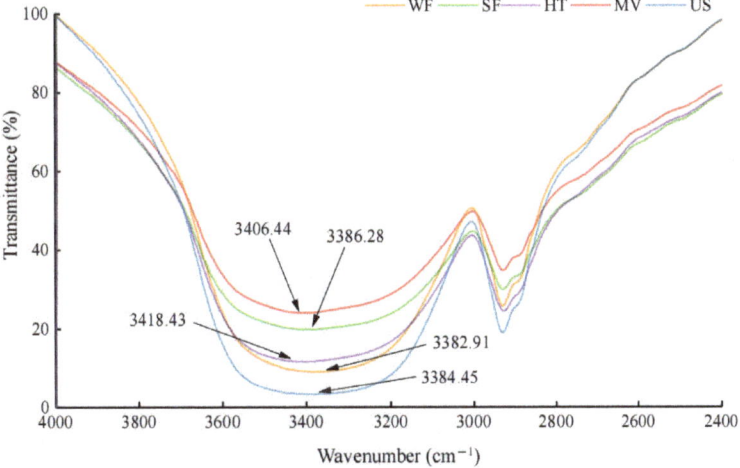

Figure 4. FTIR spectra of rice flour with different physical modifications.

In order to explore the effect of physical modification on the protein secondary structure, the change in amide I (1700–1600 cm^{-1}) was estimated from the relative area of each individual peak shown in Table 1. Hydrothermal treatment led to a significant reduction in α-helix up to 45.1% and an increase in random coil up to 34.3% in HT compared to SF. This result indicated that hydrothermal treatment could cause the unfolding and rearrangement of polymeric subunits, and eventually lead to disordered structure. There was an increase in α-helix content of US, but the β-turn content of US decreased. These observations may be explained in part by the reorganization of the loose protein structure under cavitation effect of ultrasound. These findings were inconsistent with Vela et al. [24], who reported that ultrasound decreased the content of α-helix and β-sheet in rice protein. This differing result might be related to the differences in protein and sonication conditions. In MW, the increase of β-sheet was accompanied by a decrease in β-turn, indicating the transformation between these two structures, and microwave treatment helped implement the conversion from β-turn to β-sheet. This change may improve the texture of rice bread. According to a previous study in wheat flour, the rise of β-sheet and the decline in β-turn made the gluten network of dough more ordered, which reduced the hardness of steamed bread and improved the quality [25].

2.5. Hydration Properties

Water absorption index (WAI), water solubility (WS) and swelling power (SP) of rice flour with different physical modifications are shown in Table 2. The results exhibited that hydration properties values of SF and WF in 25 °C were statistically insignificant ($p > 0.05$) and lower than ($p < 0.05$) that of modified rice flour. In contrast, WAI, WS and SP of HT was highest in all samples. The higher hydration properties values could be due to the highest DS content and the cracks and fragments of the damaged granules, which made the granules easier to absorb water and expand [26]. In addition, high values of SP in HT could be caused by the damage of the crystalline structure of rice starch (Figure 3) and the combination of water molecules with the free hydroxyl groups in amylose or amylopectin [27]. Ultrasound treatment also significantly increased the WAI and WS of rice flour in 25 °C, which may be because ultrasonic treatment destroyed the amorphous region of starch particles, resulting in amylose leaching and increased solubility.

Table 2. Hydration, thermal and pasting properties of rice flour with different physical modifications.

Sample	WF	SF	HT	MW	US
Hydration properties					
WAI (g/g)	3.05 ± 0.20c	3.00 ± 0.10c	10.54 ± 0.08a	3.81 ± 0.12b	3.74 ± 0.44b
WS (%)	2.45 ± 0.49b	2.14 ± 0.76b	4.33 ± 0.44a	1.93 ± 0.29b	3.42 ± 0.22a
SP (g/g)	3.13 ± 0.21c	3.07 ± 0.07c	11.02 ± 0.03a	3.89 ± 0.12b	3.88 ± 0.46b
Thermal properties					
To (°C)	62.47 ± 0.11a	60.78 ± 0.45b	-	60.18 ± 0.06b	62.10 ± 0.74a
Tp (°C)	69.02 ± 0.03a	67.36 ± 0.17b	-	66.76 ± 0.17b	66.75 ± 0.75b
Te (°C)	79.98 ± 0.00a	79.32 ± 0.40bc	-	76.26 ± 2.74cd	74.33 ± 0.43d
ΔH (J·g^{-1})	8.48 ± 0.02a	7.63 ± 0.19bc	-	7.85 ± 0.17b	7.33 ± 0.30c
ΔT (°C)	17.51 ± 0.11a	18.54 ± 0.36a	-	16.09 ± 2.77a	12.22 ± 0.56b
Pasting properties					
PV (cP)	4144 ± 23a	4209 ± 46a	3276 ± 78b	4067 ± 113a	2498 ± 122c
TV (cP)	2580 ± 8a	2300 ± 54b	2449 ± 60a	2506 ± 10a	1731 ± 86c
BD (cP)	1564 ± 15b	1909 ± 13a	829 ± 39c	1561 ± 103b	768 ± 36c
FV (cP)	4218 ± 7a	3853 ± 52b	3901 ± 90b	4182 ± 52a	3099 ± 95c
SB (cP)	1638 ± 1ab	1553 ± 7bc	1452 ± 81cd	1677 ± 43a	1369 ± 29e

WF: wet-milled rice flour, SF: semidry-milled rice flour, HT: semidry-milled rice flour treated by hydrothermal, US: semidry-milled rice flour treated by ultrasound, MW: semidry-milled rice flour treated by microwave. WAI: Water absorption index, WS: water solubility, SP: swelling power, To: onset temperatures, Tp: peak temperatures, Te: end temperatures, ΔH: enthalpy change, ΔT: gelatinization temperature range; PV: peak viscosity, TV: trough viscosity, BD: breakdown viscosity, FV: final viscosity, SB: setback viscosity. Different letters in the same row indicate significant differences ($p < 0.05$).

2.6. Thermal Properties

The influence of physical modifications on rice flour thermal properties was determined and the results are shown in Table 2. Higher To, Tp, Te and ΔH were observed in WF compared to SF, which was due to the complete starch granules and fine flour particles, leading to a greater barrier for energy transfer and higher heat stability [13]. To, Tp, Te and ΔH were not detected in HT since the starch granules was completely gelatinized after hydrothermal treatment. Compared with SF, no significant ($p > 0.05$) changes were observed in the value of gelatinization temperature and ΔH of MW. Nevertheless, the ΔH of US presented a decreased trend, which might be related to the higher DS content and the weakening of amorphous regions. The pores and cracks on the surface of DS (Figure 2), as well as the destruction of amorphous areas caused by cavitation effect (Section 3.3), make it easier for water to diffuse into particles and further penetrate into the crystalline area of starch. Decreased ΔH was also reported for rice starch after ultrasonication [18]. Meanwhile, the end temperatures were significantly decreased after ultrasonication (74.33 °C) compared to that of SF (79.32 °C), leading to a narrowed ΔT. This result is mainly due to the reinforcement of starch structure induced by ultrasound treatment. The ultrasonic treatment distorted amorphous regions of the starch granules (Figure 3), increased the starch homogeneity and thus strengthened the remaining crystalline structure (Table 2). As a result, the energy required for gelatinization (ΔH) was reduced and the gelatinization temperature (ΔT) was narrowed. The results are consistent with the reports of Amini, Razavi and Mortazavi [28] and Chi et al. [29].

2.7. Pasting Properties

Pasting properties of rice flour with different physical modifications are shown in Table 2. MW exhibited much higher pasting viscosity compared to those of other physical modified rice flours and close to those of WF. HT exhibited lower values of pasting viscosity than those of SF, as a consequence of the molecular destruction and gelatinization of starch granules during hydrothermal treatment. In addition, sonication led to a significant decrease in the viscosity profile, consistent with the study in corn starch by Yang et al. [30]. Cracks on the surface of US starch particles contribute to the penetration of water for hydration, causing significant decrease in viscosity. Besides, SB value of starch was related to the amylose structure, which mainly constituted the amorphous regions. Ultrasound treatment distorted the amorphous regions, resulting in the leaching out of amylose molecules and a decrease in SB. However, an opposite result was observed by Yang et al. [18], who found an increase in pasting viscosity in ultrasound treated rice starch compared to native starch. It could be explained by the secondary structure of rice protein. Ultrasound treatment may reduce proteins folding, which could increase the sites for binding water. This would reduce the possibility of other components contacting water and decrease the rate and extent of leaching out of granular components from the granule, thereby reducing the pasting viscosities [31].

2.8. Rheological Properties of Rice Batter

The viscosity and elasticity of batter are the key indicators of bread quality [32]. The mixed rice flour could form a starch-based batter structure. The function of HPMC is to simulate the structure of gluten, to improve the viscoelasticity of the paste and stabilize the gas in the mixing process [33]. The dynamic viscoelastic properties of batter added different modified rice flour were investigated in Figure 5. Frequency sweeps showed that G' exceeded G'' in the whole frequency range tested and tan δ < 1 at any point, indicating the solid-like behavior of all the batter. With the increasing frequency, G' and G'' were increased, showing that viscoelasticity was enhanced. The G' and G'' values of the modified rice flour were higher than those of SF, and the tan δ curve of the modified rice flour was lower than that of SF. These results indicated the viscoelasticity of the batter was enhanced and the gel strength of the batter was improved after adding physical modified rice flour. The rheological properties of the batter with US were the closest to those of WF, suggesting that batter with US had more viscoelastic and structured. This was probably because the amorphous regions of starch were destroyed by sonication, which promoted the leaching of amylose and the extension of the helix structure increased the association of linear chains, which was able to form a consolidated network [34]. The viscoelasticity of the batter added MW also improved compared to that of SF. However, batter added HT exhibited highest G' and G'' values and lowest tan δ, which indicated a harder texture. This was because HT with a high DS content was easier to absorb water, leading to excessive intermolecular cross-linking of the batter, which would negatively affect the development of the dough network and lead to a stiffer bread crumb [35].

2.9. Bread Quality

Table 3 and Figure 6 show the specific volume and microstructure of rice bread added physically modified rice flour. It can be found that the specific volume of rice bread with the addition of US (3.63 mL/g) and MW (3.51 mL/g) was greater than that of SF (3.14 mL/g), and comparable to that of WF (3.72 mL/g). In order to obtain a large volume of bread, the dough not only need to be soft enough to expand during proofing, but also have enough strength to bear the rapid expansion of CO_2 gas cell during baking. The higher bread volume could be related to the increase of batter viscosity as a result of ultrasound and microwave treatment. The rheological properties (Figure 5) showed that the viscoelasticity of batter with US and MW were improved compared to SF, and the elasticity and viscous modulus of the batter with US were closest to those of WF. The increase of dough viscoelasticity improved the gas retention ability of dough, and prevented gas from escaping during fermentation and baking, which could produce rice bread with higher specific volume to a certain extent [36]. Therefore, the bread with US and MW had higher volume compared to the SF formulation. However, excessive viscosity would reduce the ductility of the dough and limit the generation and maintenance of gas, resulting in a low specific volume of bread [37]. This would explain why the batter with 15% of HT, having highest elasticity and viscous modulus and lowest tan δ, led to breads with the smallest bread volume (1.18 mL/g).

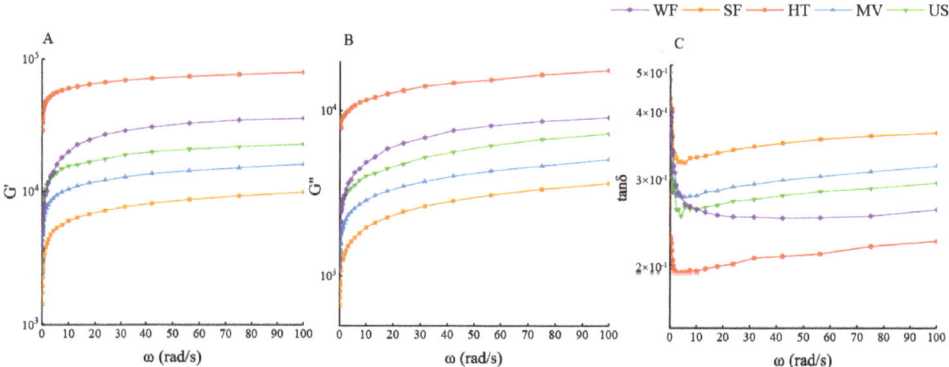

Figure 5. (**A**) The storage modulus (G'), (**B**) loss modulus (G") and (**C**) the loss tangent (tan δ) of rice flour batter added physically modified rice flour.

Table 3. Texture properties of rice bread added physically modified rice flour.

Sample	Specific Volume (mL/g)	Hardness (g)	Resilience (%)	Cohesion	Springiness (%)	Gumminess	Chewiness
WF	3.72 ± 0.12a	48.59 ± 2.8b	47.18 ± 1.6a	0.89 ± 0.01a	165.74 ± 5.5a	44.15 ± 2.8bc	68.62 ± 4.3b
SF	3.14 ± 0.10b	51.30 ± 4.7b	42.73 ± 1.7b	0.83 ± 0.01b	97.39 ± 1.8b	49.61 ± 5.1b	49.74 ± 6.2bc
HT	1.18 ± 0.07c	701.70 ± 125.5a	33.54 ± 2.1c	0.66 ± 0.02e	81.23 ± 6.2c	468.40 ± 21.3a	442.94 ± 58.9a
MW	3.51 ± 0.19a	46.06 ± 5.2b	30.73 ± 0.9d	0.74 ± 0.02d	93.78 ± 2.7b	38.81 ± 4.7cd	35.91 ± 4.7c
US	3.63 ± 0.08a	42.27 ± 4.0b	32.41 ± 1.4d	0.78 ± 0.02c	101.35 ± 6.6b	32.62 ± 3.4d	31.29 ± 2.8c

Different letters in the same row indicate significant differences ($p < 0.05$). WF: wet-milled rice flour, SF: semidry-milled rice flour, HT: semidry-milled rice flour treated by hydrothermal, US: semidry-milled rice flour treated by ultrasound, MW: semidry-milled rice flour treated by microwave.

From the microstructure of breadcrumbs shown in Figure 6B, it can be seen that the network structure formed by gelatinized starch, HPMC and denatured protein showed a smooth surface, dense and porous, and was similar to the honeycomb gluten network. Under the microscope, the gas cells of the bread added physically modified rice flour became larger. However, the cohesion of the network structure of the bread with HT was enhanced and the matrix tended to thicken. This occurred because the excessive cross-linking of the batter with HT resulted in a compact structure, which limited the expansion of the bread and was not conducive to the edible quality of the bread.

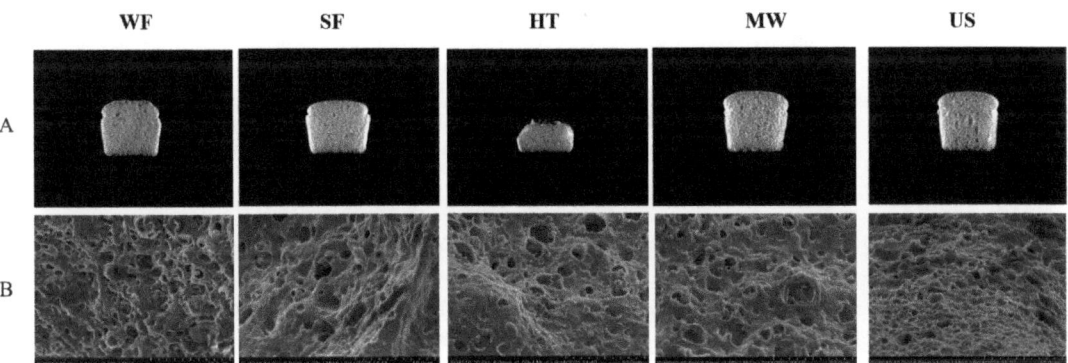

Figure 6. (A) Specific volume and (B) microstructure of rice bread added physically modified rice flour.

Texture properties of the rice bread added physically modified rice flour are shown in Table 3. Hardness is usually considered a key indicator of sensory perception, which is related to the specific volume of the rice bread. A higher specific volume means a higher amount of air retained in the crumb structure, leading to a softer texture [38]. The use of US and MW make breadcrumbs softer compared to SF, corresponding to a 17% and 10% decrease in hardness, respectively. Studies have shown that the cavitation of ultrasound can destroy the amorphous region of starch, leading to the exposition of hydrophilic groups and the increase of dough hydrophilicity, which reduce the hardness [25]. With the increase of ultrasonic time, the short-chain molecules rearranged to produce a more compact structure, and the linear molecules were more tightly bound to form a new ordered structure, which facilitated the generation and maintenance of the network structure of the dough and bread. Whereas, the microwave treatment destroyed the hydrogen bonds within the molecule and the protein secondary structure (Section 3.4), broke the long chains of starch into short chains, which may contribute to the soft texture of rice bread. The same result was previously reported by Villanueva et al. [8], who found that the partial replacement of rice flour by microwave treatment significantly increased bread specific volume. However, the addition of HT led to a harder crumb (from 51.30 g to 701.70 g). The small bread volume corresponded to less amount of air retained in the batter during proofing and baking, mainly due to the excessive dough viscoelasticity, which endorsed a stiffer crumb. Similar results were observed in gumminess and chewiness, probably because these indicators were mainly affected by hardness. However, the resilience, cohesion and springiness of the bread added different physical modified rice flour were overall lower than that of the control bread.

2.10. Principal Component Analysis (PCA)

Figure 7 shows the results of principal component analysis for rice flour and rice bread attributes, to explore the relationship of rice flour with different physical modifications and rice bread added physical modified flour. Overall, PC1 and PC2 accounted for 80.28% of the total variation, indicating the plane of PC1 and PC2 reflects, to a large extent, the main contributions of the response variables. The results show that different physical modifications led to significantly different flour properties and bread quality. In Figure 7, according to the properties of rice flour and rice bread, the samples are divided into three groups as a whole. HT was located in the positive quadrant of PC1 and PC2, which was separated from the rest of the samples since it possessed high DS content, water solubility and water absorption, and the bread with HT had a stiffer crumb texture. This result indicated that hydrothermal treatment could destroy the rice starch, increase the DS content and make the granules absorb water easier, which lead to the hard crumb texture and high chewiness of bread. SF and MW were located in the negative part of PC1 and PC2. US and WF were located in the negative quadrant of PC1 and the positive quadrant of PC2, and their bread exhibited high specific volume as well as high springiness, cohesion and resilience, which was due to the high extensibility and resistance of batter adding US and WF. In general, adding ultrasonic modified rice flour would improve the quality of bread based on semidry-milled rice flour, whereas hydrothermal treatment is not a desirable modification method.

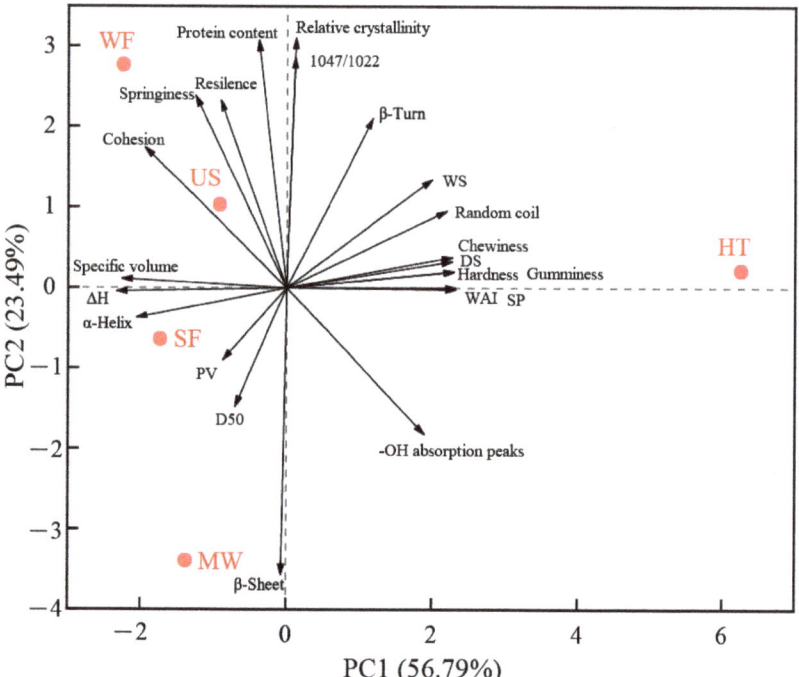

Figure 7. Principal component analysis.

3. Materials and Methods

3.1. Rice Flour

Polished japonica rice (Panjin, China; harvested in 2020) provided by Panjin Pengyue Rice Industry Co., Ltd. was used in this study. The moisture content was 11.46%, protein: 8.21% dry basis (db), total starch: 80.75% db and amylose: 18.80% db, respectively. Semidry-milled rice flour (SF) was prepared according to Qin et al. [13]. A total of 200 g of polished japonica rice was first treated by hot air in an oven at 45 °C for 1 h and subsequently tempered for 35 min with the addition of water. The added water volume was calculated as follows:

$$\text{Added water (mL)} = m \times (\omega_1 - \omega_2)/(1 - \omega_1) \qquad (1)$$

where m represented the weight of rice flour, ω_1 represented the saturated moisture content of rice grain (approximately 28% according to preliminary experiment) and ω_2 represented the moisture content of rice grain after hot air treatment. After grinding the rice with a cyclone mill (CT410, Foss Analytical Co., Ltd., Hillerød, Denmark), the obtained rice flour was dried in an oven at 45 °C until the moisture content was less than 12%.

Wet-milled rice flour (WF) was prepared as follow: 200 g of polished rice grains were soaked in 400 mL water for 2 h and then milled in a colloid mill (JMS-30A, Langtong Machinery Co., Ltd., Hebei, China). The obtained slurry was freeze-dried (SCIENTZ-10N,

Scientz Biotechnology Co., Ltd., Ningbo, China) and then sifted through a 100-mesh screen. In this study, SF and WF were set as the control.

3.2. Different Physical Modification Methods

3.2.1. Ultrasound Treatment

SF was suspended in distilled water to give a concentration of 30% (w/v) and stirred for 10 min to obtain a homogeneous slurry. An ultrasonic processer (ScientzIID, Zhejiang, China) equipped with a 6 mm titanium alloy probe was used. According to the method of Yang et al. (2019) [18] with minor modifications, the slurry was treated with ultrasound for 20 min (5 s on and 5 s off) at a constant frequency of 20 kHz with an output power of 600 W, and the probe was placed at 1.5 cm depth. Then, they were retrieved by freeze-drying, followed by sifted through a 100-mesh screen. SF treated by ultrasound was coded as US and stored in a desiccator at room temperature.

3.2.2. Microwave Treatment

A 100 g of SF was put in a glass container and then placed in the center of the microwave oven (M3-L205C, Midea, Foshan, China). As per the previous method with slight modifications [19], samples were irradiated at 300 W for 5 min. The irradiated samples were freeze dried and then sifted through a 100-mesh screen. SF treated by microwave was coded as MW and stored in a desiccator at room temperature.

3.2.3. Hydrothermal Treatment

Hydrothermally treated rice flour (HT) was prepared by referring to the method of Bourekoua et al. [39] with minor modification. SF was suspended in distilled water to obtain 20% (w/w) suspension and stirred continuously in a hot water bath until the inner temperature reached 65 °C. The rice slurry was heated in a hot water bath for 30 min. Then, the paste was cooled at room temperature and freeze-dried. The dried rice was milled into flour by a grinder and passed through a 100-mesh screen. The sample was stored in a desiccator at room temperature.

3.3. Particle Size Distribution, Damaged Starch (DS) Content and Protein Content

The particle size of all samples was determined by Malvern MasterSizer 3000 (Malvern Instrument, Ltd., Worcestershire, UK). In addition, D50 (the particle diameters at cumulative volume percentage of 50%) was calculated by the software provided by the device. The DS content (AACC Method 76–31.01) was measured by commercial assay kit (K-SDAM, Megazyme International Ltd., Wicklow, Ireland).

3.4. Scanning Electron Microscopy (SEM)

Rice flour was fixed onto the surface of a circular aluminum stud with double-sided sticky tape and then observed using a scanning electron microscopy (Hitachi S-570, Hitachi, Co., Ltd., Tokyo, Japan) under 10 kV.

3.5. X-ray Diffraction (XRD)

The crystallization characteristics of freeze-dried rice flour were determined by X-ray powder diffractometer (Bruker D8 Advance, Salbuluken, Germany). Scans were performed at diffraction angles (2θ) ranging from 5° to 45°, with a scanning rate of 10°/min and step size of 0.02°. Then, 40 kV and 40 mA with Cu-Kα radiation were set as operating conditions. The relative crystallinity (%) of all samples was calculated by MDI Jade 6.0 software (Materials Data, Inc., Livermore, CA, USA).

3.6. Water Hydration Property

Water hydration property included water absorption index (WAI), water solubility (WS) and swelling power (SP). Additionally, 100 mg of rice flour was dispersed in 10 mL distilled water and stirred for 30 min at 25 °C. The suspensions were then cooled to room temperature and centrifuged at 8500× g for 30 min. Subsequently, the sediment was weighed for calculating WAI. The supernatants were collected in an aluminum dish of a known weight, dried at 105 °C until the weight was constant [40]. WAI, WS and SP were calculated according to the previous study [13].

3.7. Short-Range Ordered Structure of Starch and Secondary Structure of Protein

The FTIR spectra of rice flour samples were obtained using the FTIR spectrometer (TENSOR 27, Borken, Germany). Samples were equilibrated at 45 °C for 24 h and mixed with KBr (1:75, w/w), followed by grounding and pressing into tablets. The wavenumber range was set in the range of 4000–600 cm^{-1} at a resolution of 4 cm^{-1} with 64 scans. The spectrum was normalized by using OMNIC (Thermo Fisher Scientific, Waltham, MA, USA). The changes in the short-range order structure of starch were calculated by the absorbance ratio 1047/1022 (R1047/1022). Amide I bands (1700–1600 cm^{-1}) were analyzed using PeakFit 4.12 (SeaSolve Software Inc., San Jose, CA, USA).

3.8. Differential Scanning Calorimetry (DSC)

The thermal characteristics of physical modified rice flour were measured by a DSC8000 analyzer (PerkinElmer, Norwalk, CT, USA). Then, 3 mg of rice flour was accurately weighed in an aluminum pan, and deionized water was added at a ratio of 1:2 (w/w). The sealed samples were balanced overnight at 4 °C and scanned at a temperature range of 30–110 °C at a rate of 10 °C/min. A sealed empty pan was used as a reference [41].

3.9. Pasting Properties

A Rapid Viscosity Analyzer (RVA-TecMaster, Perten Instruments, NSW, Australia) was adopted to measure the pasting properties of the samples. Each sample (3 g, dry basis) was mixed with 25 g distilled water before being transferred into an aluminum canister. The sample was equilibrated at 50 °C for 1 min, heated from 50 °C to 95 °C at 0.2 °C/s, held for 2.5 min at 95 °C, then cooled to 50 °C at the rate of 0.2 °C/s, and finally maintained at 50 °C for 2 min [13].

3.10. Rheological Properties of Gels

The dynamic viscoelasticity of rice batter (consisting of all ingredients except butter and yeast) were assessed by a Rheometer (Physica MCR 301, Anton Paar GmbH, Graz, Austria). The resting time of batter samples before the measurement was 5 min. Dynamic viscoelasticity measurements were done using parallel plate geometry (50 mm diameter, 1 mm gap) at 25 °C. Conduct frequency scanning of 0.1–100 rad/s with 0.01% strain [42]. Three replicates were performed for each analysis.

3.11. Breadmaking Procedure

The physical modified rice flour of US, MW and HT were prepared into bread, and mixed with SF in the ratio of 3:17 (For example, 15 g modified rice flour and 85 g SF). The gluten-free rice bread formula was based on 100 g mixed rice flour, 90 g water, 4 g HPMC, 20 g sugar, 10 g batter and 2 g yeast. Yeast was first dissolved in water at 28–30 °C to hydrate. The other dry ingredients were mixed evenly for 5 min using a Pin Mixer (JHMZ200, Beijing Dongfu Jiuheng Instrument Technology Co., Ltd., Beijing, China). Then, the hydrated yeast mixture and batter were added and mixed for 5 min. After stirring, 50 g of batter was filled in a baking tray (70 mm × 70 mm × 40 mm) and proofed in a fermentation chamber (CF-6000, Zhongshan Kashi Electric Appliance Co., Ltd., Zhongshan, China) for 40 min at 38 °C and 85% relative humidity. The batter was put in an electric oven (T7-L328E, Midea Group Co. Ltd., Foshan, China) and the baking parameters were set at 150 °C for 15 min. After baking, the bread was taken out and then cooled for 1 h. Finally, the samples were sealed and stored for determination. Bread was made three times for each formulation.

3.12. Gluten-Free Rice Bread Quality

Bread volume (mL) was measured after 1 h cooling and determined by the rapeseed displacement method 10-05 (AACC International, Saint Paul, MN, USA, 2000). The specific volume (mL/g) was calculated as the volume divided by the weight measured 1 h after baking.

After baking and cooling, the breadcrumbs were immediately frozen by liquid nitrogen and then freeze-dried. The size of 4 mm $\times$ 2 mm $\times$ 2 mm small pieces were cut from the center of the breadcrumbs, and then their microstructures were examined by SEM at magnifications $1000\times g$ with 10 kV.

Crumb texture was determined by using a Texture Analyzer (TA-XT2i, Stable Micrio System, Godalming, England). Small piece sizes of 15 mm $\times$ 15 mm $\times$ 15 mm were cut from the center of the breadcrumbs. The texture profile analysis (TPA) double compression test was performed by an aluminum probe (36 mm diameter; P/36R), which could penetrate 50% depth at 1 mm/s speed test with a 5 s delay between the first and second compression. The trigger force was 5 g. A total of 12 replicates were carried out.

4. Conclusions

Rice flour was subjected to ultrasound, microwave and hydrothermal treatments. These physical modified treatments significantly changed the particle size and structure of rice flour, which affected their hydration, pasting and rheological behavior, and further changed the rice bread quality. After hydrothermal treatment, the structure of starch granules was seriously destroyed, and the percentage of irregular curl in the protein secondary structure increased. The excessive batter viscoelasticity of HT limited the development of the batter during fermentation, resulting in a stiffer bread crumb and smaller bread volume. In addition, microwave treatment weakened the hydrogen bonding force and changed the secondary structure of the protein with the conversion from β-turn to β-sheet. The doughs made with MW showed higher viscoelasticity and more stable cell structure than that of SF. Especially, the starch structure of US was reinforced and the protein secondary structure was modified with the decrease in β-turn and increase in α-helix. These molecular changes could explain the decrease of the pasting profile and narrowing of the gelatinization range in US. Moreover, the viscoelasticity of the batter containing 15% US was significantly improved compared to the SF batter, leading to a higher bread volume and softer crumb texture close to those of bread with WF. In conclusion, ultrasound modification is an alternative to improve the morphology and physical properties of rice flour and further improve gluten-free bread quality.

Author Contributions: W.Q.: Investigation, Data curation, Writing—original draft. H.X.: Writing—review & editing. A.W.: Validation, Methodology. X.G.: Resources, Methodology. Z.C.: Software, Resources. Y.H.: Software, Methodology. L.W.: Resources. L.L.: Methodology. F.W.: Project administration. L.T.: Conceptualization, Project administration, Funding acquisition. All authors have read and agreed to the published version of the manuscript.

Funding: This work was supported by the National Natural Science Foundation of China [31972005] and the National Key Research and Development Program of China (2017YFD0401104-05).

Institutional Review Board Statement: Not applicable.

Informed Consent Statement: Not applicable.

Data Availability Statement: Not applicable.

Conflicts of Interest: The authors declare no conflict of interest.

Sample Availability: Samples of the compounds are available from the authors.

References

1. Brouns, F.; van Rooy, G.; Shewry, P.; Rustgi, S.; Jonkers, D. Adverse Reactions to Wheat or Wheat Components. *Compr. Rev. Food Sci. Food Saf.* **2019**, *18*, 1437–1452. [CrossRef] [PubMed]
2. Santos, F.G.; Aguiar, E.V.; Rosell, C.M.; Capriles, V.D. Potential of chickpea and psyllium in gluten-free breadmaking: Assessing bread's quality, sensory acceptability, and glycemic and satiety indexes. *Food Hydrocoll.* **2021**, *113*, 106487. [CrossRef]
3. Solaesa, Á.G.; Villanueva, M.; Muñoz, J.M.; Ronda, F. Dry-heat treatment vs. heat-moisture treatment assisted by microwave radiation: Techno-functional and rheological modifications of rice flour. *LWT-Food Sci. Technol.* **2021**, *141*, 110851. [CrossRef]
4. Sun, X.; Saleh, A.S.M.; Sun, Z.; Ge, X.; Shen, H.; Zhang, Q.; Yu, X.; Yuan, L.; Li, W. Modification of multi-scale structure, physicochemical properties, and digestibility of rice starch via microwave and cold plasma treatments. *LWT-Food Sci. Technol.* **2022**, *153*, 112483. [CrossRef]
5. Fonseca, L.M.; Halal, S.; Dias, A.R.G.; Zavareze, E.D.R. Physical modification of starch by heat-moisture treatment and annealing and their applications: A review. *Carbohyd. Polym.* **2021**, *274*, 118665. [CrossRef]
6. Ruiz, E.; Srikaeo, K.; Revilla, L. Effects of Heat Moisture Treatment on Physicochemical Properties and Starch Digestibility of Rice Flours Differing in Amylose Content. *Food Appl. Biosci.* **2018**, *6*, 140–153.
7. Li, Y.-D.; Xu, T.-C.; Xiao, J.-X.; Zong, A.-Z.; Qiu, B.; Jia, M.; Liu, L.-N.; Liu, W. Efficacy of potato resistant starch prepared by microwave–toughening treatment. *Carbohyd. Polym.* **2018**, *192*, 299–307. [CrossRef]
8. Villanueva, M.; Harasym, J.; Muñoz, J.M.; Ronda, F. Rice flour physically modified by microwave radiation improves viscoelastic behavior of doughs and its bread-making performance. *Food Hydrocoll.* **2019**, *90*, 472–481. [CrossRef]
9. Shabbir, M.A.; Ahmed, W.; Latif, S.; Inam-Ur-Raheem, M.; Manzoor, M.F.; Khan, M.R.; Bilal, R.M.; Aadil, R.M. The quality behavior of ultrasound extracted sunflower oil and structural computation of potato strips appertaining to deep-frying with thermic variations. *J. Food Process. Preserv.* **2020**, *44*, e14809. [CrossRef]
10. Tayyab Rashid, M.; Ahmed Jatoi, M.; Safdar, B.; Wali, A.; Muhammad Aadil, R.; Sarpong, F.; Ma, H. Modeling the drying of ultrasound and glucose pretreated sweet potatoes: The impact on phytochemical and functional groups. *Ultrason. Sonochem.* **2020**, *68*, 105226. [CrossRef]
11. Cui, R.; Zhu, F. Effect of ultrasound on structural and physicochemical properties of sweetpotato and wheat flours. *Ultrason. Sonochem.* **2020**, *66*, 105118. [CrossRef]
12. Li, S.; Li, Q.; Zhu, F.; Song, H.; Wang, C.; Guan, X. Effect of vacuum combined ultrasound treatment on the fine structure and physiochemical properties of rice starch. *Food Hydrocoll.* **2022**, *124*, 107198. [CrossRef]
13. Qiu, W.; Liu, Z.; Wang, A.; Chen, Z.; He, Y.; Wang, L.; Liu, L.; Wang, F.; Tong, L. T. Influence of particle size on the properties of rice flour and quality of gluten-free rice bread. *LWT-Food Sci. Technol.* **2021**, *151*, 112236. [CrossRef]
14. Tong, L.-T.; Zhu, R.; Zhou, X.; Zhong, K.; Wang, L.; Liu, L.; Hu, X.; Zhou, S. Soaking time of rice in semidry flour milling was shortened by increasing the grains cracks. *J. Cereal Sci.* **2017**, *74*, 121–126. [CrossRef]
15. Kaur, H.; Gill, B.S. Effect of high-intensity ultrasound treatment on nutritional, rheological and structural properties of starches obtained from different cereals. *Int. J. Biol. Macromol.* **2019**, *126*, 367–375. [CrossRef]
16. Villanueva, M.; Harasym, J.; Muñoz, J.M.; Ronda, F. Microwave absorption capacity of rice flour. Impact of the radiation on rice flour microstructure, thermal and viscometric properties. *J. Food Eng.* **2018**, *224*, 156–164. [CrossRef]
17. Zhu, L.; Zhang, H.; Wu, G.; Qi, X.; Wang, L.; Qian, H. Effect of structure evolution of starch in rice on the textural formation of cooked rice. *Food Chem.* **2020**, *342*, 128205. [CrossRef] [PubMed]
18. Yang, W.; Kong, X.; Zheng, Y.; Sun, W.; Chen, S.; Liu, D.; Zhang, H.; Fang, H.; Tian, J.; Ye, X. Controlled ultrasound treatments modify the morphology and physical properties of rice starch rather than the fine structure. *Ultrason. Sonochem.* **2019**, *59*, 104709. [CrossRef]
19. Kumar, Y.; Singh, L.; Sharanagat, V.S.; Patel, A.; Kumar, K. Effect of microwave treatment (low power and varying time) on potato starch: Microstructure, thermo-functional, pasting and rheological properties. *Int. J. Biol. Macromol.* **2020**, *155*, 27–35. [CrossRef]
20. Vela, A.J.; Villanueva, M.; Ronda, F. Low-frequency ultrasonication modulates the impact of annealing on physicochemical and functional properties of rice flour. *Food Hydrocoll.* **2021**, *120*, 106933. [CrossRef]
21. Han, Z.; Li, Y.; Luo, D.H.; Zhao, Q.; Cheng, J.H.; Wang, J.H. Structural variations of rice starch affected by constant power microwave treatment. *Food Chem.* **2021**, *359*, 129887. [CrossRef] [PubMed]
22. Zhang, J.; Luo, D.; Xiang, J.; Xu, W.; Xu, B.; Li, P.; Huang, J. Structural Variations of Wheat Proteins under ultrasound treatment. *J. Cereal Sci.* **2021**, *99*, 103219. [CrossRef]
23. Hu, H.; Wu, J.; Li-Chan, E.C.Y.; Zhu, L.; Zhang, F.; Xu, X.; Fan, G.; Wang, L.; Huang, X.; Pan, S. Effects of ultrasound on structural and physical properties of soy protein isolate (SPI) dispersions. *Food Hydrocoll.* **2013**, *30*, 647–655. [CrossRef]
24. Vela, A.J.; Villanueva, M.; Solaesa, Á.G.; Ronda, F. Impact of high-intensity ultrasound waves on structural, functional, thermal and rheological properties of rice flour and its biopolymers structural features. *Food Hydrocoll.* **2021**, *113*, 106480. [CrossRef]
25. Luo, D.; Wu, R.; Zhang, J.; Zhang, K.; Xu, B.; Li, P.; Yuan, Y.; Li, X. Effects of ultrasound assisted dough fermentation on the quality of steamed bread. *J. Cereal Sci.* **2018**, *83*, 147–152. [CrossRef]
26. Lee, Y.T.; Shim, M.J.; Goh, H.K.; Mok, C.; Puligundla, P. Effect of jet milling on the physicochemical properties, pasting properties, and in vitro starch digestibility of germinated brown rice flour. *Food Chem.* **2019**, *282*, 164–168. [CrossRef] [PubMed]
27. Sujka, M.; Jamroz, J. Ultrasound-treated starch: SEM and TEM imaging, and functional behaviour. *Food Hydrocoll.* **2013**, *31*, 413–419. [CrossRef]

28. Mohammad Amini, A.; Razavi, S.M.; Mortazavi, S.A. Morphological, physicochemical, and viscoelastic properties of sonicated corn starch. *Carbohyd. Polym.* **2015**, *122*, 282–292. [CrossRef]
29. Chi, C.; Li, X.; Lu, P.; Miao, S.; Zhang, Y.; Chen, L. Dry heating and annealing treatment synergistically modulate starch structure and digestibility. *Int. J. Biol. Macromol.* **2019**, *137*, 554–561. [CrossRef]
30. Yang, Q.Y.; Lu, X.X.; Chen, Y.Z.; Luo, Z.G.; Xiao, Z.G. Fine structure, crystalline and physicochemical properties of waxy corn starch treated by ultrasound irradiation. *Ultrason. Sonochem.* **2019**, *51*, 350–358. [CrossRef]
31. Meadows, Pasting Process in Rice Flour Using Rapid Visco Analyser Curves and First Derivatives. *Cereal Chem.* **2002**, *79*, 559–562. [CrossRef]
32. Qin, W.; Lin, Z.; Wang, A.; Xiao, T.; He, Y.; Chen, Z.; Wang, L.; Liu, L.; Wang, F.; Tong, L.-T. Influence of damaged starch on the properties of rice flour and quality attributes of gluten-free rice bread. *J. Cereal Sci.* **2021**, *101*, 103296. [CrossRef]
33. Renzetti, S.; Rosell, C.M. Role of enzymes in improving the functionality of proteins in non-wheat dough systems. *J. Cereal Sci.* **2016**, *67*, 35–45. [CrossRef]
34. Monroy, Y.; Rivero, S.; García, M.A. Microstructural and techno-functional properties of cassava starch modified by ultrasound. *Ultrason. Sonochem.* **2018**, *42*, 795–804. [CrossRef]
35. Crockett, R.; Ie, P.; Vodovotz, Y. How do xanthan and hydroxypropyl methylcellulose individually affect the physicochemical properties in a model gluten-free dough? *J. Food Sci.* **2011**, *76*, E274–E282. [CrossRef]
36. Marston, K.; Khouryieh, H.; Aramouni, F. Effect of heat treatment of sorghum flour on the functional properties of gluten-free bread and cake. *LWT-Food Sci. Technol.* **2016**, *65*, 637–644. [CrossRef]
37. Ronda, F.; Pérez-Quirce, S.; Villanueva, M. Rheological Properties of Gluten-Free Bread Doughs: Relationship with Bread Quality. In *Advances in Food Rheology and Its Applications*; Ahmed, J., Ptaszek, P., Basu, S., Eds.; Woodhead Publishing: Cambridge, UK, 2017; Chapter 12; pp. 297–334.
38. Roman, L.; Reguilon, M.P.; Gomez, M.; Martinez, M.M. Intermediate length amylose increases the crumb hardness of rice flour gluten-free breads. *Food Hydrocoll.* **2020**, *100*, 105451. [CrossRef]
39. Bourekoua, H.; Benatallah, L.; Zidoune, M.N.; Rosell, C.M. Developing gluten free bakery improvers by hydrothermal treatment of rice and corn flours. *LWT-Food Sci. Technol.* **2016**, *73*, 342–350. [CrossRef]
40. Kim, M.; Oh, I.; Jeong, S.; Lee, S. Particle size effect of rice flour in a rice-zein noodle system for gluten-free noodles slit from sheeted doughs. *J. Cereal Sci.* **2019**, *86*, 48–53. [CrossRef]
41. Boulemkahel, S.; Betoret, E.; Benatallah, L.; Rosell, C.M. Effect of low pressures homogenization on the physico-chemical and functional properties of rice flour. *Food Hydrocoll.* **2021**, *112*, 106373. [CrossRef]
42. Pongjaruvat, W.; Methacanon, P.; Seetapan, N.; Fuongfuchat, A.; Gamonpilas, C. Influence of pregelatinised tapioca starch and transglutaminase on dough rheology and quality of gluten-free jasmine rice breads. *Food Hydrocoll.* **2014**, *36*, 143–150. [CrossRef]

Article

Effects of Annealing on the Properties of Gamma-Irradiated Sago Starch

Jau-Shya Lee [1,*], Jahurul Haque Akanda [2], Soon Loong Fong [3], Chee Kiong Siew [1] and Ai Ling Ho [1]

[1] Faculty of Food Science and Nutrition, University Malaysia Sabah, Jalan UMS, Kota Kinabalu 88400, Sabah, Malaysia; cksiew@ums.edu.my (C.K.S.); alho@ums.edu.my (A.L.H.)
[2] Department of Agriculture, School of Agriculture, University of Arkansas, 1200 North University Drive, M/S 4913, Pine Bluff, AR 71601, USA; akandam@uapb.edu
[3] ITS Nutriscience Sdn Bhd, 2, Jalan Sg. Kayu Ara 32/38, Berjaya Industrial Park, Shah Alam 40460, Selangor, Malaysia; slfong@its-nutriscience.com
* Correspondence: jslee@ums.edu.my

Abstract: The increase in health and safety concerns regarding chemical modification in recent years has caused a growing research interest in the modification of starch by physical techniques. There has been a growing trend toward using a combination of treatments in starch modification in producing desirable functional properties to widen the application of a specific starch. In this study, a novel combination of gamma irradiation and annealing (ANN) was used to modify sago starch (*Metroxylon sagu*). The starch was subjected to gamma irradiation (5, 10, 25, 50 kGy) prior to ANN at 5 °C (T_o-5) and 10 °C (T_o-10) below the gelatinization temperature. Determination of amylose content, pH, carboxyl content, FTIR (Fourier Transform Infrared) intensity ratio ($R_{1047/1022}$), swelling power and solubility, thermal behavior, pasting properties, and morphology were carried out. Annealing irradiated starch at T_o-5 promoted more crystalline perfection as compared to T_o-10, particularly when combined with 25 and 50 kGy, whereby a synergistic effect was observed. Dual-modified sago starch exhibited lower swelling power, improved gel firmness, and thermal stability with an intact granular structure. Results suggested the potential of gamma irradiation and annealing to induce some novel characteristics in sago starch for extended applications.

Keywords: sago starch; gamma irradiation; annealing; DSC; pasting; swelling power; crystalline order; gel firmness

1. Introduction

Apart from being the most abundant carbohydrate source for human staple food, the low cost, easy tailoring, biocompatibility, renewability, and extensive sources of starch have made it a most promising biodegradable natural polymer for food and nonfood industries [1]. In its native form, starch has limited industrial application due to its low solubility, low transparency, poor heat, shear and acid stability, fast retrogradation, and poor refrigerated and frozen storage stability [2,3]. The unique molecular and granular structure of starch [3] renders it versatile for various kinds of modifications, either by enzymatic, chemical, or physical means, to achieve the desired functional properties for industrial applications. The tailored modification of starch is very essential to obtain specific functional properties for innumerous industrial applications. Even though chemical modifications provide more options for the functionalization of starch [4], the food and pharmaceutical industries prefer starches without chemical modification for safety considerations [5].

The ionizing radiation processing is a progressive, fast, and environmentally friendly way to produce changes in the structural and functional properties of starch [6]. The irradiation of starch also requires minimal sample preparation, does not induce a significant increase in temperature, and is free of chemical residues [7–10]. After decades of study, food irradiation was proven to be safe as a processing treatment for foods [11]. Gamma radiation alters the properties of starch by generating free radicals that break the glycoside bonds and cause the decomposition of the macromolecules, while it may also promote the cross-linking of starch molecules [8]. The advantages of ionizing radiation have attracted great research interest to investigate the effects of irradiation on various types of conventional and nonconventional starches such as arrowroot starch [12], mung bean starch [13], and kithul starch [14].

In recent years, there has been a growing trend toward treating starch with hydrothermal methods, including annealing. Annealing (ANN) is carried out by incubating starch samples with moisture content higher than 65% (w/w) at a temperature that is above the glass transition temperature (T_g) but lower than the onset temperature (T_o) for gelatinization [15]. Though simple, annealing was reported able to improve the functional properties of starch for industrial application. Annealing contributes to improving the stability of starch paste to heat and shear, increasing the paste clarity and low gelation concentration, as well as the lower digestibility in Prata banana [16]. Annealing has been interpreted as a 'sliding diffusion', which entails the movement of complete molecular sequences within a crystalline lattice in starch, and/or a 'complete or partial fusion' of crystals and the subsequent recrystallization of the melted materials at the annealing temperature [3]. The interactions between starch chains improve crystalline perfection and change the physicochemical properties of the starches. According to Zavareze and Dias [17], the extent of these changes are influenced by the starch composition and source, the ratio of amylose to amylopectin, and the treatment conditions (temperature, time, and moisture levels).

Due to the potential for the further improvement and modulation of the functional properties of starch, a combination of modification methods for starches has started to attract research attention in recent years. The combination of dual step ANN and hydroxypropylation was found to significantly improve the properties of native barley starch [18]. Hu et al. [19] reported that dual modification by enzymatic debranching and hydroxypropylation can improve the mechanical properties of normal maize starch film. Dual modification of corn starch by annealing and succinylation was found to further improve the stability of starch paste against heat and shearing stress [20]. Other starch modifications by combined treatments reported in the literature include the combination of hydroxypropylation and acid hydrolysis for potato starch [21], hydrothermal treatment with sonication for Carioca bean starch [22], the combination of ultrasound and ANN for glutinous rice [23], the combination of gamma radiation and acetylation for wheat starch [24], the combination of sonication and gamma radiation for lentil starch [25], the combination of pullulanase debranching and microwave irradiation [26], and many others. In brief, these combined modification treatments produced starches with new properties that differed from the single modification.

Malaysia is one of the main producers of sago starch, which is extracted from the pith of a palm plant (*Metroxylon sagu*). It is considered an underutilized starch in food applications due to its poor functional properties [27]; hence, modification to improve its properties would potentially widen its utilization. In line with the ever-increasing demand for modified starch produced using a safe, fast, and economically viable method, the combination of gamma irradiation and ANN was used to modify the sago starch. To our knowledge, no work has been carried out to investigate starch modification using a combination of these two physical treatments. Our earlier preliminary work explored the effects of ANN on the pH, the degree of the short-range crystalline order, pasting, and the thermal properties of the sago starch treated with different doses of gamma radiation [28]. We found that increasing radiation doses improved the crystalline perfection by annealing, thus modifying the thermal and pasting properties of the sago starch. It is known that the extent of the crystal growth/perfection by ANN is dependent on the mobility of the glucan chains in the amorphous regions of the starch, which can be manipulated by the incubation temperature or the amount of plasticizer (water, in this case) [3]. We hypothesized that by increasing the annealing temperature, the extent of double helices formation in the irradiated starch would be further enhanced. In the previous study, the annealing temperature was at 10 °C below the onset of gelatinization. In the present study, we investigated the effect of higher annealing temperature (5 °C below the onset of gelatinization) on the irradiated sago starch and examined further the changes in the amylose content, carboxyl content, swelling power, solubility, gel firmness, and the morphology of the starch on top of the pH, the degree of short-range crystalline order, pasting, and the thermal behavior.

2. Results and Discussions

2.1. Apparent Amylose Content

The apparent amylose content of the native sago starch was found to be 31.38% (Table 1), slightly higher than the content reported by the starch manufacturer, 30.8%. This is because all the starch samples were defatted prior to the determination of the amylose content, and the removal of the fatty acids from the central hydrophobic cavity of the amylose molecules [17] made them more available to form complexes with the iodine reagent. Compared to the native sago, the amylose content of the irradiated counterparts was lower ($p \leq 0.05$) and gradually decreased as the dose of the irradiation was increased (Table 1). Similar results were previously reported for lotus seed starch [2], brown rice starch [29], tapioca starch [7], and corn starch [30]. Increased irradiation doses induced more damage to the conformation of the amylose molecules, hence lowering the iodine-binding ability. Othman et al. [31] observed a reduction of the amylose content in gamma-irradiated sago starch at 6 kGy; however, the apparent amylose content was found increased at higher doses of 10 and 25 kGy. They attributed this to the degradation of amylopectin that released more linear chains of amylose to form a complex with iodine. This phenomenon was, however, not seen in the present study.

Table 1. The apparent amylose content, pH, apparent carboxyl content, and IR intensity ratio of 1047/1022 cm^{-1} ($R_{1047/1022}$) for native and modified sago starches.

Sample	Apparent Amylose Content (%)	pH	Apparent Carboxyl Content (%)	$R_{1047/1022}$
Native	31.38 ± 0.13 c	5.09 ± 0.02 h**	N.D. *	0.57 ± 0.01 ef**
5 kGy	24.21 ± 1.41 b	4.91 ± 0.02 g**	0.033 ± 0.006 a	0.43 ± 0.01 c**
10 kGy	24.45 ± 0.77 b	4.12 ± 0.01 e**	0.079 ± 0.002 b	0.44 ± 0.02 c**
25 kGy	23.64 ± 1.33 ab	3.72 ± 0.01 c**	0.105 ± 0.004 c	0.36 ± 0.01 b**
50 kGy	21.09 ± 0.19 a	3.31 ± 0.01 a**	0.181 ± 0.009 e	0.31 ± 0.01 a**
ANN(T$_o$-5)	37.00 ± 0.32 def	5.10 ± 0.05 h	N.D. *	0.67 ± 0.00 i
ANN(T$_o$-10)	39.89 ± 0.33 g	5.25 ± 0.02 i	N.D. *	0.62 ± 0.01 gh
5kGyANN(T$_o$-5)	36.37 ± 0.24 de	4.60 ± 0.03 f	N.D. *	0.63 ± 0.01 h
10kGyANN(T$_o$-5)	36.12 ± 0.49 d	4.14 ± 0.02 e	N.D. *	0.59 ± 0.01 fg
25kGyANN(T$_o$-5)	37.98 ± 2.08 defg	3.80 ± 0.00 d	0.111 ± 0.014 c	0.72 ± 0.01 j
50kGyANN(T$_o$-5)	36.67 ± 0.25 de	3.43 ± 0.01 b	0.168 ± 0.009 d	0.71 ± 0.01 j
5kGyANN(To-10)	38.79 ± 0.70 efg	4.57 ± 0.01 f**	N.D. *	0.48 ± 0.01 d**
10kGyANN(To-10)	39.32 ± 0.61 fg	4.17 ± 0.00 e**	N.D. *	0.44 ± 0.01 c**
25kGyANN(To-10)	37.37 ± 0.17 defg	3.81 ± 0.01 d**	0.114 ± 0.003 cd	0.54 ±0.01 e**
50kGyANN(To-10)	39.84 ± 1.12 g	3.43 ±0.01 b**	0.173 ±0.008 e	0.62 ± 0.01 gh**

$^{a-j}$ Means with different lowercase letters within the same column are significantly different ($p \leq 0.05$). * N.D. indicates 'Not Detectable'. ** Data reported in Lee et al. [28].

On the contrary, the annealed samples had significantly higher amylose content than the native starch with the ANN(T$_o$-10) and showed higher amylose content than the ANN(T$_o$-5) ($p \leq 0.05$). A similar observation was reported by Babu et al. [32] for foxtail millet starch. One of the many possible alterations of the starch internal structure by annealing is the formation of double helices [17]. The more organized molecular configuration of the starch acquired by annealing may be responsible for this observation. All dual-modified starches displayed higher amylose content ($p \leq 0.05$) than their irradiated counterparts, but insignificant differences when compared to the annealed starch samples ($p > 0.05$). Depolymerization by gamma radiation may most probably facilitate the reorganization of the starch molecules during annealing to acquire more 3-D helicoidal structures that are suitable to fix in the iodine. More amylose-like molecules were formed by the reorganization of the dual-modified starch molecules.

2.2. PH and Apparent Carboxyl Content

A substantial drop in pH was observed with an increase in the irradiation dose (Table 1), suggesting that oxidation had taken place with the formation of acid groups that increased the acidity of the samples. The irradiation process contributed to the oxidation of the hydroxyl groups of the starch samples, especially into the carbonyl and carboxyl groups, as well as the formation of different acids such as acetic, formic, pyruvic, and glucuronic acid [13]. This result is in corroboration with the apparent carboxyl content of the samples. Previous studies also reported similar results [7,13,14]. It is interesting to note that the pH of the dual-modified samples differed from the irradiated counterparts in a radiation dose-dependent manner: where 5kGyANN(T_o-5) and 5kGyAnn(T_o-10) were lower than 5kGy ($p \leq 0.05$); insignificant differences were found between 10kGy, 10kGyANN(T_o-5) and 10kGyAnn(T_o-10) ($p \leq 0.05$); whereas 25kGyANN(T_o-5), 25kGyAnn(T_o-10), 50kGyANN(T_o-5), and 50kGyAnn(T_o-10) were reported higher than 25kGy and 50kGy, respectively. The incubation temperature had no influence in this case ($p \leq 0.05$). When compared to 25kGy and 50kGy, the dual-modified counterparts (25kGyANN(T_o-5), 25kGyAnn(T_o-10), 50kGyANN(T_o-5), and 50kGyAnn(T_o-10)) turned out to be less acidic, indicating the potential involvement of the acid groups in the structural reconfiguration of the starch macromolecules during ANN.

Since no hydrolysis of the glycosidic bonds was expected, no carboxyl content was detectable in the native and annealed samples, even though the ANN(T_o-10) exhibited a slightly higher pH than the native counterpart ($p \leq 0.05$). As the carboxyl content was mainly attributed to the starch degradation by irradiation, ANN did not remarkably alter the carboxyl content of the dual-modified starch, regardless of the ANN temperature.

2.3. Infrared Spectra Analysis

Starch samples were analyzed by FTIR spectroscopy to confirm the breakdown of the glycosidic bonds and the changes to the short-range crystalline order (double helices) in the crystalline region and the amorphous region near the granule surface [33]. The degradation effect of gamma irradiation on the starch short-range crystalline order is seen with the increasing treatment dosage (Table 1), where the $R_{1047/1022}$ dropped gradually until the lowest value was obtained at 50 kGy ($p \leq 0.05$). In contrast, the $R_{1047/1022}$ increased after ANN because the treatment provided conditions enabling the rearrangement of the starch molecules for crystalline perfection. Compared to the ANN(T_o-10), the higher annealing temperature of ANN(T_o-5) accelerated the rate of hydration and increased the kinetic energy of the glucan chains to take part in the molecular rearrangement. Therefore, the more pronounced effect of annealing on the structural change of the starch was obtained at a temperature closer to its gelatinization temperature. The recrystallisation-promoting effect of the higher annealing temperature of ANN(To-5) was also notable when combined with gamma irradiation. In the dual-modified samples, a synergistic effect in promoting crystalline perfection was observed, in which 25kGyANN(T_o-5) and 50kGyANN(T_o-5) achieved the highest $R_{1047/1022}$ ($p \leq 0.05$), indicating the higher molecular order of the double helix short-range in the starch granules [5]. It is postulated that these higher irradiation doses produced more new shorter molecules [30] as the starting materials to participate in the recrystallisation process, resembling the melted materials induced by ANN [3].

2.4. Swelling Power and Solubility

Upon irradiation, the swelling power of the sago starch was noticeably reduced with the increasing irradiation dose (Table 2). A reduction of almost 77.5% of the swelling ability was reported when the sago starch was irradiated up to 50 kGy. This indicates the severe depolymerization of some of the amylose and amylopectin molecules [34] and the destabilization of the hydrogen bonds within the double helices of the starch [35] under high-dose treatment. Gamma irradiation also significantly increased the amount of soluble fraction with the increasing dose. The soluble fractions are the leaching of the degraded amylose and/or amylopectin after maximum swelling; therefore, the samples that experienced more severe radiation-induced damage (contained smaller starch fractions) tended to be more soluble. The increase in the solubility was due to the increase in the polarity because of chain scission under irradiation and the decrease in interchain hydrogen bonds [36]. As expected, both of the annealed samples also showed reduced swelling power and solubility ($p \leq 0.05$). In corroboration to the results of the FTIR, the ANN(T_o-5), which experienced more extensive reorganization of the structure, had lower swelling power and solubility than the ANN(T_o-10) ($p < 0.05$). Compared to gamma radiation, the suppression of solubility by ANN was more apparent than the suppression of the swelling power. According to Zavareze and Dias [17], the interplay between the extent of crystalline perfection and the amylose–amylose and/or amylose–amylopectin interactions decrease the number of available water binding sites and hence suppress the swelling of the granules and the leaching of the soluble. The swelling behavior of the dual-modified samples closely resembled the irradiated counterparts ($p > 0.05$), except for 50kGyANN(T_o-5), which displayed the lowest swelling ability ($p \leq 0.05$), approximately 40% lower than 50 kGy. The ANN caused a solubility suppression effect on the irradiated sago starch with a more profound outcome by ANN at (T_o-5) than (T_o-10). The reduction in the solubility was in proportion with the increase in the irradiation dose, where a reduction from 59% to 89% was observed for the dual-modified samples as compared to the irradiated counterparts. In brief, the dual-modified samples behaved like irradiated samples in terms of the swelling power, but the solubility was lower than the irradiated samples due to the suppression effect by ANN.

Table 2. The swelling power and solubility (in excess water at 85 °C) and gel firmness for native and modified sago starches.

Sample	Swelling Power	Solubility (%)	Gel Firmness (g)
Native	35.92 ± 0.83 [i]	24.50 ± 1.24 [d]	30.67 ± 2.08 [bc]
5 kGy	11.18 ± 0.42 [def]	39.29 ± 0.66 [ef]	15.33 ± 1.53 [a]
10 kGy	10.18 ± 1.32 [cde]	42.23 ± 2.32 [f]	14.00 ± 1.00 [a]
25 kGy	9.30 ± 0.38 [cde]	65.11 ± 1.59 [g]	15.33 ± 2.08 [a]
50 kGy	8.09 ± 0.72 [bcd]	85.40 ± 0.53 [h]	15.10 ± 2.65 [a]
ANN(T_o-5)	17.80 ± 0.47 [g]	8.69 ± 1.02 [a]	68.67 ± 4.16 [f]
ANN(T_o-10)	24.37 ± 2.84 [h]	15.27 ± 1.46 [bc]	46.04 ± 3.46 [de]
5kGyANN(T_o-5)	13.65 ± 1.07 [f]	16.21 ± 0.25 [bc]	50.33 ± 6.66 [de]
10kGyANN(T_o-5)	11.77 ± 1.57 [ef]	16.61 ± 5.08 [bc]	53.02 ± 0.98 [de]
25kGyANN(T_o-5)	8.24 ± 0.12 [bcd]	11.21 ± 1.87 [ab]	30.33 ± 6.11 [bc]
50kGyANN(T_o-5)	4.86 ± 0.06 [a]	9.55 ± 0.18 [a]	11.35 ± 1.53 [a]
5kGyANN(T_o-10)	9.33 ± 0.04 [cde]	22.10 ± 0.34 [cd]	42.67 ± 5.51 [cd]
10kGyANN(T_o-10)	10.13 ± 0.17 [cde]	43.56 ± 0.72 [f]	60.11 ± 11.00 [ef]
25kGyANN(T_o-10)	7.45 ± 0.87 [abc]	45.27 ± 3.93 [f]	44.02 ± 4.58 [cd]
50kGyANN(T_o-10)	5.79 ± 0.13 [ab]	33.76 ± 3.91 [e]	25.10 ± 6.56 [ab]

[a–j] Means with different lowercase letters within the same column are significantly different ($p \leq 0.05$).

2.5. Gel Firmness

Table 2 shows that the native sago starch gel was able to withstand a higher deformation force than the irradiated counterparts ($p \leq 0.05$). The formation of carboxyl groups by the radiolytic degradation of the starch molecules caused the electrostatic repulsion of the molecular association during gel formation by retrogradation [37]. As the gel is a water entrapment system by three-dimensional networks, the low intra- and intermolecular associations made the gel weaker. Gamma irradiation caused about 50% of reduction in the gel strength, independent of the radiation doses ($p > 0.05$). Depending on the starch origin and the radiation dose, the irradiated starch gel firmness/strength was found to be either improved [13] or reduced [38]. Polesi et al. [38] attributed the reduction in rice gel firmness to the excessive breakdown of the amylose molecules during irradiation, which hampered the association of molecules during retrogradation. In accordance with the reduction of the swelling power and solubility, the volume of the annealed starch gels was lowered [39] due to the lower water retention ability; hence, becoming less elastic and firmer. Concomitantly with the lowest swelling power and solubility, ANN(T_o-5) exhibited the highest gel firmness among all tested samples ($p \leq 0.05$). The results obtained show that ANN may improve the gel firmness of irradiated starch without being affected by the incubation temperature ($p > 0.05$). The beneficial effect was, however, only reported at lower radiation doses of 5 kGy, 10 kGy, and 25 kGy. At 50 kGy, all the irradiated samples and dual-modified samples showed insignificant differences ($p < 0.05$).

2.6. Thermal Properties

The gelatinization temperature of the irradiated sago starch was gradually decreased with the increasing dose of gamma radiation (Table 3), and a significant reduction was observed for T_o (onset temperature), T_p (peak temperature), and T_c (conclusion temperature) at 50 kGy ($p \leq 0.05$). This observation agrees with the results of the swelling power and solubility (Table 2), whereby the depolymerization of the starch macromolecules into shorter chain molecules after irradiation weakened the associative forces in the granules and consequently eased the initiation of the phase transition and subsequent cooperative melting. Comparatively high irradiation doses, such as 50 kGy, may cause the disruption of the crystalline domain in starch granules, as well as the disruption of the double-helical order [33], hence lowering the gelatinization temperatures. Liu et al. [36] found the decreases in the gelatinization temperature and the enthalpy (ΔH) of irradiated maize starch were not statistically significant from 0 to 20 kGy, but a significant decrease in T_o, T_p, and ΔH was observed from 20 to 50 kGy. Chung et al. [30] reported insignificant differences for the gelatinization temperature of irradiated normal corn starch from the native counterpart at 1, 5, and 10 kGy, while a significant decrease was found at 25 and 50 kGy. They related the reduced gelatinization temperature to the weaker starch granules resulting from the cleavage of glycosidic bonds by irradiation. Other inconsistent observations were also reported for the alteration of the gelatinization behavior of irradiated starch. Othman et al. [31] revealed that a small increase in the gelatinization temperature was observed for sago starch irradiated at 10 and 25 kGy. They related this observation to the presence of small molecular products (monosaccharides, small chain polysaccharides) resulting from the molecular degradation by gamma ray. According to Chung and Liu [39], gelatinization temperatures reflect the stability of starch crystallites. A decrease in the gelatinization temperature by gamma irradiation was due to the production of a defective crystalline structure and an increase in the proportion of short chains in amylopectin. On the other hand, an increase in the gelatinization temperature was attributed to the destruction of weak crystalline structures by gamma irradiation, leaving behind a more stable structure that required a higher temperature for gelatinization. In the present study, gamma irradiation did not cause any change in ΔH (enthalpy). The ΔH for irradiated wheat starch was also reported to be insignificantly different from that of the native starch until 50 kGy, which was ascribed to the unchanged crystallinity [34]. It is therefore very likely that the crystallinity of the sago starch was not severely disrupted by the irradiation condition used in the present study.

Annealing at (T_o-5) and (T_o-10) brought about an increase in the T_o and T_p of the sago starch ($p < 0.05$). The increase in the gelatinization temperature by ANN has been shown to be most pronounced for T_o and least for T_c [3]. The effect of ANN on the gelatinization characteristics is well established, where there tends to be an increase in T_o and T_p, a decrease in the gelatinization range, and either no change or an increase in the gelatinization enthalpy [17]. The ANN promoted bond strengthening and hence a higher temperature will be required to gelatinize the starch granules [40]. The increased ΔH of the annealed sago starch implies an improved starch reinforcement within the granules and the attainment of a higher configuration stability [17]. Enthalpy represents the energy needed for the dissociation of the double-helical order; therefore, the increase in the order of the double helices (as suggested by the results of the FTIR analysis) may most probably explain the higher ΔH of the annealed starches. The granular reinforcement effect of ANN was remarkably enhanced after gamma irradiation, along with a significant raise in the gelatinization temperatures ($p \leq 0.05$). The narrowing of the gelatinization temperature range, which may imply the higher crystallite homogeneity (double-helical structures) was also observed, particularly for ANN(T_o-5) and the corresponding dual-modified samples. Even though the values of T_c for the two annealed samples are high, they are statistically insignificant from the native and irradiated samples (Table 3). This may most likely be due to the high standard deviation of 50 kGyANN(T_o-5), 25kGyANN(T_o-5), and 50kGyANN(T_o-10). The high standard deviations indicate that the T_c values (the ending point of gelatinization) are spread out over a wider range, suggesting that these three samples contained more heterogeneous crystallites (induced by irradiation) as compared to the annealed counterparts.

Table 3. Gelatinization parameters of native and modified sago starch.

Sample	Onset Temperature, T_o (°C)	Peak Temperature, T_p (°C)	Conclusion Temperature, T_c (°C)	Melting Temperature Range, ΔT (°C)	Enthalpy ΔH (J/g)
Native **	69.06 ± 0.02 [ab]	74.69 ± 0.39 [b]	79.96 ± 0.41 [abc]	10.90 ± 0.43 [ab]	5.43 ± 0.73 [ab]
5 kGy **	69.39 ± 0.33 [b]	74.08 ± 0.39 [b]	78.80 ± 0.93 [ab]	9.41± 0.91 [ab]	4.54 ± 1.29 [ab]
10 kGy **	69.50 ± 0.51 [b]	74.17 ± 0.18 [b]	78.39 ± 0.20 [ab]	8.89 ± 0.57 [ab]	4.21 ± 0.61 [ab]
25 kGy **	69.23 ± 0.19 [b]	73.99 ± 0.17 [b]	78.78 ± 0.40 [ab]	9.55± 0.20 [ab]	6.20 ± 0.91 [ab]
50 kGy **	66.12 ± 0.47 [a]	72.44 ± 0.10 [a]	77.03 ± 0.35 [a]	10.91± 0.24 [ab]	7.75 ± 1.96 [ab]
ANN(T_o-5)	75.81 ± 0.16 [d]	77.88 ± 0.17 [d]	81.86 ± 0.26 [abcd]	6.05 ±0.12 [a]	10.29 ± 0.69 [b]
ANN(T_o-10)	72.22 ±0.27 [c]	75.57 ± 0.42 [c]	80.54 ± 0.90 [abc]	8.32 ± 0.63 [ab]	10.15 ± 1.77 [b]
5kGyANN(T_o-5)	79.49 ± 0.14 [f]	81.25 ± 0.17 [gh]	85.22 ± 0.35 [cdef]	5.73 ±0.22 [a]	7.17 ± 2.30 [ab]
10kGyANN(T_o-5)	80.02 ± 0.06 [fg]	81.98 ± 0.19 [h]	86.82 ± 0.73 [def]	6.80 ± 0.70 [ab]	8.23 ± 2.46 [ah]
25kGyANN(T_o-5)	80.64 ± 0.41 [g]	83.06 ± 0.20 [i]	88.57 ± 0.54 [ef]	7.93 ± 0.94 [ab]	6.21 ± 4.17 [ab]
50kGyANN(T_o-5)	80.50 ± 0.50 [fg]	83.74 ± 0.10 [i]	89.41 ± 2.84 [f]	8.91 ± 3.33 [ab]	3.18 ± 4.26 [ab]
5kGyANN(T_o-10) **	76.07 ± 0.04 [d]	78.32 ± 0.10 [d]	81.99 ± 0.10 [abcd]	5.92 ± 0.08 [a]	5.85 ± 2.16 [ab]
10kGyANN(T_o-10) **	76.68 ±0.51 [de]	79.38 ± 0.29 [e]	83.61 ± 0.74 [bcde]	6.93 ± 0.09 [ab]	6.03 ± 3.77 [ab]
25kGyANN(T_o-10) **	77.34 ± 0.55 [e]	80.39 ± 0.34 [f]	89.61 ± 5.96 [f]	12.27 ± 5.60 [b]	6.35 ± 4.81 [ab]
50kGyANN(T_o-10) **	77.75 ± 0.65 [e]	81.02 ± 0.35 [fg]	86.53 ± 4.45 [def]	8.78 ± 1.88 [ab]	8.22 ± 1.26 [ab]

[a–i] Means with different lowercase letters within the same column are significantly different ($p \leq 0.05$). ** Data for these samples are adopted from Lee et al. [28].

Owing to the starch reinforcement by ANN, a notable increase in T_o, T_p, and T_c was depicted ($p \leq 0.05$) in the dual-modified samples, with a more substantial change in the ANN(T_o-5)-corresponding samples, whereby these samples appeared to have the highest gelatinization temperatures. This observation agreed well with the results obtained for the $R_{1047/1022}$, swelling power, and solubility. By comparing the effect of ANN on high-amylose wheat (HAWS) and maize starch (HAMS) with a similar apparent amylose content, Li et al. [41] concluded that chain mobility is the key contributing factor to support greater structural rearrangement by ANN and the formation of a thermostable molecular order in the HAWS. As suggested earlier, the availability of irradiation-induced new shorter molecules and a high incubation temperature at T_o-5 provided the required chain mobility for successive structural rearrangement in the samples. In the present study, an inconclusive change in ΔH was reported for all the dual-modified starches.

2.7. Pasting Properties

Investigation of the pasting profile plays a vital role to elucidate the technological functionalities that encompasses the gelatinization, pasting, and retrogradation characteristics of starch. The results obtained (Table 4) show that the influences of the modification treatments on the pasting temperature are in accordance with the changes seen in the gelatinization temperatures (Table 3) in the ascending order of irradiated starches < annealed starches < dual-modified starches. The enhancement of the pasting temperature by the synergistic effects of gamma radiation and ANN was more remarkable in ANN(T_o-5)-corresponding samples.

Table 4. Pasting properties of native and modified sago starch.

Sample	Pasting Temperature (°C)	Peak Viscosity (cP)	Breakdown (cP)	Final Viscosity (cP)	Setback (cP)
Native **	74.38 ± 0.08 [c]	397.92 ± 2.17 [n]	256.17 ± 1.64 [k]	193.14 ± 4.51 [h]	51.39 ± 4.84 [f]
5 kGy **	73.88 ± 0.20 [b]	179.00 ± 2.49 [g]	131.42 ± 0.55 [e]	60.64 ± 1.48 [e]	13.05 ± 0.86 [c]
10 kGy **	73.60 ± 0.05 [b]	171.69 ± 0.79 [f]	158.89 ± 0.85 [f]	16.78 ± 0.90 [c]	3.97 ± 0.57 [ab]
25 kGy **	73.63 ± 0.03 [b]	128.14 ± 0.56 [e]	128.06 ± 0.70 [e]	2.39 ± 0.26 [b]	2.31 ± 0.32 [ab]
50 kGy **	72.37 ± 0.03 [a]	53.77 ± 0.42 [c]	54.89 ± 0.54 [c]	0.29 ± 0.04 [a]	1.42 ± 0.09 [ab]
ANN(T_o-5)	77.85 ± 0.26 [e]	387.83 ± 3.03 [m]	167.08 ± 3.75 [g]	274.72 ± 1.61 [j]	53.97 ± 0.80 [f]
ANN(T_o-10)	75.87 ± 0.28 [d]	370.33 ± 6.96 [l]	198.42 ± 3.56 [i]	224.86 ± 3.76 [i]	52.94 ± 0.34 [f]
5kGyANN(T_o-5)	80.73 ± 0.03 [i]	243.60 ± 0.80 [j]	240.80 ± 0.90 [j]	91.88 ± 0.84 [g]	89.08 ± 0.77 [g]
10kGyANN(T_o-5)	81.55 ± 0.05 [j]	191.81 ± 1.40 [h]	193.81 ± 1.25 [h]	26.29 ± 0.07 [d]	28.29 ± 0.18 [e]
25kGyANN(T_o-5)	82.78 ± 0.03 [k]	99.44 ± 0.21 [d]	102.11 ± 0.21 [d]	2.44 ± 0.42 [b]	5.10 ± 0.41 [b]
50kGyANN(T_o-5)	82.67 ± 0.08 [k]	21.78 ± 0.68 [a]	22.06 ± 0.10 [a]	0.52 ± 0.45 [a]	0.79 ± 1.09 [a]
5kGyANN(T_o-10) **	77.93 ± 0.03 [e]	251.65 ± 1.40 [k]	193.45 ± 0.40 [h]	78.43 ± 0.25 [f]	20.23 ± 1.44 [d]
10kGyANN(T_o-10) **	78.70 ± 0.05 [f]	207.91 ± 0.09 [i]	189.55 ± 0.04 [h]	22.41 ± 0.36 [d]	4.04 ± 0.41 [ab]
25kGyANN(T_o-10) **	79.57 ± 0.03 [g]	127.39 ± 0.25 [e]	126.89 ± 0.25 [d]	2.53 ± 0.46 [b]	2.03 ± 0.46 [ab]
50kGyANN(T_o-10) **	79.93 ± 0.03 [h]	34.34 ± 0.86 [b]	35.39 ± 0.97 [b]	0.22 ± 0.08 [a]	1.12 ± 0.11 [ab]

[a–m] Means with different lowercase letters within the same column are significantly different ($p \leq 0.05$). ** Data for these samples are adopted from Lee et al. [28].

Starch pasting involves the process of viscosity development that occurs after heating starch above the gelatinization temperature [42]. Native sago displayed the highest peak viscosity ($p \leq 0.05$); moreover, in parallel with the effects of irradiation and ANN on the swelling capacity of the starch (Table 2), a reduction in the peak viscosities was observed for these samples. It should be emphasized that granular swelling is not the only contributing factor to the perceived paste viscosities under continuous heating and stirring during the measurement; the amylose leaching, granular rigidity and integrity, and molecular size of the irradiated-induced fractions contribute collectively to the consistency of the pastes. The change in the paste viscosity for the dual-modified samples was dose-dependent, whereby the peak viscosities for the dual-modified samples at 5 and 10 kGy were higher than the irradiated counterparts; on the other hand, the peak viscosities of the dual-modified samples at 25 and 50 kGy were lower than the irradiated counterparts. In general, the peak viscosities for the ANN(T_o-10)-corresponding samples were reported higher than the ANN(T_o-5)-corresponding samples ($p < 0.05$). The alteration in viscosity was in corroboration to the swelling power and solubility behaviors of these samples (Table 2). It is interesting to note that ANN may enhance the peak viscosity of the irradiated sago starch at low irradiation doses (5 and 10 kGy). Combining ANN with low irradiation doses may potentially amplify the granule rigidity and resistance to shear [43], and hence a higher viscosity. The incubation temperature for ANN also had a significant influence on the peak viscosities of the dual-modified samples, whereby at a higher incubation temperature (T_o-5), more extensive molecular interactions took place to restrict the starch swelling and solubility, and subsequently a lower peak viscosity.

As expected, the breakdown decreased in proportion to the irradiation dose applied ($p \leq 0.05$) because a higher irradiation dose brought about more severe molecular destruction and weaker starch granules, and hence a lower ability to withstand shearing at a high temperature. The final viscosity and setback also decreased with the increasing radiation dose. The setback and final viscosity are attributed to a reordering or polymerization of the leached amylose and long linear amylopectin [44] upon the cooling of the paste. With the increase of the radiation dose, the tendency for retrogradation was diminished, indicating the degradation of the starch macromolecules, the reduction of the crystallization capacity, and the disruption of the molecular structures [45]. The results obtained showed that the sago starch irradiated at 25 and 50 kGy contained highly fragile granules that underwent utter granular rupture with nearly negligible final viscosities and setback. Barroso and del Mastro [12] pointed out that the observed changes to the pasting properties may also be affected by the radiolytic effects of oxidation when the irradiation was carried out in the presence of air, as in the present study. The ANN brought about a decrease in the breakdown and an increase in the final viscosity of the sago starch ($p \leq 0.05$) without affecting the setback ($p > 0.05$). The starch bond strengthening by ANN mentioned in the earlier session had improved the granular stability against collapse by continuous shearing at the maximum swelling volume. In general, all the dual-modified samples were found to exhibit a relatively high breakdown and low final viscosities and setback, indicating the irradiation exerted a more pronounced effect than ANN in the pasting properties of the dual-modified samples.

2.8. Starch Morphology

Figure 1 shows the microstructure of the native and irradiated starch. Native sago granules are predominantly ovoid, with some having a spherical shape, and the presence of the typical characteristic of a truncated end [31]. Irradiation up to 50 kGy did not cause any physical damage to the sago granules. The starch granules remained smooth and intact. No significant granular size increment can be detected. Othman et al. [31] also reported that the gamma irradiation of sago starch up to 25 kGy did not affect the granule size and shape. Depending on the starch origin and irradiation dose, some researchers reported surface disruptions such as cracking, fissures, and pores [13,46]. By using gel permeation chromatography (GPC), Castanha et al. [13] found that the molecular size of mung bean starch became smaller due to the hydrolysis of the glycosidic bonds.

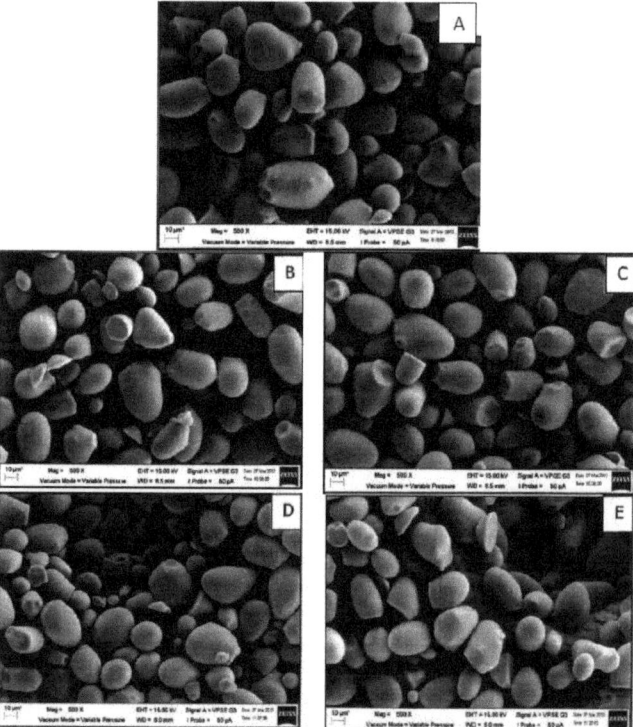

Figure 1. Scanning electron micrographs of (**A**) native sago; (**B**) 5 kGy; (**C**) 10 kGy; (**D**) 25 kGy; (**E**) 50 kGy (500× magnification).

Dent surfaces and the development of pores on the starch granules (indicated by the arrows) are found on the ANN(T_o-5) and the ANN(T_o-10), as depicted in Figure 2. Xu et al. [47] reported most of the potato starch granules retained their original appearance, whereby only a small number of starch granules appeared with grooves and dents on their granule surfaces after annealing. Rocha et al. [48] found an increase of the pores on the surface of the granules for both the normal and waxy con starches and attributed this to the hydrolysis activity of the endogenous amylase under the annealing condition. No pronounced change occurred in the granule morphology when ANN was applied to green banana flour [49]. Waduge et al. [50] found that annealed starch exhibited increased granular size due to the ingress of moisture through the amorphous regions of the starch during annealing, which was not seen in this study.

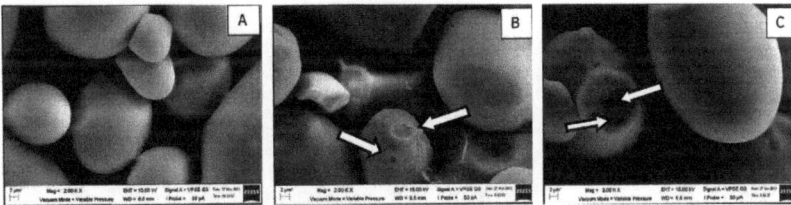

Figure 2. Scanning electron micrographs for (**A**) native sago; (**B**) ANN(T_o-5); (**C**) ANN(T_o-10) (2000× magnification).

Partial gelatinization had taken place in the annealed and dual-modified starches, as shown by the agglomeration of the starch granules (Figures 3 and 4). No morphological differentiation can be found in these single-modified and dual-modified samples. Since the ANN temperature for the samples in Figure 3 (T_o-5) is closer to the onset gelatinization temperature of the sago starch, more extensive agglomerations are observed compared to those annealed at (T_o-10) (Figure 4). Differing from this study, even though Rocha et al. [48] used an ANN temperature much closer to the melting of corn starches (3 °C below T_o), no agglomeration of the starch granules was observed by the researchers.

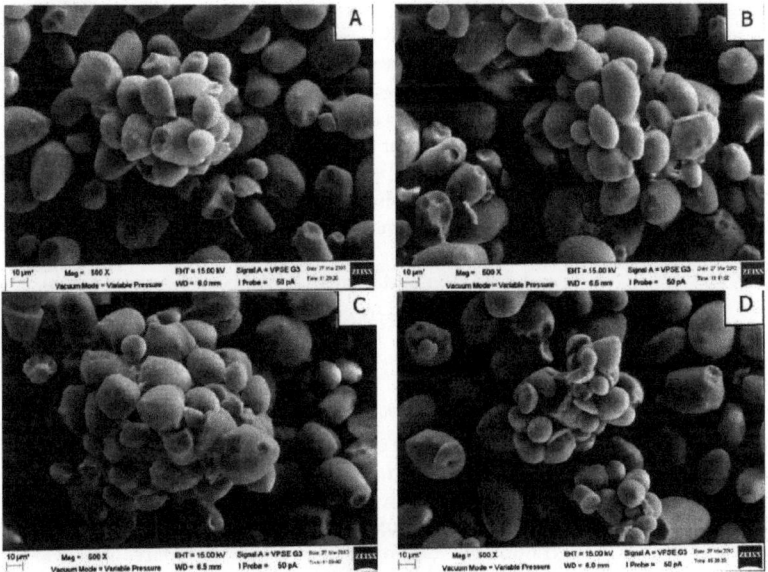

Figure 3. Scanning electron micrographs for (**A**) 5kGyANN(T_o-5); (**B**) 10kGyANN(T_o-5); (**C**) 25kGyANN(T_o-5); (**D**) 50kGyANN(T_o-5) (500× magnification).

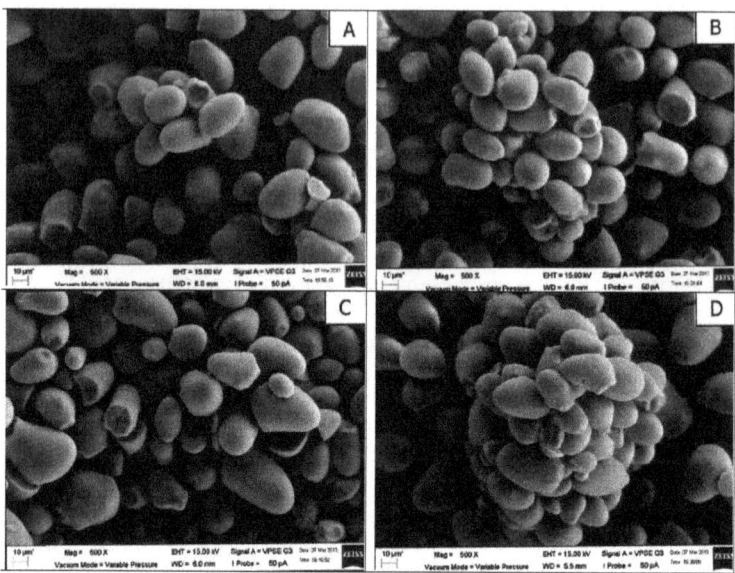

Figure 4. Scanning electron micrographs for (**A**) 5kGyANN(T$_o$-10); (**B**) 10kGyANN(T$_o$-10); (**C**) 25kGyANN(T$_o$-10); (**D**) 50kGyANN(T$_o$-10) (500× magnification).

3. Material and Methods

3.1. Sample Preparation

Food-grade sago (*Metroxylon sagu*) starch (12.0% moisture content, 0.12% fat content, 0.19% protein content) was bought from Nee Seng Ngeng and Sons Sago Industries Sdn. Bhd. (Malaysia). Sago starch was packed in a polyethylene bag and irradiated with a ^{60}Co gamma source at the Malaysian Nuclear Agency (J.L. Shepherd Gammacell, Model 109 Irradiator). The starch samples were subjected to four doses of irradiation (5, 10, 25, 50 kGy) as described by Chung et al. [30]. The irradiation was carried out at an ambient temperature and the dose rate was 9.08 kGy/h. The dosimetry was performed using a Harwell Amber Perspex dosimeter (Type 3042 Batch H). After irradiation, annealing was performed by subjected starch slurries (starch (dry basis):water, 1:2) at 5 °C and 10 °C below the onset of gelatinization temperature in a water bath for 24 h. The onset of the gelatinization temperature of the sago starch was determined using a Differential Scanning Calorimeter (Diamond DSC, Perkin Elmer, Waltham, MA, USA) prior to the annealing. After incubation, the starch samples were centrifuged using a bench-top centrifuge (Thermo Scientific, Waltham, MA, USA) for 10 min (2000 g), decanted, washed with deionized water, and oven-dried (Memmert, Schwabach, Germany) at 40 °C to achieve a uniform moisture content (~12%). The samples were ground to pass through a sieve of 150 μm (Retsch, Haan, Germany). The samples were sealed in a polyethylene bag and kept at 4 °C until further analysis.

3.2. Apparent Amylose Content

The amylose content was determined using the iodine-binding spectrophotometry method modified from Williams et al. [51]. An amount of 20 mg of defatted starch (soxhlet extraction for 24 h with 95% ethanol) was added with 8 mL of 90% dimethyl sulfoxide (DMSO$_4$) and mixed in a 25 mL volumetric flask. The content was heated for 15 min with continuous swirling at 85 °C until the starch sample was fully dissolved, and the content was topped up to 25 mL using distilled water. After that, 1 mL of the starch solution was added with 40 mL of distilled water and 10 mL of iodine solution, and the solution was incubated for 15 min before the measurement. The absorbance was measured (Lambda

35 UV/Vis Spectrophotometer, Perkin Elmer) at 625 nm against a reagent blank as the reference. A standard curve for mixtures of pure potato amylose and amylopectin was plotted and the regression equation of the standard curve was used to calculate the total amylose content of the samples (y = 0.210x + 0.021; R^2 = 0.989).

3.3. PH and Apparent Carboxyl Content

An amount of 4 g of starch was suspended in 50 mL of distilled water and the pH value was recorded using a pH meter (Eutech pH 700, Singapore). The method of Chattopadhyay et al. [52] was used to determine the apparent carboxyl content of the starch samples. An amount of 1 g of starch was added with 25 mL of 0.1 N HCl and continuously stirred for 30 min, followed by filtration and washing with distilled water until the sample was free of chlorine. After that, 300 mL of distilled water was added, and the starch solution was boiled for 10 min to gelatinize the sample. Titration was carried out using 0.1 N NaOH with phenolphthalein as the indicator.

$$\text{Apparent \% carboxyl} = \frac{(\text{sample} - \text{blank}) \text{ ml} \times \text{normality of NaOH}}{\text{sample weight (g in dry basis)}} \times 0.045 \times 100$$

3.4. Infrared Spectra Analysis

The infrared spectra of the starch samples were obtained on KBr pellets (starch:KBr, 1:100) of the samples using a Spectrum 100 Spectrometer (Perkin Elmer), and the intensity ratio from 1047 to 1022 cm^{-1} was calculated [33] to evaluate the starch short-range orders. Spectra were averaged over 64 scans.

3.5. Swelling Power and Solubility

The swelling power and solubility were determined as described by Schoch [53], with slight modification. The starch suspension (3 g of starch with 180 g of distilled water) was heated with continuous stirring at 85 °C for 30 min. After that, 20 g of distilled water was added to make up the total water to be 200 g. After centrifugation at 2200 rpm for 15 min, 50 mL of the supernatant was withdrawn, and oven dried at 105 °C for 24 h. The following calculations were employed:

$$\% \text{ Solubles (dry basis)} = \frac{\text{weight of soluble starch} \times 400}{\text{weight of sample on dry basis}}$$

$$\text{Swelling power} = \frac{\text{weight of sedimented paste} \times 100}{\text{weight of sample on dry basis} \times (100 - \% \text{ solubles, dry basis})}$$

3.6. Thermal Properties

Thermal analysis was performed using a Differential Scanning Colorimeter (PYRIS^{TM} Diamond, Perkin Elmer) with slight modification from Chung et al. [30]. An amount of 2 mg (dry basis) of the starch sample was added with 2 mg of distilled water and kept at room temperature to equilibrate for 24 h. The sample was scanned from 30 °C to 150 °C (10 °C/min) to obtain the gelatinization parameters. Indium and tin were used as calibration standards.

3.7. Pasting Properties

Pasting properties of the starch were measured using a Rapid Visco Analyzer (RVA-4, Newport Scientific, Warriewood, Australia). An amount of 3.0 g of starch (dry basis) was adjusted with distilled water to achieve constant weight of 28.0 g in the aluminum canister. The heating and cooling profile used was according to Ng et al. [54]. The temperature was first held at 50 °C for 1 min and raised to 95 °C (12 °C/min) and held for another 2.5 min before cooling down to 50 °C (12 °C/min). The starch paste was held at 50 °C for 1.5 min before the pasting parameters (pasting temperature, peak viscosity, breakdown, final viscosity, and setback) were recorded.

3.8. Gel Firmness

Gel firmness was measured using a texture analyzer (TA.XT Plus, Stable Micro Systems, Godalming, UK) with a load cell of 25 N. The cylindrical (23 mm diameter × 20 mm height) starch gel (10% w/v d.b.) was prepared according to Adawiyah et al. [55]. An amount of 5 g of the starch sample was added with 50 mL of distilled water and cooked for 30 min (80 °C) with continuous stirring. The paste was then poured into the cylindrical mold and cooled to form a gel. The gel was stored at 4 °C for 24 h prior to the measurement. Starch gel was compressed at a speed of 1 mm/s with a compression plate (40 mm diameter) until the strain reached 90% of its original height. The maximum force recorded during compression was defined as the firmness of the gel [56].

3.9. Starch Morphology

The granules were observed by Scanning Electron Microscopy (EVO MA10, Carl Zeiss Microscopy, Jena, Germany) with 500× and 2000× magnification. Starch sample was suspended in a carbon tape and coated with a fin layer of gold–palladium using a sputter coater (Polaron SC500, Quorum Technologies, Lewes, UK).

3.10. Statistical Analysis

Data reported are the average of at least triplicate determinations (from three batches of samples prepared independently). SPSS (Statistical Package for the Social Sciences) version 20 was used for statistical analysis. One-way analysis of variance (ANOVA) with Tukey's HSD test was used to compare the means (95% level of significance).

4. Conclusions

Results obtained support the hypothesis that a higher annealing (ANN) temperature (T_o-5) enhanced the crystalline perfection by increasing the short-range orders of the irradiated sago starch as compared to a lower ANN temperature (T_o-10). Synergism in molecular reorganization was observed when the modification was carried out by combining a high radiation dose (25 kGy and 50 kGy) with the ANN temperature of T_o-5. Among the properties investigated, the annealing temperature significantly influenced the swelling power, the gelatinization behavior, and the pasting profile of the irradiated sago starch. The dual-modified sago starch recorded a lower swelling power and a higher thermal stability by exhibiting higher gelatinization temperatures as well as a narrower melting temperature range when compared to the native starch or the single-modified starch. The gel firmness was also significantly improved. The physical modification methods employed also offer a sago starch with new properties without destroying the granular structure, which might be important in extending its applications. Future work investigating in-depth structural changes and starch digestibility are needed to provide a better insight of the modification techniques employed.

Author Contributions: Conceptualization, J.-S.L.; methodology, J.-S.L. and S.L.F.; validation, J.H.A., C.K.S. and A.L.H.; formal analysis, S.L.F.; investigation, S.L.F.; resources, J.H.A.; writing—original draft preparation, J.-S.L.; writing—review and editing, C.K.S. and A.L.H.; visualization, A.L.H.; supervision, J.-S.L.; project administration, J.-S.L.; funding acquisition, J.-S.L. All authors have read and agreed to the published version of the manuscript.

Funding: This research was funded by the Ministry of Higher Education, Malaysia, grant number FRGS0190-ST-1/2010. The APC was funded by Universiti Malaysia Sabah.

Institutional Review Board Statement: Not applicable.

Informed Consent Statement: Not applicable.

Data Availability Statement: Not applicable.

Conflicts of Interest: The authors declare no conflict of interest. The funders had no role in the design of the study; in the collection, analyses, or interpretation of data; in the writing of the manuscript, or in the decision to publish the results.

Sample Availability: Modified starches are not available from the authors.

References

1. Wang, X.; Huang, L.; Zhang, C.; Deng, Y.; Xie, P.; Liu, L.; Cheng, J. Research advances in chemical modifications of starch for hydrophobicity and its applications: A review. *Carbohydr. Polym.* **2020**, *240*, 116292. [CrossRef] [PubMed]
2. Punia, S.; Dhull, S.B.; Kunner, P.; Rohilla, S. Effect of γ-radiation on physic-chemical, morphological and thermal characteristics of lotus sees (*Nelumbo nucifera*) starch. *Int. J. Biol. Macromol.* **2020**, *157*, 584–590. [CrossRef] [PubMed]
3. Jayakody, L.; Hoover, R. Effect of annealing on the molecular structure and physicochemical properties of starches from different botanical origins—A review. *Carbohydr. Polym.* **2008**, *74*, 691–703. [CrossRef]
4. Fan, Y.; Picchioni, F. Modification of starch: A review on the application of "green" solvents and controlled functionalization. *Carbohydr. Polym.* **2020**, *214*, 116350. [CrossRef] [PubMed]
5. Chi, C.; Li, X.; Lu, P.; Miao, S.; Zhang, Y.; Chen, L. Dry heating and annealing treatment synergistically modulate starch structure and digestibility. *Int. J. Biol. Macromol.* **2019**, *137*, 554–561. [CrossRef]
6. Brașoveanu, M.; Nemțanu, M.R. Aspects on starches modified by ionizing radiation processing. In *Applications of Modified Starches*; Huicochea, E.F., Rendon, R., Eds.; IntechOpen: Rijeka, Croatia, 2018; pp. 49–68. [CrossRef]
7. Kanatt, S.R. Irradiation as a tool for modifying tapioca starch and development of an active food packaging film with irradiated starch. *Radiat. Phys. Chem.* **2020**, *173*, 108873. [CrossRef]
8. Bhat, R.; Karim, A.A. Impact of radiation processing on starch. *Compr. Rev. Food Sci. Food Saf.* **2009**, *8*, 44–58. [CrossRef]
9. Al-Kaisey, M.T.; Mohammed, M.A.; Alwan, A.K.H.; Mohammed, M.H. The effect of gamma irradiation on the viscosity of two barley cultivars for broiler chicks. *Radiat. Phys. Chem.* **2002**, *63*, 295–297. [CrossRef]
10. Diehl, J.F. Food irradiation—past, present and future. *Radiat. Phys. Chem.* **2002**, *63*, 211–215. [CrossRef]
11. Roberts, P.B. Food irradiation is safe: Half a century of studies. *Radiat. Phys. Chem.* **2014**, *105*, 78–82. [CrossRef]
12. Barroso, A.G.; del Mastro, N.L. Physicochemical characterization of irradiated arrowroot starch. *Radiat. Phys. Chem.* **2019**, *158*, 194–198. [CrossRef]
13. Castanha, N.; Miano, A.C.; Sabadoti, V.D.; Augusto, P.E.D. Irradiaiton of mung beans (*Vigna radiate*): A prospective study correlating the properties of starch and grains. *Int. J. Biol. Macromol.* **2019**, *129*, 460–470. [CrossRef] [PubMed]
14. Sudheesh, C.; Sunooj, K.V.; George, J. Kithul palm (*Caryota urens*) as a new source of starch: Effect of single, dual chemical modifications and annealing on the physicochemical properties and in vitro digestibility. *Int. J. Biol. Macromol.* **2019**, *125*, 1084–1092. [CrossRef]
15. Tester, R.F.; Debon, S.J.J. Annealing of starch—A review. *Int. J. Biol. Macromol.* **2000**, *27*, 1–12. [CrossRef]
16. Almeida, M.C.B.D.M.; Costa, S.D.S.; Cavalcanti, M.T.; Almeida, E.L. Characterization of Prata Banana (musa AAB-Prata) starch: Native and modified by annealing. *Starch-Stärke* **2020**, *72*, 1900137. [CrossRef]
17. Zavareze, E.d.R.; Dias, A.R.G. Impact of heat-moisture treatment and annealing in starches: A review. *Carbohydr. Polym.* **2011**, *83*, 317–328. [CrossRef]
18. Devi, R.; Sit, N. Effect of single and dual steps annealing in combination with hydroxypropylation on physicochemical, functional and rheological properties of barley starch. *Int. J. Biol. Macromol.* **2019**, *129*, 1006–1014. [CrossRef]
19. Hu, X.; Jia, X.; Zhi, C.; Jin, Z.; Miao, M. Improving properties of normal maize starch films using dual-modification: Combination treatment of debranching and hydroxypropylation. *Int. J. Biol. Macromol.* **2019**, *130*, 197–202. [CrossRef]
20. Ariyantoro, A.R.; Katsuno, N.; Nishizu, T. Effects of dual modification with succinylation and annealing on physicochemical, thermal and morphological properties of corn starch. *Foods* **2018**, *7*, 133. [CrossRef]
21. Li, L.; Hong, Y.; Gu, Z.; Cheng, L.; Li, Z.; Li, C. Effect of a dual modification by hydroxypropylation and acid hydrolysis on the structure and rheological properties of potato starch. *Food Hydrocoll.* **2018**, *77*, 825–833. [CrossRef]
22. Zanella, P.V.; Goncalves, D.V.; Moomand, K.; Levien, V.N.; Pilatti-Riccio, D.; da Roza Zavareze, E. Characteristics of modified Carioca bean starch upon single and dual annealing, heat-moisutre-treatment, and sonication. *Starch-Stärke* **2019**, *71*, 1800173. [CrossRef]
23. Kunyanee, K.; Luangsakul, N. The effects of dual modification with ultrasound and annealing treatments on the properties and glycemic index of the thani glutinous rice cultivar 'RD6'. *Int. J. Agric. Technol.* **2019**, *15*, 933–946.
24. Kong, X.; Zhou, X.; Sui, Z.; Bao, J. Effects of gamma irradiation on physicochemical properties of native and acetylated wheat starches. *Int. J. Biol. Macromol.* **2016**, *91*, 1141–1150. [CrossRef] [PubMed]
25. Majeed, T.; Wani, I.A.; Hussain, P.R. Effect of dual modification of sonication and γ-irradiation on physicochemical and functional properties of lentil (*Lens culinaris* L.) starch. *Int. J. Biol. Macromol.* **2017**, *101*, 358–365. [CrossRef] [PubMed]
26. Mutlu, S.; Kahraman, K.; Severcan, S.; Öztürk, S. Modelling the effects of debranching and microwave irradiation treatments on the properties of high amylose corn starch by using Response Surface Methodology. *Food Biophys.* **2018**, *13*, 263–273. [CrossRef]
27. Zailani, M.A.; Kamilah, H.; Husaini, A.; Seruji, A.Z.R.A.; Sarbini, S.R. Functional and digestibility properties of sago (*Metroxylon sagu*) starch modified by microwave heat treatment. *Food Hydrocoll.* **2022**, *122*, 107042. [CrossRef]
28. Lee, J.-S.; Jahurul, M.H.A.; Fong, S.-L. Dual modification of sago starch by gamma irradiation and annealing. *IOP Conf. Ser. Mater. Sci. Eng.* **2021**, *1092*, 012091. [CrossRef]
29. Kumar, P.; Prakash, K.S.; Jan, K.; Swer, T.L.; Jan, S.; Verma, R.; Bashir, K. Effects of gamma irradiation on starch granule structure and physiochemical properties of brown rice starch. *J. Cereal Sci.* **2017**, *77*, 194–200. [CrossRef]

30. Chung, K.-H.; Othman, Z.; Lee, J.-S. Gamma irradiation of corn starches with different amylose-to-amylopectin ratio. *J. Food Sci. Technol.* **2015**, *52*, 6218–6229. [CrossRef] [PubMed]
31. Othman, Z.; Hassan, O.; Hashim, K. Physicochemical and thermal properties of gamma-irradiated sago (*Metroxylon sagu*) starch. *Radiat. Phys. Chem.* **2015**, *109*, 48–53. [CrossRef]
32. Babu, A.S.; Mohan, R.J.; Primalavalli, R. Effect of single and dual-modifications on stability and structural characteristics of foxtail millet starch. *Food Chem.* **2019**, *271*, 457–465. [CrossRef] [PubMed]
33. Chung, H.-J.; Liu, Q.; Hoover, R. Effect of single and dual hydrothermal treatments on the crystalline structure, thermal properties and nutritional fractions of pea, lentil, and navy bean starches. *Food Res. Int.* **2010**, *43*, 501–508. [CrossRef]
34. Atrous, H.; Benbettaieb, N.; Hosni, F.; Danthine, S.; Blecker, C.; Attia, H.; Ghorbel, D. Effect of γ-radiation on free radicals formation, structural changes and functional properties of wheat starch. *Int. J. Biol. Macromol.* **2015**, *80*, 64–76. [CrossRef]
35. Bao, J.; Ao, Z.; Jane, J.L. Characterization of physical properties of flour and starch obtained from gamma-irradiated white rice. *Starch-Stärke* **2005**, *57*, 480–487. [CrossRef]
36. Liu, T.; Ma, Y.; Xue, S.; Shi, J. Modifications of structure and physicochemical properties of maize starch by γ-irradiation treatments. *LWT-Food Sci. Technol.* **2012**, *46*, 156–163. [CrossRef]
37. Xie, S.X.; Liu, Q.; Cui, S.W. Starch modification and applications. In *Food Carbohydrates: Chemistry, Physical Properties and Applications*, 1st ed.; Cui, S.W., Ed.; CRC Press: Boca Raton, FL, USA, 2005; pp. 357–406.
38. Polesi, L.F.; Sarmento, S.B.S.; Ganniatti-Brazaca, S.G. Starch digestibility and functional properties of rice starch subjected to gamma radiation. *Rice Sci.* **2018**, *25*, 42–51. [CrossRef]
39. Chung, H.-J.; Liu, Q. Molecular structure and physicochemical properties of potato and bean starches as affected by gamma-irradiation. *Int. J. Biol. Macromol.* **2010**, *47*, 214–222. [CrossRef] [PubMed]
40. Gomes, A.M.M.; Silva, C.E.M.; Ricardo, N.M.P.S.; Sasaki, J.M.; Germani, R. Impact of annealing on the physicochemical properties of unfermented cassava starch (polvilho doce). *Starch-Stärke* **2004**, *56*, 419–423. [CrossRef]
41. Li, H.; Dhital, S.; Flanagan, B.M.; Mata, J.; Gilbert, E.P.; Gidley, M.J. High-amylose wheat and maize starches have distinctly different granule organization and annealing behaviour: A key role for chain mobility. *Food Hydrocoll.* **2020**, *105*, 105820. [CrossRef]
42. Ai, Y.; Jane, J.-l. Gelatinization and rheological properties of starch. *Starch-Starke* **2015**, *67*, 213–224. [CrossRef]
43. Jacobs, H.; Eerlingen, R.C.; Clauwaert, W.; Delcour, J.A. Influence of annealing on the pasting properties of starches from varying botanical sources. *Cereal Chem.* **1995**, *72*, 480–487.
44. Chung, H.J.; Liu, Q. Effect of gamma irradiation on molecular structure and physicochemical properties of corn starch. *J. Food Sci.* **2009**, *74*, C353–C361. [CrossRef]
45. Valencia, G.A.; Henao, A.C.A.; Zapata, R.A.V. Comparative study and characterization of starches isolated from unconventional tuber sources. *J. Polym. Eng.* **2012**, *32*, 531–537. [CrossRef]
46. Gani, A.; Nazia, S.; Rather, S.A.; Wani, S.M.; Shah, A.; Bashir, M.; Masoodi, F.A.; Gani, A. Effect of γ-irradiation on granule structure and physicochemical properties of starch extracted from two types of potatoes grown in Jammu & Kashmir, India. *LWT-Food Sci. Technol.* **2014**, *58*, 239–246.
47. Xu, M.; Saleh, A.S.M.; Gong, B.; Li, B.; Jing, L.; Gou, M.; Jiang, H.; Li, W. The effect of repeated versus continuous annealing on structural physicochemical, and digestive properties of potato starch. *Food Res. Int.* **2018**, *111*, 324–333. [CrossRef] [PubMed]
48. Rocha, T.S.; Felizardo, S.G.; Jane, J.-L.; Franco, C.M.L. Effect of annealing on the semicrystalline structure of normal and waxy corn starches. *Food Hydrocoll.* **2012**, *29*, 93–99. [CrossRef]
49. Cahyana, Y.; Wijaya, E.; Halimah, T.S.; Marta, H.; Suryadi, E.; Kurniati, D. The effect of different thermal modifications on slowly digestible starch and physicochemical properties of green banana flour (*Musa acuminate colla*). *Food Chem.* **2019**, *274*, 274–280. [CrossRef]
50. Waduge, R.N.; Hoover, R.; Vasanthan, T.; Gao, J.; Li, J. Effect of annealing on the structure and physicochemical properties of barley starches of varying amylose content. *Food Res. Int.* **2006**, *39*, 59–77. [CrossRef]
51. Williams, P.C.; Kuzina, F.D.; Hlynka, I. Rapid colorimetric procedure for estimating the amylose content of starches and flours. *Cereal Chem.* **1970**, *47*, 411–420.
52. Chanttopadhyay, S.; Singhal, R.S.; Kulkarni, P.R. Optimisation of conditions of synthesis of oxidised starch form corn and amaranth for use in film-forming applications. *Carbohydr. Polym.* **1997**, *34*, 203–212. [CrossRef]
53. Schoch, T.J. Swelling power and solubility of granular starches. In *Methods in Carbohydrate Chemistry*; Whistler, R.I., Smith, R.J., BeMiller, J.N., Eds.; Academic Press: New York, NY, USA, 1964; Volume IV, pp. 106–108.
54. Ng, J.-Q.; Siew, C.K.; Mamat, H.; Matanjun, P.; Lee, J.-S. Effect of acid methanol treatment and heat moisture treatment on in vitro digestibility and estimated glycemic index of raw and gelatinized sago (*Metroxylon sagu*) starch. *Starch-Stärke* **2018**, *70*, 1700198. [CrossRef]
55. Adawiyah, D.R.; Sasaki, T.; Kohyama, K. Characterization of arenga starch in comparison with sago starch. *Carbohydr. Polym.* **2013**, *92*, 2306–2313. [CrossRef] [PubMed]
56. Liu, S.; Yuan, T.Z.; Wang, X.; Reimer, M.; Isaak, C.; Ai, Y. Behaviors of starches evaluated at high heating temperatures using a new model of Rapid Visco Analyzer—RVA 4800. *Food Hydrocoll.* **2019**, *94*, 217–228. [CrossRef]

Article

Understanding the Palatability, Flavor, Starch Functional Properties and Storability of Indica-Japonica Hybrid Rice

Xue Gong [1,†], Lin Zhu [2,†], Aixia Wang [1], Huihan Xi [1], Mengzi Nie [1], Zhiying Chen [1], Yue He [1], Yu Tian [1], Fengzhong Wang [1,*] and Litao Tong [1,*]

1. Key Laboratory of Agro-Products Processing Ministry of Agriculture, Institute of Food Science and Technology, Chinese Academy of Agricultural Sciences, Beijing 100193, China; xuegong1214@163.com (X.G.); zhk960656147@163.com (A.W.); 18437906881@163.com (H.X.); niemengzi315@163.com (M.N.); 15989109538@163.com (Z.C.); hy1257585181@163.com (Y.H.); tianyu152302@163.com (Y.T.)
2. Key Laboratory of Preservation Engineering of Agricultural Products, Ningbo Academy of Agricultural Sciences, Institute of Agricultural Products Processing, Ningbo 315040, China; zhulin0822@163.com
* Correspondence: wangfengzhong@sina.com (F.W.); tonglitao@caas.cn (L.T.); Tel./Fax: +86-10-6281-7417 (L.T.)
† These authors contributed equally to this work.

Abstract: The rice quality and starch functional properties, as well as the storability of three YY-IJHR cultivars, which included YY12 (biased japonica type YY-IJHR), YY1540 (intermedius type YY-IJHR) and YY15 (biased indica type YY-IJHR), were studied and compared to N84 (conventional japonica rice). The study results suggested that the three YY-IJHR varieties all had greater cooking and eating quality than N84, as they had lower amylose and protein content. The starch of YY-IJHR has a higher pasting viscosity and digestibility, and there was a significant difference among the three YY-IJHR cultivars. Rice aroma components were revealed by GC-IMS, which indicated that the content of alcohols vola-tile components of YY-IJHR were generally lower, whereas the content of some aldehydes and esters were higher than N84. In addition, YY-IJHR cultivars' FFA and MDA contents were lower, which demonstrated that YY-IJHR had a higher palatability and storability than those of N84 in fresh rice and rice stored for 12 months. In conclusion, this study suggested that YY-IJHR had better rice quality and storability than N84. PCA indicated that the grain quality and storability of YY12 and YY15 were similar and performed better than YY1540, while the aroma components and starch functional properties of YY-IJHR cultivars all had significant differences.

Keywords: starch-based food; indica–japonica hybrid rice; palatability; aroma components; starch functionality; storability

1. Introduction

Rice (*Oryza sativa*) with abundant starch, is a globally important starch-based food that provides calories to over 60% of the world's population [1]. To meet the rice demand of large populations worldwide, breeders around the world make use of the strong heterosis between indica and japonica rice to improve the rice yield. In recent years, the breeding of indica–japonica hybrid rice has developed very quickly and some cultivars with high yield have successfully been released. Yongyou indica–japonica hybrid rice (YY-IJHR), a new hybrid rice with high yield, is widely cultivated in southern China [2]. At the city of Ningbo in Zhejiang Province, China, the mean grain yield of Yongyou11, Yongyou12, Yongyou13, Yongyou15, Yongyou538, Yongyou1538, Yongyou1540 and Yongyou4540 were all over 12.0 t/ha, which was significantly higher than conventional rice [3,4]. Most studies on YY-IJHR focus on cultivation techniques and yield [4–7]. In recent years, there have been preliminary studies on the characteristics of indica–japonica hybrid rice starch [8]. With the continuous expansion of planting area, the rice quality and storage characteristics of YY-IJHR have attracted increasing attention from rice industrialists, physiologists and consumers.

Rice quality is reflected by the nutritional components and eating quality, as well as the aroma compounds. Recently, the report of Bian et al. [9] suggested that different panicle types of indica–japonica hybrid cultivars had a different yield and grain quality. Starch, as the major component of rice, as well as its functional properties, play an important role in determining rice's edible quality and end-uses [10]. The study of Wani et al. [11] suggested that the starch determines the eating and cooking properties of rice grains, with the amylose content particularly affecting the taste value of cooked rice. Zhu et al. [8] indicated that the structural and physicochemical properties of indica–japonica hybrid rice starches were significantly different. In addition to making starch-based food, rice starch is also used in various processed foods as an adhesive, thickener, extending agent, and inflating agent, according to its properties [12]. However, the rice starch research is mainly focused on conventional or hybrid indica, japonica or wild rice [13,14]. The knowledge of YY-IJHR's starch properties is limited [8]; there was no information regarding comparative studies on the starch functionality of YY-IJHR before and after storage. Furthermore, the type and quantity of aroma compounds were also important indicators of rice quality, which made an important contribution to the taste value of rice. The gas chromatography–ion mobility spectrometry (GC-IMS), as a sensitive and efficient approach, had been increasingly applied to investigate the volatiles in rice, such as Guangxi fragrant rice [15]. However, as a new cultivar, little work has been performed on the aroma volatiles of YY-IJHR, especially the study determining the YY-IJHR volatiles by GC-IMS. In this study, we determined the aroma components of YY-IJHR using GC-IMS and analyzed the characteristic aromas of different cultivars by principal component analysis (PCA). To comprehensively describe the rice quality of YY-IJHR, we evaluated and comparatively analyzed three YY-IJHR cultivars' characteristic aroma volatiles, eating quality and starch functional properties.

As is well known, the storage of rice inevitably leads to changes in cooking quality, pasting properties and biochemical indicators. The study of Faruq, et al. [16] showed that ageing could influence kernel expansion, water absorption and gelatinization temperature, which ultimately affect the internal structure and cooking quality of rice grains. Another study indicated that storing rice for from 3 to 4 months after harvesting would have a great impact on the cooking quality of grains, maturing the rice and making it taste better [17]. However, to our knowledge, there was little information about the changes in the major cooking quality and pasting property parameters of YY-IJHR under 12 months' storage. Furthermore, a study has shown that the biochemical indicators regarding the storability of rice stored for one year obviously changed, such as an increase in FFA content and a decrease in antioxidant enzymes' activity [18]. However, rice storability is a complex trait, which is influenced by many aspects, such as seed genotype, endogenous enzymes activity and storage conditions [19]. Knowledge of rice storability also plays a guiding role in rice uses and consumer choice. As a new rice variety, the storability of YY-IJHR cultivars is not clear.

An understanding of rice quality, the functional properties of rice starch and rice storability is of critical importance to optimize industrial applications and allow for consumers to select suitable rice varieties [12]. To the best of our knowledge, there was no information systematically evaluating the rice properties of YY-IJHR. Therefore, the objectives of this study were: (i) determine the rice quality, including edible quality and aroma components of YY-IJHR; (ii) examine the functional properties of rice starches before and after 12 months' storage; (iii) explore the storability of YY-IJHR. Such information may not only contribute to the understanding of rice quality, starch functional properties and the storability of YY-IJHR, but also provide necessary knowledge for consumers, the future work of rice breeding and better applications in the rice industry.

2. Results and Discussion

2.1. Rice Quality

2.1.1. Rice Nutritional and Eating Quality

Amylose determines the eating and cooking properties of rice grains and protein is typically the second-highest component in many rice-starch based foods, at about 4–20% by weight [10]. The breeding of high-eating-quality rice has been focused on obtaining low amylose and protein content [11]. To evaluate the nutritional and eating quality of YY-IJHR, amylose and protein content were determined. As shown in Figure 1A, the amylose content was higher in N84 (21.65%) compared with the YY-IJHR cultivars (16.45–19.2%). The protein content of N84 was 8.85%, while the protein content of YY-IJHR cultivars was in the range 8.05~8.30% (Figure 1B). The nutritional quality of hybrid rice substantially differed from that of conventional japonica rice, because hybrid rice was an F2-segregated population derived from hybrid F1. Theoretically, there was a greater separation in the indica–japonica hybrid rice (IJHR) of the hybrid F2 population. For YY-IJHR varieties, both amylose and protein contents were decreased compared to those of its parents, which might be explained by the character segregation brought by indica–japonica hybrid genes in the YY series.

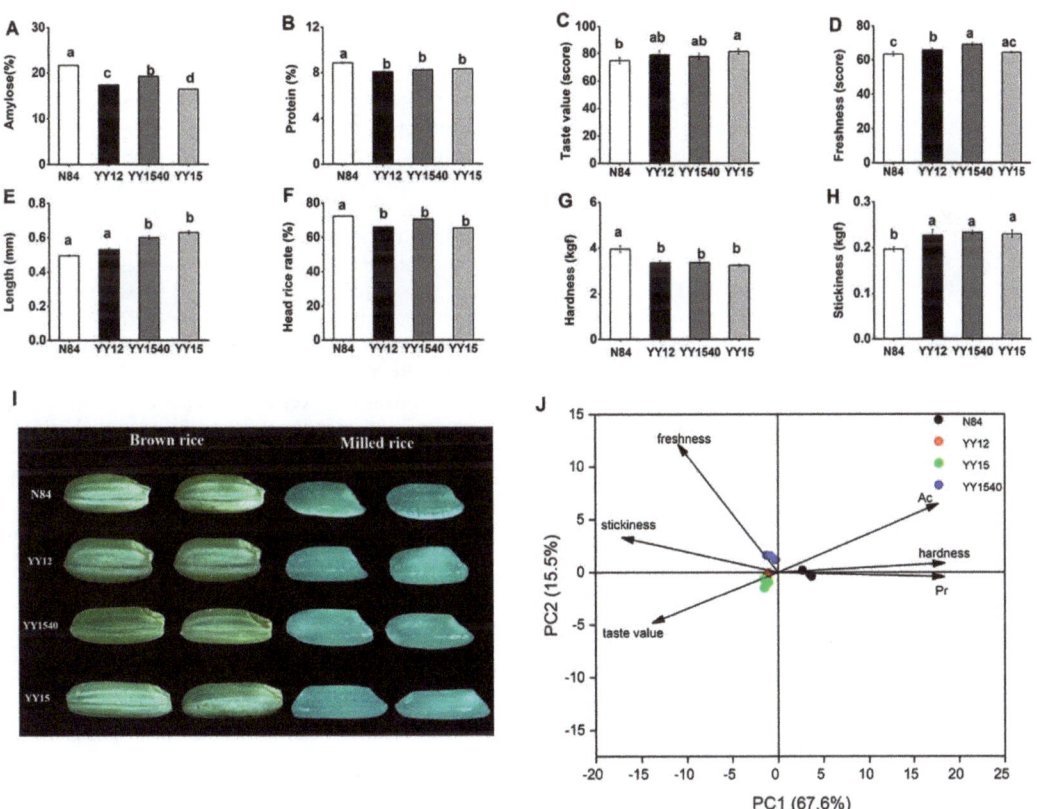

Figure 1. Rice comprehensive quality of YY-IJHR. The amylose content (**A**), protein content (**B**), taste value (**C**), freshness (**D**), length (**E**), head rice rate (**F**), hardness (**G**), and stickiness (**H**), and stereomicrograph (**I**), as well as principal component analysis (PCA) of grain quality (**J**) of N84 and YY-IJHR cultivars. Different lowercase letters indicate significant differences at $p < 0.05$ (Ducan's test).

The edible quality of rice was mainly reflected by its taste value and freshness [12]. As shown in Figure 1C, the taste values of YY12, YY1540, and YY15 were 79.00, 77.76, 81.36, respectively; which were also higher than that of N84 (74.76). Previous studies showed the rice amylose and protein contents were negatively correlated with sensory quality [13]; these data may explain the higher eating quality of YY-IJHR cultivars compared to N84. Freshness was an important palatability characteristic of rice grain. As shown in Figure 1D, the freshness scores of YY12, YY1540, and YY15 were 65.66, 69.00, 64.33, respectively; all of these were higher than that of N84 (63.3). This result indicated that YY-IJHR had a better palatability than N84.

Taken together, these data suggest that YY-IJHR varieties have an advantage over N84 in terms of edible quality and YY15 had the best taste value, while YY1540 had the best freshness among the three YY-IJHR cultivars.

2.1.2. The Appearance and Textural Characteristic

The grain length of YY-IJHR cultivars was measured and compared with that of N84. As shown in Figure 1E, the average lengths of YY12, YY1540, and YY15 were 5.3 mm, 5.7 mm and 6.3 mm, respectively, all of which were longer than that of N84 (5.0 mm). Generally, the shape of common indica rice was long and thin, whereas the shape of conventional japonica rice was short and circular. YY-IJHR was a cultivar containing an indica–japonica hybrid genotype, so its length was longer than that of conventional japonica (Figure 1I). The head rice rate, an important factor affecting the sensory evaluation of rice, was considered a major determinant of quality grades in some developing countries [20]. As shown in Figure 1F, the head rice rate of N84 was significantly higher than that of the YY-IJHR cultivars. Therefore, improving the head rice rate might be a problem for the next YY series of indica–japonica hybrid rice breeding.

Texture is a multi-faceted sensory property, with hardness and stickiness being the most common determined and discriminable textural properties of cooked rice [21]. As shown in Figure 1G,H, there were significant differences in terms of hardness and stickiness between N84 and YY-IJHR cultivars. The hardness of N84 (3.95 kgf) was higher than YY-IJHR cultivars in the range 3.24 kgf~3.36 kgf, while the stickiness of YY-IJHR varieties (0.22 kgf~0.23 kgf) was higher compared with N84 (0.19 kgf). There was no significant difference in hardness and stickiness among YY-IJHR cultivars. The stickiness of cooked rice is important for eating quality and consumer acceptance [21]. Rice stickiness had a significant correlation with leached amylopectin amount, which increased with the decrease in amylose content [21]. Since the hardness was also positively correlated with the amylose content, the low hardness and high stickiness could be explained by the lower amylose content of YY-IJHR. Furthermore, the study of Peng et al. [22] indicated that the taste value of the hybrid rice was negatively correlated with hardness, while it was positively correlated with stickiness. These results were consistent with the results regarding edible quality in this study.

2.1.3. PCA of Rice Quality

The rice-quality characteristics among the four rice varieties was summarized in the scores plot (Figure 1J). The PCA of grain-quality parameters and samples with different rice varieties suggested that PC1 and PC2 accounted for 87.8% of the total variation. Therefore, the plane of PC1–PC2 could reflect the main contributions of the response variables. The similarities and differences between the varieties are determined by the locations of the dots, and the distances of the dots from the arrows can explain the relationship between the two in a biplot. In Figure 1J, all the samples could be divided into two groups according to their corresponding grain quality. Group 1 was the control group (N84), which exhibited high amylose content, high protein content and high hardness values. This result indicated that the high amylose and protein contents could lead to the low taste value and hard texture of cooked rice, which was consistent with the previous study [16]. As shown in Figure 1J, YY12, YY1540, and YY15 formed group 2, which possessed high stickiness and freshness as well as good taste values. This result suggested that YY-IJHR varieties had a better grain-quality performance than N84, the conventional japonica rice.

2.2. Analysis of Aroma Components of YY-IJHR

2.2.1. Odor Description of Detected Aroma Components

Previous studies have demonstrated that flavor fingerprints by GC-IMS provide useful information and, therefore, could be used to evaluate the characteristic aroma of grain products [23,24]. The identified visual spots were presented in Gallery Plot. As shown in Figure 2A, the row represented sample and the column represented substance. The color and brightness of the signal peak represents the concentration of the substance, which could directly show the complete volatile compounds' information of each sample and the difference in volatile compounds between samples. Among the 65 detected peaks (P), a total of 49 volatile components and 16 unknown substances were identified, including alcohols (9P), esters (6P), aldehydes (26P), ketones (7P) and heterocyclic substance (1P), among which 18 pairs were monomers and dimers (the chemical formula and CAS number were the same, but the morphology was different). There were 16 aroma substances that could not be characterized, which resulted from the imperfect information in the flavor composition database. These volatile components all occurred in the YY-IJHR and N84, while the different concentrations of volatiles brought variations in aroma. Among the experimentally detected volatile components, alcohols and esters were the characteristic substances in rice aroma components, and 1-octen-3-ol, n-hexanol, n-pentanol, 2-methylbutanol, and isoamyl alcohol were detected in all rice samples. These alcohols were C4–C8 low-carbon chain alcohols with an alcohol aroma, such as 1-octene-3-alcohol with mushroom and hay aroma, n-hexanol with an apple-like fruit aroma, etc. The detected esters included ethyl acetate, butyl acetate and butyl propionate, which were volatile components with a fruit aroma.

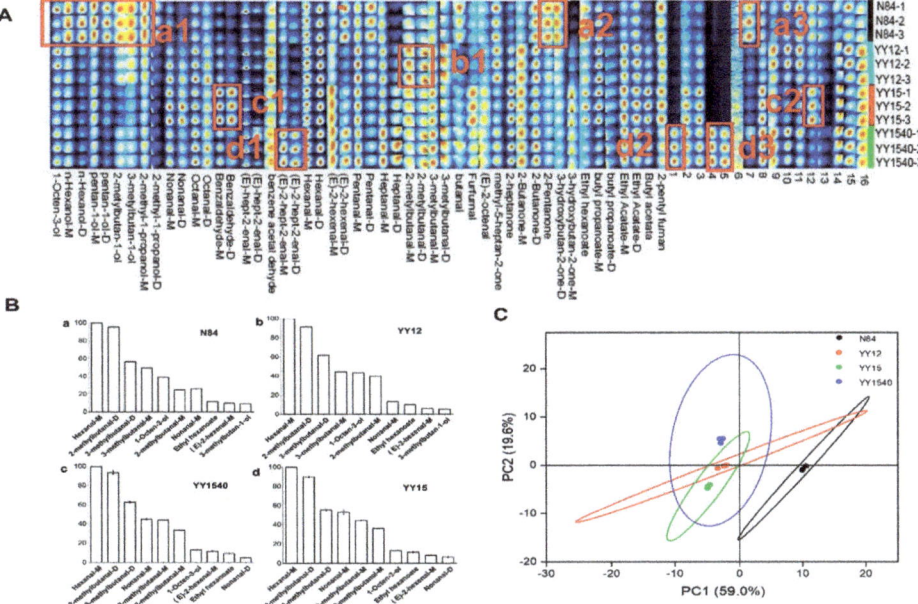

Figure 2. Characteristic of YY-IJHR aroma components. (**A**): Gallery plot fingerprint of volatile components in N84 and YY-IJHR varieties, the characteristics of volatiles among the different rice samples with circled volatile compounds (a1, a2, a3, b1, c1, c2, d1, d2, d3) are presented; (**B**): Fingerprints of the different volatile components and bar graph of the main volatile compounds (ROAV > 1) with the top 10 in four rice varieties; (**C**): PCA of flavor components of N84 and YY-IJHR.

Aldehydes and ketones containing carbonyl groups were important components of rice volatiles. Most of the ketones detected in this study were low-carbon ketones, of which C5–C8 ketones had a special aroma. For example, methyl heptanone had a lemongrass aroma and 2-heptanone had a lemongrass aroma. This had a pear fragrance, while ketones of C4 and below had no fragrance. For example, 2-butanone had a similar smell to acetone. Aldehydes were intermediates that convert alcohols into acids. Aldehydes with a low molecular weight (below C7) have an unpleasant pungent odor, such as butyraldehyde, glutaraldehyde and furfural. With the increase in molecular weight, C8–C10 aldehydes have an obvious floral or fruit aroma, such as nonanal, with a rose and citrus aroma; Octanal had a strong orange aroma. The other detected organic compounds were mainly 2-pentylfuran, which were oxygenated heterocyclic compounds with fruit and grass aromas [25].

2.2.2. Differential Aroma Compounds

As shown in Figure 2A, the characteristics of volatiles among the different rice samples with circled volatile compounds (a1, a2, a3, b1, c1, c2, d1, d2, d3) are presented. The dark red spots indicated the obviously higher concentrations of volatile compounds than the reference. It could be seen that the relative content of alcohols in the YY-IJHR varieties were significantly lower compared with N84, such as hexanol, amyl alcohol, 2-methylbutanol, and isoamyl alcohol. However, the levels of some aldehydes and fruity esters, such as benzaldehyde, furfural, 2-methylbutanal, 3-methylbutanal, ethyl acetate, and butyl acetate, were much higher in YY-IJHR than those in the control group, especially benzaldehyde and butyl acetate.

There were also significant differences in volatile flavor components among different varieties of YY-IJHR. The volatile flavor components of YY15 were mainly aldehydes and esters, whereas the YY12 were mainly aldehyde and ketone. Differential aroma compounds are landmark substances, which distinguish them from other samples [26]. Combined with ROAV (Figure 2D), 3-methylbutan-1-ol was referred to as the differential in N84 and YY12, while nonanal-D were referred to as differential volatiles in YY1540 and YY15 among the volatile compounds.

2.2.3. Key Aroma Compounds

The relative concentration of volatile flavor components was an important factor in distinguishing rice aroma. The contribution of each compound to the fresh rice flavor was evaluated by calculating the relative odor activity value (ROAV). An ROAV compound of greater than 1 indicated that compounds made a significant contribution to the flavor, and a value < 1 meant a small contribution to the flavor. As shown in Figure 2B, the top 10 volatile compounds of the ROAV value (ROAV > 1) of fresh rice grains were referred to as the main volatile compounds. This meant that the detected volatile compounds made a significant contribution to the aroma. Hexanal-M, 2-methylbutanal-D, 3-methylbutanal-D, nonanal-M, 3-methylbutanal-M, 2-methylbutanal-M, 1-Octen-3-ol, ethyl-hexanoate, (E)-2-hexenal-M were the common main volatile compounds of the four samples.

The vital compounds that formed the flavor of fresh rice comprised differential volatile compounds and main volatile compounds. Therefore, the key aroma compounds were defined by these two major components [27]. Hexanal-M, 2-methylbutanal-D, 3-methylbutanal-D, nonanal-M, 3-methylbutanal-M, 2-methylbutanal-M, 1-Octen-3-ol, ethyl-hexanoate, (E)-2-hexenal-M, 3-methylbutan-1-ol and nonanal-D were regarded as key aroma compounds in YY-IJHR.

2.2.4. PCA of Aroma Compounds

A PCA performed on the rice aroma compounds and samples of different rice varieties showed that PC1 and PC2 accounted for 78.6% of the total variation. As shown in Figure 2C, the distance between YY-IJHR group and the control group(N84) was relatively high, indicating that the difference in aroma properties between the YY-IJHR group and control group (N84) was significant. The control group was positive with PC1, whereas the YY-IJHR group was negative. This result could be attributed to the relative, higher concentration of aldehydes, and the ROAV of alcohols was lower in all YY-IJHR samples (Figure 2B).

The volatile flavor components of YY15 were mainly aldehydes and esters, while those of YY12 were mainly aldehydes and ketones (Figure 2A). As shown in Figure 2A, these aroma compounds' concentrations were relatively low. Notably, the aroma compound concentration of YY1540 was relatively high. Therefore, the cultivars of YY12 and YY15 were negative with PC2, while YY1540 was positive (Figure 2C). These results suggested that there was a significant difference among three YY-IJHR cultivars and N84, which was revealed by the PCA plot.

2.3. Starch Functional Properties of YY-IJHR

Rice starch is used in various processed foods as an adhesive, thickener, extending agent, and inflating agent. The functional properties of rice starch are influenced by the crystalline structure, the amylose–amylopectin ratio and the fine structure of amylopectin [12]. This is important for the rice-starch-based food quality and other end-use utilizations. The digestive properties of cooked rice are affected by the composition of starch, such as rapidly digestible starch and slowly digestible starch, whereas the eating and cooking quality of rice is influenced by the pasting properties of starch [28].

2.3.1. Starch Digestive Property

The digestive properties of rice starch have important potential implications for human health. Starch can be grouped into rapidly digestible starch (RDS), slowly digestible starch (SDS) and resistant starch (RS) [29]. Rapidly digestible starch (RDS) rapidly hydrolyses into sugar molecules after consumption and provides the energy needed by the body. Slowly digestible starch (SDS) is that fraction of starch that is slowly hydrolyzed to glucose molecules in small intestine for the maintenance of normal blood sugar levels [29]. Resistant starches (RS) are those that, by localization, physical, or chemical causes, are unavailable for enzymatic attack, acting as dietary fiber in our organism [30].

RS was positively correlated with amylose content in rice starch, while SDS and RDS decreased as amylose content increased [31]. The study of Noda et al. [32] indicated that the amylose content in rice starch was negatively correlated with digestibility because amylose in starch granules was not easily digested by enzymes. As shown in Figure 1A, the amylose content was significantly higher in N84 (21.65%) compared with the YY-IJHR cultivars (16.45–19.2%). This result indicated that the cultivar of N84 had more RS and lower digestibility compared to YY-IJHR.

2.3.2. The Pasting Properties

The characteristic pasting parameter values of rice have a great impact on the properties (e.g., cooking and eating quality, texture) of many rice-starch-based food matrices [10]. The pasting parameters, including peak viscosity (PV), trough viscosity (TV), breakdown (BD), final viscosity (FV), setback (SB) and pasting temperature (PT), are displayed in Table 1. PV reflects the water-holding capacity or swelling degree of the starch granules and is often related to the final product quality, because swollen and collapsed starch granules can affect the texture of the product [11]. The retrograded viscosity (BD) characterizes the thermal and shear resistance of the starch pastes, while the SB shows the retrogradation trend [13]. High BD values indicate high cooking quality because rice degrades to a small extent after cooling [13]. As shown in Table 1, PV, FV, BD of YY-IJHR were all significantly higher than those of N84. The higher PV of YY-IJHR represented the higher swelling capacity because of its lower amylose content, which is consistent with the study of Kong et.al. [13]. The underlying origin of the differences in viscosity was the composition of starch and the fine starch structure in the rice grain. The relatively higher FV of the three YY-IJHR cultivars could be explained by the fact that, with decreasing amylose content and the proportion of short branches of amylopectin, leading to a higher FV [21]. Moreover, studies have indicated that the low amylose content and high PV and BD were characteristics of a high eating quality [33]. Combining the results of the previous research [8,33], it can be concluded that YY-IJHR possessed a higher pasting viscosity and better eating quality than N84. In addition, YY15 had a relatively better eating quality among the three YY-IJHR varieties, in accordance with the results of a lower amylose content and higher taste value.

Table 1. Pasting properties of fresh rice and rice stored for 12 months in N84 and YY-IJHR.

Rice Varieties	N84	YY12	YY1540	YY15
		Fresh Rice		
PV (cp)	2428 ± 39.3 d	3356 ± 37.6 b	3238 ± 26.1 c	3574 ± 63.9 a
TV (cp)	1549 ± 35.6 b	1677 ± 43.5 b	2040 ± 58.5 a	1660 ± 86.9 b
BD (cp)	878 ± 10.8 d	1678 ± 14.5 b	1199 ± 73.8 c	1915 ± 52.8 a
FV (cp)	2484 ± 35.1 c	2670 ± 41.1 b	3287 ± 68.6 a	2608 ± 73.3 bc
SB (cp)	934 ± 4.1 c	993 ± 2.6 b	1247 ± 13.1 a	948 ± 14.2 c
PT (°C)	75.5 ± 0.1 d	80.3 ± 0.1 b	74.3 ± 0.1 c	81.2 ± 0.2 a
		Rice Stored for 12 Months		
PV (cp)	2930 ± 42.0 b	3589 ± 52.9 a	2819 ± 72.5 b	3655 ± 9.7 a
TV (cp)	1690 ± 76.0 ab	1640 ± 10.6 ab	1763 ± 61.5 a	1581 ± 38.7 b
BD (cp)	1240 ± 40.9 c	1948 ± 56.0 b	1056 ± 31.7 d	2074 ± 40.8 a
FV (cp)	2926 ± 55.5 c	2918 ± 6.9 b	3286 ± 68.3 a	2784 ± 40.4 b
SB (cp)	1236 ± 20.5 c	1278 ± 16.7 b	1523 ± 8.2 a	1202 ± 8.7 c
PT (°C)	71.1 ± 0.1 c	80.7 ± 0.1 b	86.1 ± 0.4 a	80.2 ± 0.8 b

PV, peak viscosity; FV, final viscosity; TV, trough viscosity; BD, breakdown viscosity; SB, setback viscosity; PT, pasting temperature. cp, centipoises. Means values ± standard deviation; the error bars represent the SD, and different lowercase letters indicate significant differences at $p < 0.05$ (Ducan's test).

During storage, the changes in pasting property parameters occurred due to the variation in granule size distribution, amylose content, and crystallinity. As shown in Figure 3A, a greater change in pasting properties (PV, BD, FV, SB) was exhibited in N84 than YY-IJHR after 12 months of storage, which indicated that YY-IJHR has a stronger starch function retention than N84.

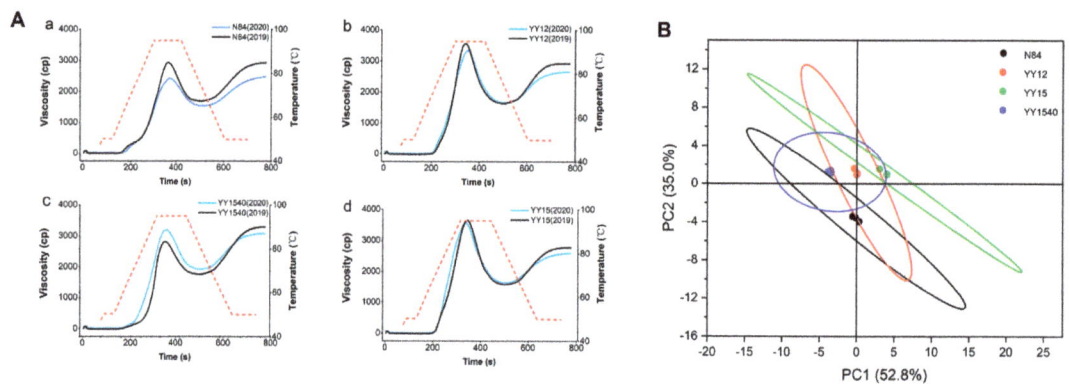

Figure 3. The pasting properties of YY-IJHR. (A) Comparison of RVA curves in new rice and rice stored for one year from N84 (a) and YY12 (b), YY1540 (c), YY15 (d) as well as PCA of pasting properties (B) of N84 and YY-IJHR cultivars.

Interestingly, a decreasing trend of PV was shown in YY1540, whereas an increasing trend was shown in the other three varieties during storage. PV represents the water-binding capacity of starch granules [8] and the higher amylose content in rice starch decreased PV [34]. Therefore, the decreasing trend of PV in YY1540 may be due to the increase in amylose during storage, which inhibited the swelling of starch granules, resulting in the decrease in PV. This phenomenon suggested that the stored rice of YY1540 was of relatively lower quality in rich-starch food or swelling agents [12].

2.3.3. PCA of Pasting Properties

To comprehensively analyze the pasting properties, data including PV, TV, BD, FV, SB, PT of fresh rice and rice stored for 12 months among the four cultivars were loaded on PCA and are summarized in the scores plot (Figure 3B). The points in the confidence ellipses were all statistically significant. The PCA of pasting property parameters (Table 2) and different rice cultivars indicated that PC1 and PC2 accounted for 86.3% of the total variation. The samples could be divided into two groups according to their contribution to the principal components. Group 1 was the conventional japonica rice (N84), which was significantly negative with the PC2, which provided the lowest pasting viscosity, in accordance with Figure 2A. Group 2 was consisted of YY12, YY1540 and YY15. As shown in Figure 2A, the PV of YY15 was the highest in both fresh rice and stored for 12 months, while the mean rice starch of YY15 has the best swelling capacity among the four cultivars. The PV of stored rice in YY1540 was significantly decreased, which meant that the starch of YY1540 had a lower retention capacity regarding starch's functional properties. As shown in Figure 2B, YY15 was significantly positive with PC2 and made the greatest contribution to PC1, while YY1540 was negatively correlated with PC1. Three YY-IJHR cultivars showed significant variation in pasting properties, which was in accordance with the previous report [8]. YY15 possessed a higher pasting viscosity, followed by YY12 and YY1540; N84 had the lowest pasting viscosity.

Table 2. Fresh rice and rice stored for 12 months' cooking characteristics for N84 and YY-IJHR.

Rice Varieties	N84	YY12	YY1540	YY15
		Fresh Rice		
Cooking expansion rate (%)	268 ± 5.7 c	355 ± 1.4 a	364 ± 5.7 a	322 ± 5.7 b
Cooking elongation rate (%)	62.4 ± 0.5 b	77.5 ± 1.4 a	80.2 ± 2.8 a	55.5 ± 7.8 b
Heating water absorption rate (%)	141.6 ± 0.4 b	145.9 ± 1.2 b	143.4 ± 2.8 b	151.8 ± 2.3 a
Soaked water absorption rate (%)	16.1 ± 0.1 c	23.9 ± 0.4 b	26 ± 4.4 ab	30.9 ± 0.4 a
		Rice Stored for 12 Months		
Cooking expansion rate (%)	460 ± 0.4 a	338 ± 2.8 b	361 ± 1.4 b	367 ± 1.4 b
Cooking elongation rate (%)	53.1 ± 0.1 b	90 ± 2.8 a	67.5 ± 10.1 b	56.6 ± 3.5 b
Heating water absorption rate (%)	344.3 ± 0.2 a	146.4 ± 2.6 b	146.6 ± 0.6 b	148.3 ± 1.8 b
Soaked water absorption rate (%)	14.7 ± 0.1 b	26.0 ± 1.3 a	27.3 ± 4.1 a	30.2 ± 2.8 a

Means values ± standard deviation, $n = 3$. Different lowercase letters indicate significant differences at $p < 0.05$ (Ducan's test).

2.4. The Cooking Quality

2.4.1. The Cooking Quality of Fresh Rice

The cooking quality of rice was an important indicator of consumer acceptance. To evaluate the cooking quality of fresh rice in YY-IJHR, the cooking expansion rate, cooking elongation rate, soaked water absorption rate and heating water absorption rate were measured and compared with N84. As shown in Table 2, the cooking expansion rate (322–364%), heating water absorption rate (143.40–151.75%) and soaked water absorption rate (23.87–38.60%) of fresh rice in YY-IJHR cultivars were greater than those of N84 (cooking expansion rate (268%), heating water absorption rate (141.62%), soaked water absorption rate (16.08%). The hydration of rice was influenced by amylose content in rice grain. Varieties with a low amylose content tended to be sticky and cohesive when cooked, absorbed more water, and thus had a greater volume after cooking [35]. Furthermore, the study of Zhu et al. [36] reported that protein could compete with starch to absorb water and further restrict the swelling of starch granules, which may affect the rice expansion and extension rate when heated or soaked with water. Therefore, YY-IJHR cultivars with low amylose and protein content had a greater cooking expansion rate and heating water absorption rate. The greatest heating and soaked water absorption rate (151.75% and 30.86%, respectively) occurred in YY15 among the three YY-IJHR cultivars, which was consistent with its lower amylose content.

The swelling ability of starch granules when heated with water to cook rice could be reflected by the PV [9]. The YY1540 cultivar has the greatest cooking expansion rate (364%) and elongation rate (80.16%) among the three YY-IJHR cultivars, in accordance with the relatively higher PV in YY1540. To improve consumer satisfaction, the information provided by this study could be used by the rice breeders and industrialists to use these rice varieties' characteristics as evidence to breed new rice with a greater cooking quality and better rice starch-based foods.

2.4.2. Comparison of the Changes in Cooking Quality during Storage

Ageing also enhances the volume expansion and water absorption of rice upon cooking, resulting in a product with a firmer texture and less stickiness [6]. In this study, the magnitude of changes was also different in the rice cultivars after 12 months of storage. It could be found that the range of changes in cooking quality (cooking expansion rate (3–45%) and heating water absorption rate (1–3%)) of YY-IJHR was smaller than that of N84 (cooking expansion rate (198%), heating water absorption rate (203%)) during storage. The steric hindrance effect of protein and starch–protein interactions during storage might also influence the cooking quality. However, the surrounding structure of protein or the bond between starch and protein were destroyed by the protease, allowing for more water to seep into the starch granules. The higher the protease content, the greater the expansion and extension of rice molecules [36]. Therefore, this greater change magnitude in the cooking quality of N84 could be attributed to the relatively higher protein and amylose content, as well as the high protease activity in N84. Moreover, the range of changes in the four cooking indexes in YY1540 was the smallest after storage. In conclusion, the results of this study suggested that YY-IJHR cultivars had a better cooking-quality retention capacity than N84, especially the YY1540 cultivar.

2.5. The Rice Storability

2.5.1. The Content of FFA and MDA

The free fatty acid (FFA) content, a sensitivity indicator of the quality changes, is often employed as a measure of the deterioration of stored grain. As shown in Figure 4A, for the fresh rice, N84 (12.00 mg/100 g) had a similar amount of FFA to YY12 (12.89 mg/100 g), and both were significantly higher than YY1540 (8.80 mg/100 g) and YY15 (8.70 mg/100 g). In addition, the FFA content of N84 (15.98 mg/100 g) significantly increased, becoming higher than that of YY-IJHR after one year of storage. The increase in FFA content in stored grain could be attributed to the hydrolysis of lipase during storage. This can impact the physical properties of rice in terms of its eating quality, flavor, compositions. With a prolongation of storage time, FFA content increased and the quality of stored grain decreased simultaneously [37]. The FFA content was an effective indicator of a deterioration in palatability in the old rice [38]. The lower FFA content of YY-IJHR meant that the quality and palatability of YY-IJHR were better than these of N84 after 12-months storage.

Malondialdehyde (MDA) is an end-product of seed lipid peroxidation and its content may represent the degree of rice seed cell damage [39]. To explore the storability of YY-IJHR, we determined the MDA content of rice grain in YY-IJHR. As shown in Figure 4B, for fresh rice, the MDA content of YY1540 (21.36 nmol/g) was the highest among the four varieties. The MDA content of YY12 showed a decreasing trend, while the other two YY-IJHR varieties showed an increasing trend after 12 months of storage. This phenomenon occurred because cells began to die, enzyme activity decreased with the extension of storage time, and FFA oxidation significantly reduced, resulting in a decline in the MDA content [39]. Moreover, the MDA content of N84 substantially increased to 46.31 nmol/g, which was significantly higher than that of YY12 (9.80 nmol/g), YY15 (19.00 nmol/g) and YY1540 (29.00 nmol/g) after 12-months' storage (Figure 4B). The increase in MDA level indicated that the degree of membrane lipid peroxidation in rice increased and rice quality decreased. This result suggested that the storability of YY-IJHR cultivars was higher than N84, which was in accordance with the result of the lower FFA content in YY-IJHR.

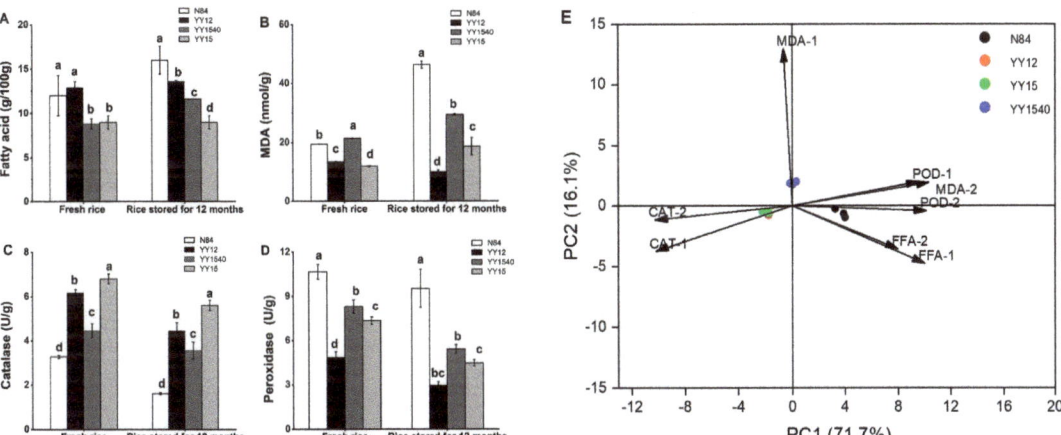

Figure 4. The storability of YY-IJHR. The FFA content (**A**), MDA content (**B**), catalase activity (**C**) and peroxidase activity (**D**) characteristics of YY-IJHR cultivars and N84.In the biplot of PCA of storability (**E**). FFA-1 represented the free fatty acid content of fresh rice, FFA-2 represented the free fatty acid content of rice stored for 12 months; MDA-1 represented the MDA content of fresh rice, MDA-2 represented the MDA content of rice stored for 12 months; CAT-1 represented the CAT activity of fresh rice, CAT-2 represented the CAT activity of rice stored for 12 months; POD-1 represented the POD activity of fresh rice, POD-2 represented the POD activity of rice stored for 12 months. Different lowercase letters indicate significant differences at $p < 0.05$ (Ducan's test).

In conclusion, both YY-IJHR cultivars' FFA and MDA contents were lower, which indicated that YY-IJHR had a lower deterioration in palatability and higher storability than N84 over the same storage time.

2.5.2. The Activity of CAT and POD

Catalase (CAT) and peroxide (POD) are rice-seed-viability-protecting enzymes that scavenge hydrogen peroxide in rice seeds. To explore the rice-seed-protecting enzymes' activity, we determined the CAT and POD activity of fresh rice and rice stored for a year. As shown in Figure 4C,D, for fresh rice, the CAT activity of YY-IJHR in the range 4.15 U/g~6.80 U/g was higher than N84 (3.28 U/g). This was consistent with the results of lower FFA and MDA content in YY-IJHR in this study. After 12 months of natural storage, the CAT and POD activities of all four rice varieties showed a decreasing trend. The decline in the activities of CAT and POD, resulting in a reduced ability to break down the peroxides that are hazardous to cells in the rice grain, thus accelerating the deterioration in rice quality. The decreasing degree of enzyme activity of the four rice varieties was similar, so we judged the storability by the PCA.

2.5.3. PCA of Storability

The relationship between rice cultivars with different aging indexes and antioxidant enzymes was summarized in the biplot (Figure 4E). The PCA of storability indexes and different rice varieties suggested that PC1 and PC2 accounted for 87.8% of the total variation. The similarities and differences between varieties were determined by the point locations and the distances of the points from the arrows may explain the relationship between the two in the biplot. As shown in Figure 4E, the storability of N84 and YY-IJHR were clearly distinguished, as the FFA content of N84 was high before and after storage, and its higher MDA content after storage indicated the inferior storability of N84. In addition, YY12 and YY15 showed similar storage characteristics and had a higher CAT activity in grain compared to N84, which led to the higher free-radical scavenging capacity and lower palatability deterioration.

3. Materials and Methods

3.1. Materials

Rice seeds were obtained from Shipu Town, Xiangshan County, Ningbo City, Zhejiang Province, China. N84 seeds were used as the high-quality rice control. YY15 (biased indica-type indica–japonica hybrid rice,) and YY1540 (intermedius-type indica–japonica hybrid rice) were harvested from the same field at the end of October. YY12 (biased japonica-type indica–japonica hybrid rice) was harvested around 15 days later due, to its late-maturing character. Fresh rice in the current season were dried to a moisture content of 14–15% after threshing. The rice was packaged into a universal transparent preparation bag and stored in the dark at 4 °C before the formal start of the experiment. The samples were stored at room temperature (26 °C) at the formal start of the experiment.

3.2. Preparation of Rice Flour

The harvested mature rice grains were dehulled with a rice-huller (Yao jiang JLGJ 2.5, China) and brown rice was milled into fine rice with a rice mill (Osaka MB-RC52 Japan). Finally, they were ground into powder by a cyclone grinder (CT410, FOSS Scino (Suzhou) Co., Ltd., Suzhou, China) and passed through an 80-mesh standard examination sieve to produce rice powder. On a dry basis, the chemical composition of rice flour consists of carbohydrates (about 80 g/100 g, starch (50–60 g/100 g), moisture (about 10–15 g/100 g) proteins (about 7–10 g/100 g), ash (about 3 g/100 g), fats (about 1.5 g/100 g), and pigments [40].

3.3. Determination of Rice Amylose and Protein Content

Protein contents (46–11.02) were measured in triplicate according to AACC-approved methods. The amylose content (AACC Method 76–31.01) was determined three times by commercial assay kit (K-AMYL, Megazyme International Ltd., Wicklow, Ireland).

3.4. Determination of Eating Quality, Length and Head Rice Rate of Rice Grain

A rice-grain taste analyzer (SATAKE STA1B, Hiroshima, Japan) was used to evaluate the taste value of cooked rice and every sample was determined three times. Freshness score was determined in triplicate by the fresh tester (SATAKE RFDM1B, Hiroshima, Japan). Ten rice grains were randomly picked and measured with a right-angle measurement (Shinwai Sokutei, 12416, Niigata, Japan), and the average was subsequently calculated. Head rice rate was determined in triplicate according to the Chinese National Standard method (GB/T 21719-2008). Pictures of brown rice and milled rice from four varieties were taken by a stereomicroscope (Nikon SMZ800N, Tokyo, Japan); Three parallel runs were performed for each sample.

3.5. Determination of Rice Hardness and Stickiness

The hardness and stickiness of N84 and YY-IJHR varieties were measured by a hardness and viscosity tester (SATAKE RHS1A, Hiroshima, Japan). The values of hardness (kgf) and stickiness (kgf) were derived from the instrument software (SATAKE RHS1A, Hiroshima, Japan). The hardness and stickiness unit was kgf, where 1 kgf unit is equal to 9.8 Newtons (N).

3.6. Determination of Rice Characteristic Aroma Components

Rice samples (5 g, dry basis) from each cultivar were collected in 20-mL headspace bottles and incubated at 80 °C for 15 min before headspace injection, with an injection volume of 500 µL. The injection needle temperature was 85 °C and incubation speed was 500 R/min. The carrier gas was high-purity nitrogen. With flavourspec®, the flavor analyzer developed GC-IMS analysis, and each sample was assayed 3 times in parallel. The test conditions of GC-IMS were in accordance with the method of Zhao and Shen [41] and each sample was detected three times. Qualitative analysis of volatile organic components was performed based on the built-in database of GC × IMS Library Search. Quantitative analysis of the corresponding peak area of volatile organic components was normalized with auto-scaling in GC-IMS data.

The contribution of each compound to the fresh rice flavor was evaluated by calculating the relative odor activity value (ROAV). The present study adopted the ROAV method, where the ROAV compound greater than 1 indicated that compounds made a significant contribution to the flavor, and a value of <1 meant a small contribution to the flavor. The component with the greatest contribution to the odor of all samples was defined as $ROAV_{max}$ and given a value of 100, with the other volatile components being calculated according to Equation (1):

$$ROAV_i = 100 \times C_i/C_{max} \times (T_{max}/T_i). \tag{1}$$

where C_i represented relative concentration (the percentage of each compound peak intensity to total peak intensity of all compounds), and T_i indicated the odor threshold of the target volatile components. C_{max} and T_{max} referred to compounds with the maximum odor activity value [26].

3.7. Determination of Rice Pasting Properties

The pasting properties of rice flour were determined by a Rapid Visco Analyzer (RVA-TecMaster, Perten Instruments, Warriewood, NSW, Australia). The rice flour samples (3.00 g, based on 14% moisture) were stirred with distilled water (25 mL) in an aluminum vessel. The test parameters were set according to the descriptions of Geng, et al. [42]. The characteristic parameters such as peak viscosity, final viscosity and pasting temperature were obtained by TCW analysis. The unit of characteristic parameters was centipoise (cp), which indicates viscosity. Each sample experiment was measured three times to obtain the end results.

3.8. Determination of Rice Cooking Quality

3.8.1. Rice Cooking Expansion and Elongation Rate Measurement

Rice kernels (10 g, dry basis) were accurately weighed and cooked in a small aluminum vessel with 20 mL of water, from which 5 g of cooked rice was loaded into a 10-mL measuring cylinder to determine the volume. The volume of raw rice V0 was measured according to the same method. The cooking expansion rate (%) was calculated according to Equation (2):

$$\text{Cooking expansion rate (\%)} = 100\% \times (V_3 - V_0)/V_0. \tag{2}$$

where W was the total weight of cooked rice; V_1 was the volume measured immediately after injecting 5 mL of water, V_2 ($V_2 = V_1 - 5$) was the rice volume; V_3 ($V_3 = V_2 \times (W/5)$) was the volume of total rice. The volume of the raw rice V_0 was measured according to the same method.

Actual elongation was measured by dividing the average length of 10 cooked kernels from that of 10 raw rice kernels. Intact rice grains (10 grains) were randomly taken from each sample, and the average length (L_0) was determined. Ten rice grains were placed into a 20-mL test tube, and allowed to soak for 30 min after the addition of 10 mL distilled water. These soaked rice were subsequently boiled in boiling water for 10 min and the rice grains were removed and placed on filter paper for 60 min to determine the average length (L_1). Cooking elongation rate was measured according to Equation (3)

$$\text{Cooking elongation rate (\%)} = L_1/L_0. \tag{3}$$

where L_1 represented the average length of cooked rice kernels and L_0 represented the average length of raw rice kernels [16].

3.8.2. Rice Water Absorption Rate Measurement

Rice grains (10 g, dry basis) were accurately weighed, and the samples were cooked for 30 min in a small aluminum vessel with 20 mL of water in a water bath. The cooked rice kernels were collected and placed on filter paper to drain the surface water. The heating water absorption rate (%) was calculated according to Equation (4)

$$\text{Heating water absorption rate (\%)} = 100\% \times (W_1 - W_0)/W_0. \tag{4}$$

where, W_0 represents the weight of 10 g unheated rice kernel and W_1 represents the weight of cooked rice [16].

To measure the soaked water absorption rate, 3 g of rice were accurately weighed and the samples were soaked for 60 min in a small aluminum vessel with 20 mL of water in a water bath and the aluminum vessel was kept in a constant-temperature water bath of 25 °C. The soaked rice kernels were collected and placed on filter paper to drain the surface water. The soaked water absorption rate (%) was calculated according to Equation (5)

$$\text{Soaked water absorption rate (\%)} = 100\% \times (W_1 - W_0)/W_0. \tag{5}$$

where W_0 represented the weight of 3-g unsoaked rice kernels and W_1 represented the weight of soaked rice.

3.9. Determination of Rice FFA and MDA Content

The FFA content of the samples was measured in triplicate according to the Chinese National Standard method (GB/T 29405-2012). In brief, 5 g of rice was added to a conical flask and shaken with 30 mL of ethanol for 1h, the extract was centrifuged, and the supernatant was added to a solution of 1% phenolphthalein ethanol (95%) and titrated with potassium hydroxide ethanol solution to the endpoint. The FFA content was calculated according to the volume of potassium hydroxide ethanol that was consumed by rice grains. The MDA content was determined three times by the kit (A003-1-2, Nanjing Jian Cheng Research Ltd., Nanjing, China).

3.10. Determination of Rice Catalase (CAT) and Peroxidase (POD) Activity

The CAT activity was determined in triplicate by a commercial assay kit (A007-1-1, Nanjing Jian Cheng Research Ltd., Nanjing, China). Every milligram of tissue protein that breaks down 1 μmol H_2O_2 every second was defined as a unit of catalase activity (U/mg protein). The POD activity was measured in triplicate by commercial assay kit (A084-3-1, Nanjing Jian Cheng Research Ltd., Nanjing, China), and this method used the change in absorbance at 420 nm using the principle that POD catalyzes the reaction of hydrogen peroxide. The amount of enzyme catalyzing 1 mg of substrate per minute per milligram of tissue protein was defined as one unit of enzyme activity.

3.11. Statistical Analysis

Data processing was conducted with Microsoft Excel software (Microsoft, Washington, DC, USA) and reported as mean ± standard deviation (SD). One-way ANOVA with Ducan's tests were performed by IBM SPSS statistical software (version 26.0) at 0.05 significance level. The principal component analysis (PCA) was performed with Origin software (version Pro 2021).

4. Conclusions

As an important global starch-based food, rice contributes to nearly 40% of the world's total caloric intake. YY-IJHR, as a new variety with a high yield and wide planting area, along with its rice quality, starch functionality and storage characteristics, have been attracting increasing attention. The previous study mainly focused on cultivation techniques and yield. However, the rice quality and starch's functional properties, as well as the storability of YY-IJHR, are important for consumers and industrial applications. Here, three typical YY-IJHR varieties, which included YY12 (biased japonica type YY-IJHR), YY1540 (intermedius type YY-IJHR) and YY15 (biased indica type YY-IJHR), were studied and comparatively analyzed. They were compared with N84 (high-quality conventional japonica rice) in terms of comprehensive rice quality and the functional properties of starch, as well as storability.

The results of this study suggested that the three YY-IJHR varieties all had a greater cooking and eating quality than N84, as they had lower protein and amylose contents. Rice aroma components were revealed by GC-IMS. The alcohol contents in the volatile flavor components of YY-IJHR were generally lower, whereas the content of some aldehydes and esters were higher compared with N84. The volatile aroma components of YY15 were mainly aldehydes and esters, whereas YY12 were mainly aldehydes and ketones. The data from the current study provided the knowledge that YY-IJHR starch has a higher pasting viscosity and digestibility, which may be helpful to common consumers. In addition, both YY-IJHR cultivars' FFA and MDA contents were lower, which demonstrated that YY-IJHR had higher palatability and storability than N84 in fresh rice and rice stored for 12 months.

Therefore, YY-IJHR not only has a high yield, but also exhibited a superior grain quality and storability compared with the N84. PCA indicated that grain quality and storability of YY12 and YY15 were similar and performed better than YY1540, while the aroma components and starch functional properties of the YY-IJHR cultivars all showed significant differences. This study will provide valuable information for common consumers when selecting suitable rice varieties and necessary knowledge for physiologists to maximize the yield potential and optimize the application of rice.

Author Contributions: X.G.: investigation, data curation, writing original draft; L.Z.: methodology, investigation; A.W.: supervision, methodology; H.X.: resources, methodology; M.N.: software, resources; Z.C.: software, methodology; Y.H.: resources, software; Y.T.: methodology; F.W.: supervision, resources, project administration, funding acquisition; L.T.: conceptualization, validation, project administration, funding acquisition. All authors have read and agreed to the published version of the manuscript.

Funding: This work was supported by Fundings: Science and Technology Cooperative Innovation Project between Ningbo Academy of Agricultural Sciences and Chinese Academy of Agricultural Sciences (Grant No. 2019CXGC005) and the National Key Technologies R&D Program of China (Grant No. 2017YFD0401104-05).

Institutional Review Board Statement: Not Applicable.

Informed Consent Statement: Not Applicable.

Data Availability Statement: Not Applicable.

Conflicts of Interest: The authors declare no conflict of interest.

Sample Availability: Samples of the compounds are available from Lin Zhu (Key Laboratory of Preservation Engineering of Agricultural Products, Ningbo Academy of Agricultural Sciences, Institute of Agricultural Products Processing, Ningbo 315040, China; zhulin0822@163.com).

References

1. Qu, J.; Wang, M.; Liu, Z.; Jiang, S.; Xia, X.; Cao, J.; Lin, Q.; Wang, L. Preliminary study on quality and storability of giant hybrid rice grain. *J. Cereal Sci.* **2020**, *95*, 103078. [CrossRef]
2. Meng, T.; Xu, J.; Shao, Z.; Ge, M.; Zhang, H.; Wei, H.; Dai, Q.; Huo, Z.; Xu, K.; Guo, B.; et al. Advantages and their formation characteristics of the highest population productivity of nitrogen fertilization in Japonica/Indica hybrid rice of Yongyou series. *Acta Agron. Sin.* **2015**, *41*, 1711–1725. [CrossRef]
3. Wei, H.-H.; Li, C.; Xing, Z.-P.; Wang, W.-T.; Dai, Q.-G.; Zhou, G.-S.; Wang, L.; Xu, K.; Huo, Z.-Y.; Guo, B.-W.; et al. Suitable growing zone and yield potential for late-maturity type of Yongyou japonica/indica hybrid rice in the lower reaches of Yangtze River, China. *J. Integr. Agric.* **2016**, *15*, 50–62. [CrossRef]

4. Xu, D.; Zhu, Y.; Chen, Z.-F.; Han, C.; Hu, L.; Qiu, S.; Wu, P.; Liu, G.-D.; Wei, H.-Y.; Zhang, H.-C. Yield characteristics of japonica/indica hybrids rice in the middle and lower reaches of the Yangtze River in China. *J. Integr. Agric.* **2020**, *19*, 2394–2406. [CrossRef]
5. Shao, Z.; Zhao, Y.; Zhang, Y.; Wang, Y.; Wang, Y.; Yang, L. Effect of ozone stress on yield characteristics of indica-japonica hybrid rice Yongyou 538 in two consecutive growing seasons. *Environ. Exp. Bot.* **2021**, *186*, 104447. [CrossRef]
6. Xu, X.-M.; Liang, K.-J.; Zhang, S.-G.; Shang, W.; Zhang, Y.-Y.; Wei, X.-Y.; Ke, B. Analysis of Indica-Japonica Differentiation in Rice Parents and Derived Lines Using ILP Markers. *Agric. Sci. China* **2009**, *8*, 1409–1418. [CrossRef]
7. Zhou, T.Y.; Li, Z.K.; Li, E.P.; Wang, W.L.; Yuan, L.M.; Zhang, H.; Liu, L.J.; Wang, Z.Q.; Gu, J.F.; Yang, J.C. Optimization of nitrogen fertilization improves rice quality by affecting the structure and physicochemical properties of starch at high yield levels. *J. Integr. Agric.* **2022**, *21*, 1576–1592. [CrossRef]
8. Zhu, D.; Zhang, H.; Guo, B.; Xu, K.; Dai, Q.; Wei, C.; Zhou, G.; Huo, Z. Physicochemical properties of indica-japonica hybrid rice starch from Chinese varieties. *Food Hydrocoll.* **2017**, *63*, 356–363. [CrossRef]
9. Bian, J.-L.; Ren, G.-L.; Han, C.; Xu, F.-F.; Qiu, S.; Tang, J.-H.; Zhang, H.-C.; Wei, H.-Y.; Gao, H. Comparative analysis on grain quality and yield of different panicle weight indica-japonica hybrid rice (*Oryza sativa* L.) cultivars. *J. Integr. Agric.* **2020**, *19*, 999–1009. [CrossRef]
10. Zhu, D.; Fang, C.; Qian, Z.; Guo, B.; Huo, Z. Differences in starch structure, physicochemical properties and texture characteristics in superior and inferior grains of rice varieties with different amylose contents. *Food Hydrocoll.* **2021**, *110*, 106170. [CrossRef]
11. Wani, A.A.; Singh, P.; Shah, M.A.; Schweiggert-Weisz, U.; Gul, K.; Wani, I.A. Rice Starch Diversity: Effects on Structural, Morphological, Thermal, and Physicochemical Properties-A Review. *Compr. Rev. Food Sci. Food Saf.* **2012**, *11*, 417–436. [CrossRef]
12. You, S.Y.; Oh, S.K.; Kim, H.S.; Chung, H.J. Influence of molecular structure on physicochemical properties and digestibility of normal rice starches. *Int. J. Biol. Macromol.* **2015**, *77*, 375–382. [CrossRef] [PubMed]
13. Kong, X.; Zhu, P.; Sui, Z.; Bao, J. Physicochemical properties of starches from diverse rice cultivars varying in apparent amylose content and gelatinisation temperature combinations. *Food Chem.* **2015**, *172*, 433–440. [CrossRef] [PubMed]
14. Peng, Y.; Mao, B.; Zhang, C.; Shao, Y.; Wu, T.; Hu, L.; Hu, Y.; Tang, L.; Li, Y.; Tang, W.; et al. Influence of physicochemical properties and starch fine structure on the eating quality of hybrid rice with similar apparent amylose content. *Food Chem.* **2021**, *353*, 129461. [CrossRef] [PubMed]
15. Chen, T.; Chen, X.; Meng, L.; Wei, Z.; Chen, B.; Wang, Y.; Chen, H.; Cheng, Q. Characteristic Fingerprint Analysis of the Moldy Odor in Guangxi Fragrant Rice by Gas Chromatography-Ion Mobility Spectrometry (GC-IMS). *Anal. Lett.* **2022**, 1–13. [CrossRef]
16. Faruq, G.; Prodhan, Z.H.; Nezhadahmadi, A. Effects of Ageing on Selected Cooking Quality Parameters of Rice. *Int. J. Food Prop.* **2015**, *18*, 922–933. [CrossRef]
17. Faruq, G.; Mohamad, O.; Hadzim, M.; Meisner, C.A. Optimization of Aging Time and Temperature for Four Malaysian Rice Cultivars. *Pak. J. Nutr.* **2003**, *2*, 125–131.
18. Zhou, Z.; Robards, K.; Helliwell, S.; Blanchard, C. Ageing of Stored Rice: Changes in Chemical and Physical Attributes. *J. Cereal Sci.* **2002**, *35*, 65–78. [CrossRef]
19. Park, C.-E.; Kim, Y.-S.; Park, K.-J.; Kim, B.-K. Changes in physicochemical characteristics of rice during storage at different temperatures. *J. Stored Prod. Res.* **2012**, *48*, 25–29. [CrossRef]
20. Custodio, M.C.; Cuevas, R.P.; Ynion, J.; Laborte, A.G.; Velasco, M.L.; Demont, M. Rice quality: How is it defined by consumers, industry, food scientists, and geneticists? *Trends Food Sci. Technol.* **2019**, *92*, 122–137. [CrossRef]
21. Li, H.; Fitzgerald, M.A.; Prakash, S.; Nicholson, T.M.; Gilbert, R.G. The molecular structural features controlling stickiness in cooked rice, a major palatability determinant. *Sci. Rep.* **2017**, *7*, 43713. [CrossRef] [PubMed]
22. Peng, Y.; Mao, B.; Zhang, C.; Shao, Y.; Wu, T.; Hu, L.; Hu, Y.; Tang, L.; Li, Y.; Zhao, B.; et al. Correlations between Parental Lines and Indica Hybrid Rice in Terms of Eating Quality Traits. *Front. Nutr.* **2020**, *7*, 583997. [CrossRef]
23. Liu, Q.; Wu, H.; Luo, J.; Liu, J.; Zhao, S.; Hu, Q.; Ding, C. Effect of dielectric barrier discharge cold plasma treatments on flavor fingerprints of brown rice. *Food Chem.* **2021**, *352*, 129402. [CrossRef]
24. Gu, H.; Chen, T.; Chen, M.; Lu, D.; Chen, B. Application of Gas Chromatography—Ion Migration Spectrometry (GC—IMS) to Evaluate the Degree of Mildew in Rice. *J. Chin. Cereals Oils Assoc.* **2019**, *34*, 118–124.
25. Burdock, G.A. *Fenaroli's Handbook of Flavor Ingredients*, 6th ed.; CRC Press: Boca Raton, FL, USA, 2009.
26. Wang, A.X.; Yi, C.; Xiao, T.; Qin, W.; Chen, Z.; He, Y.; Wang, L.; Liu, L.; Wang, F.; Tong, L.T. Volatile compounds, bacteria compositions and physicochemical properties of 10 fresh fermented rice noodles from southern China. *Food Res. Int.* **2021**, *150 Pt A*, 110787. [CrossRef]
27. Yu, M.; Xiang, X.; Tan, H.; Zhang, Q.; Shan, Y.; Yang, H. Potential correlation between volatiles and microbiome of Xiang xi sausages from four different regions. *Food Res. Int.* **2021**, *139*, 109943. [CrossRef] [PubMed]
28. Bao, J. Rice starch. In *Rice*; AACC International Press: Saint Paul, MN, USA, 2019; pp. 55–108.
29. Englyst, H.N.; Kingman, S.M.; Cummings, J.H. Classification and measurement of nutritionally important starch fractions. *Eur. J. Clin. Nutr.* **1992**, *46* (Suppl. 2), S33–S50.
30. Arp, C.G.; Correa, M.J.; Ferrero, C. Resistant starches: A smart alternative for the development of functional bread and other starch-based foods. *Food Hydrocoll.* **2021**, *121*, 106949. [CrossRef]
31. Zhou, X.; Ying, Y.; Hu, B.; Pang, Y.; Bao, J. Physicochemical properties and digestibility of endosperm starches in four indica rice mutants. *Carbohydr. Polym.* **2018**, *195*, 1–8. [CrossRef] [PubMed]

32. Noda, T.; Nishiba, Y.; Sato, T.; Suda, I. Properties of Starches from Several Low-Amylose Rice Cultivars. *Cereal Chem.* **2003**, *80*, 193–197. [CrossRef]
33. Hu, Y.; Cong, S.; Zhang, H. Comparison of the Grain Quality and Starch Physicochemical Properties between Japonica Rice Cultivars with Different Contents of Amylose, as Affected by Nitrogen Fertilization. *Agriculture* **2021**, *11*, 616. [CrossRef]
34. Jane, J.L.; Chen, Y.Y.; Lee, L.F.; McPherson, A.E.; Wong, K.S.; Radosavljevic, M.; Kasemsuwan, T. Effects of amylopectin branch chain length and amylose content on the gelatinization and pasting properties of starch. *Cereal Chem.* **1999**, *76*, 629–637. [CrossRef]
35. Won, J.G.; Yoshida, T.; Uchimura, Y. Genetic Effect on Amylose and Protein Contents in the Crossed Rice Seeds. *Plant Prod. Sci.* **2002**, *5*, 17–21. [CrossRef]
36. Zhu, L.; Wu, G.; Cheng, L.; Zhang, H.; Wang, L.; Qian, H.; Qi, X. Investigation on molecular and morphology changes of protein and starch in rice kernel during cooking. *Food Chem.* **2020**, *316*, 126262. [CrossRef] [PubMed]
37. Wang, Q.; Han, F.; Wu, Z.; Lan, T.; Wu, W. Estimation of Free Fatty Acids in Stored Paddy Rice Using Multiple-Kernel Support Vector Regression. *Appl. Sci.* **2020**, *10*, 6555. [CrossRef]
38. Matsue, Y.; Uchimura, Y.; Sato, H.; Ogata, T. An Efficient Method for Evaluating the Palatability Deterioration during Storage in Rice. *Plant Prod. Sci.* **2015**, *6*, 107–111. [CrossRef]
39. Zhou, X.; Zhu, F.; Zhang, Y.; Peng, C. Analysis of the Storage Property, Physiological, Biochemical Indicators Parameters and the Pasting Characteristics of Rice in Different Storage Time. *J. Chin. Cereals Oils Assoc.* **2020**, *35*, 108–114, 124.
40. Wang, X.; Jin, Y.; Cheng, L.; Li, Z.; Li, C.; Ban, X.; Gu, Z.; Hong, Y. Pasting properties and multi-scale structures of Spirodela starch and its comparison with normal corn and rice starch. *Food Hydrocoll.* **2022**, *132*, 107865. [CrossRef]
41. Zhao, Q.; Shen, Q. Application of GC-IMS Technology Combined with Chemometrics Method in Classification of Hulless Barley. *J. Chin. Cereals Oils Assoc.* **2020**, *35*, 165–169.
42. Geng, D.-H.; Zhou, S.; Wang, L.; Zhou, X.; Liu, L.; Lin, Z.; Qin, W.; Liu, L.; Tong, L.-T. Effects of slight milling combined with cellulase enzymatic treatment on the textural and nutritional properties of brown rice noodles. *LWT* **2020**, *128*, 109520. [CrossRef]

Article

Effect of Microwave Irradiation on Acid Hydrolysis of Faba Bean Starch: Physicochemical Changes of the Starch Granules

Mayra Esthela González-Mendoza [1], Fernando Martínez-Bustos [2], Eduardo Castaño-Tostado [1] and Silvia Lorena Amaya-Llano [1,*]

[1] Facultad de Química, Universidad Autónoma de Querétaro, Cerro de las Campanas S/N, Querétaro 76010, Mexico; gonzalez.mendoza.mayra@gmail.com (M.E.G.-M.); ecastano@uaq.mx (E.C.-T.)
[2] Centro de Investigación y Estudios Avanzados del Instituto Politécnico Nacional, Unidad Querétaro, Libramiento Norponiente 2000, Real de Juriquilla, Querétaro 76230, Mexico; fmartinez@cinvestav.mx
* Correspondence: samayal@uaq.mx

Abstract: Starch is the most abundant carbohydrate in legumes (22–45 g/100 g), with distinctive properties such as high amylose and resistant starch content, longer branch chains of amylopectin, and a C-type pattern arrangement in the granules. The present study concentrated on the investigation of hydrolyzed faba bean starch using acid, assisted by microwave energy, to obtain a possible food-grade coating material. For evaluation, the physicochemical, morphological, pasting, and structural properties were analyzed. Hydrolyzed starches developed by microwave energy in an acid medium had low viscosity, high solubility indexes, diverse amylose contents, resistant starch, and desirable thermal and structural properties to be used as a coating material. The severe conditions (moisture, 40%; pure hydrochloric acid, 4 mL/100 mL; time, 60 s; and power level, 6) of microwave-treated starches resulted in low viscosity values, high amylose content and high solubility, as well as high absorption indexes, and reducing sugars. These hydrolyzed starches have the potential to produce matrices with thermo-protectants to formulate prebiotic/probiotic (symbiotic) combinations and amylose-based inclusion complexes for functional compound delivery. This emergent technology, a dry hydrolysis route, uses much less energy consumption in a shorter reaction time and without effluents to the environment compared to conventional hydrolysis.

Keywords: microwave energy; pasting properties; legume starch; microstructure

Citation: González-Mendoza, M.E.; Martínez-Bustos, F.; Castaño-Tostado, E.; Amaya-Llano, S.L. Effect of Microwave Irradiation on Acid Hydrolysis of Faba Bean Starch: Physicochemical Changes of the Starch Granules. *Molecules* **2022**, *27*, 3528. https://doi.org/10.3390/molecules27113528

Academic Editors: Litao Tong and Lili Wang

Received: 3 May 2022
Accepted: 27 May 2022
Published: 31 May 2022

Publisher's Note: MDPI stays neutral with regard to jurisdictional claims in published maps and institutional affiliations.

Copyright: © 2022 by the authors. Licensee MDPI, Basel, Switzerland. This article is an open access article distributed under the terms and conditions of the Creative Commons Attribution (CC BY) license (https://creativecommons.org/licenses/by/4.0/).

1. Introduction

Pulses contain a large amount of slow-release carbohydrates, they have a high level of protein (18 to 25 g/100 g) and contain minerals and vitamins. Some of the most important legumes are beans (*Phaseolus vulgaris* L.), lentils (*Lens culinaris* L.), peas (*Pisum sativum* L.), faba beans (*Vicia faba* L.), and chickpeas (*Cicer arietinum* L.). Since they are a high-starch source, their extraction and use are of great interest [1]. Starch is principally composed of two anhydroglucose polymers (amylopectin and amylose) linked by α-(1,4) bonds in linear segments. Additionally, the first has branching points connected to the main chain by an α-(1,6) link. Amylopectin molecules possess a unique structure that confers a substantial degree of crystallinity to the molecules [2]. The crystal structure is divided into three types, A, B, and C, which differ within the compaction limit. In legume starches, type C is the most common and the most resistant to digestion, whereas A is characteristic for cereals [3]. Moreover, a type C crystalline structure implies a mixture of types A and B. Nevertheless, the percentage of each one can vary depending on the source [4] and, consequently, the amount of RS produced during retrogradation. RS refers to the amount of starch and products of starch degradation that have not been absorbed in the gut of healthy humans [5]. Moreover, this fraction passes through the colon, where the microbiota ferments it and mainly produces short-chain fatty acids. Because of this fact, RS has positive effects on diverse diseases [6,7]. Compared to most used starches, pea, lentil, and faba bean starches have relatively high amylose contents, longer branch chains of amylopectin, and the characteristic C-type polymorphous arrangement in the granules. Legume starches also contain a lot of resistant starch (RS) compared to cereal starches [1]. Related to bean starch, faba bean starch has a higher proportion of carbohydrates [8] and differs in its pasting properties [9].

Native starches are modified by diverse methods (physical, chemical, or enzymatic methods) and they are extensively used in the food industry. Starch modified by acid hydrolysis alters the granular structure of starches, resulting in a different behavior upon heating in water and produce pastes with lower intrinsic viscosity values, increased water solubility, and good film-forming properties [10]. These types of starches are used in industry for various applications, for example, thickeners in the food industry or as a wall material to encapsulate diverse food ingredients and active compounds. This latter application is of great interest and mainly modified starches are used. Starches that contain high levels of RS or amylose, have the capability to produce a structural network that facilitates the formation of round-shape microcapsules after spray drying [11]. In addition, starches with high amylose can form amylose-based inclusion complexes of great interest in a wide variety of fields, including drugs, microorganisms, functional compounds delivery, food science, and separation industries. There are currently increasing investigations of RS matrices with protectants to formulate symbiotic (prebiotic/probiotic) combinations for microencapsulation and the delivery of targeted probiotic bacteria into the colon [12].

The functional properties of pulse starches can be modified using emerging technologies such as microwave energy, which has several advantages, including energy-saving, high conversion, rapidity, and no effluents to the environment [13]. Microwave energy accelerates the depolymerization of starch and has been used to modify the physicochemical properties of starch [14–18]. The susceptibility of different starches to microwave irradiation depended not only on their crystalline structure but also on their amylose content [19]. Microwave irradiation creates heat inside the processed materials because of rapid alterations of the electromagnetic field at high frequencies [20]. Hence, microwave irradiation has an effect in a shorter process time, with higher yield, and better quality of products than those obtained by conventional processing techniques [21]. Using microwave irradiation, the starch was completely hydrolyzed within 5 min in a suspension of starch (10 g/100 mL) in hydrochloric acid (0.5 M) [22]. Warrand et al. [23] hydrolyzed pure amylose under acidic conditions by microwave irradiation and conventional heating. Microwave was shown to be more efficient than conventional heating. In addition, acid microwave heating directly transformed starch and fiber into depolymerized products and resulted in simpler sugars than autoclaving [24]. These investigations showed that the presence of acid can enhance the effects of microwave energy, while coupling accelerates the hydrolysis.

Faba bean, due to its features mentioned before, has the potential to be used as a food-grade material. Thus, the focus of the current investigation was to prepare hydrolyzed faba bean starch using acid assisted by microwave energy. Hydrolysis using microwave energy is an environmentally friendly process because it has low energy consumption, a shorter reaction time, and no generation of effluents.

2. Results and Discussion
2.1. Starch Isolation

The yield of starch extraction from faba bean seeds (*Vicia faba* var. major) was 30–33 g/100 g, similar to the one reported by Ambigaipalan et al. [25]. These authors reported mean values of 34.6 of pulse starches isolated from different cultivars of faba bean. Due to the existence of insoluble proteins, fiber, and minerals within starch granules, it is difficult to obtain pure legume starch [26]. The purity of starch is influenced by the extraction method. In this work, the wet method was used, which has a lower yield. However, starches with a higher degree of purity are achieved [27]. NFBS showed $8.3 \pm 0.14\%$ moisture content, within the range (7–15%) obtained for native legume starches [28]. The ash content was 0.07 ± 0.029 g/100 g, similar to that reported by Zhang et al. [26] and

Piecyk and Domian [1] for faba bean seed. Regarding the protein content, the value (0.85 ± 0.012 g/100 g) was higher than the one reported by Ambigaipalan et al. [25] and Zhang et al. [26], which was 0.38 and 0.30 g/100 g, respectively, for faba bean starch. Other authors, such as Hoover et al. [28], reported values similar to those obtained in this work for the same legume. These differences may be due to the extraction process used. Concerning the lipid content, this was 2.0 ± 0.03 g/100 g, being higher than the one reported by Hoover et al. [28]. The purity of starch can be qualified by the composition of ash and protein [27]. Thus, the results indicate that the starch obtained has a high degree of purity.

2.2. Characterization of Native and Hydrolyzed Faba Bean Starches

The characterization of native faba bean starch (NFBS), conventional hydrolysis faba bean starch (CFBS), and hydrolyzed faba bean starches (HFBS) is presented in Table 1. Results for NFBS agreed with those previously reported [1,26,27]. Amylose and RS contents showed values of 38.3 and 12.8 g/100 g, respectively. Lower values (from 25.8 to 33.6 g/100 g) were reported for amylose and higher for RS (from 8.1 to 15.0 g/100 g) for pulse starches isolated from different cultivars of faba bean [27]. All the assays processed with microwave energy (MWE) and hydrochloric acid (HA), including CFBS, showed lower values of amylose and RS than native faba bean starch. The amylose content that resulted after the hydrolysis of starch varied according to the starch source and hydrolysis conditions. Existing knowledge on microwave treatment of starches with completely different amylose contents suggest that the MWE is preferably transmitted to the amorphous region of the starch granule, and the crystalline regions are affected afterwards [10,29]. The hydrolysis conditions of faba bean starch slightly decreased the liberation of amylose, although the application of severe conditions of moisture (M), HA, and MWE (HFBS8 treatment) resulted in the highest amylose content and the lowest RS. The acid hydrolysis decreased viscosity and increased solubility. In addition, when the reaction time of the hydrolysis of maize starch was increased, the crystallinity of the starch increased while its amylose content decreased [30]. Microwave treatment reduced the degree of amylopectin branching on lotus seed starch leading to the degradation of linear chains and the reduction of crystal regions, which further promoted the formation of RS during cooling [21]. Amaya-Llano et al. [31] reported that HA concentration from 1–3 mL/100 mL reduced amylose of jicama and maize starches, nevertheless a great increase (3–5) raised the amylose content of both starches. This increase of amylose was attributed to the fast degree of depolymerization of amylopectin. Aaliya et al. [32] reported that use of MWE showed a significant difference in the swelling and solubility of talipot starches. It can be seen that there are no significant differences between NFBS and CFBS in solubility and absorption indexes, similar to that reported by Cruz-Benítez et al. [33] for cassava starch. HFBS8 showed major differences due to the clear depolymerization of the granules. Therefore, the water solubility index (WSI) increased from 2.5 to 49.5%, which can also be corroborated by the rise in reducing sugars up to 4.7 g/L. Such behavior was similar to that reported by González et al. [34] in lentil starch modified by microwave irradiation. The HFBS8 treatment was able to retain water inside its structure, maybe because of the recrystallization process. However, another possibility could be that this intense treatment caused structural changes in the amylopectin molecules, resulting in the formation of longer amylopectin chains due to fusion of the granules and exposed more hydroxyl groups, thereby increasing the swelling power [35]. To understand the mechanism of the microwave acid hydrolysis phenomenon, two-way interaction effects on each response variable were interpreted. All two-way interactions effects were statistically significant for each response variable, except for RS and the water absorption index (WAI) in the interaction M with pure hydrochloric acid (A) (Figure 1(B1,D1)). The interactions effects on amylose concentration (Figure 1(A1–A3)), WSI (Figure 1(C1–C3)) and reducing sugars (Figure 1(E1–E3)) show that only having low acid concentration, short time, and low power level (P), when the moisture is modified from 20 to 40%, does not produce a statistical change. The explanation of the behavior could be that at these levels, the molecules do not reach a sufficient state of excitation to modify the starch structure to a great extent [14]. While for high levels of each factor, it was observed that hydrolysis occurs and, therefore, an increase in amylose, solubility and reducing sugars. For the effects of the significant interactions on RS (Figure 1(B2,B3)), it can be observed that M affects RS only when Time = 60 or Power = 2. It has been reported that RS content decreases depending on the degree of acid hydrolysis and its conditions, namely, temperature, time, type of acid, and concentration [7,36,37]. The non-significant interaction between M and A (Figure 1(B1)) shows that RS decreases when M augments. This reduction is independent of the value of A, probably because both factors can modify starch. M improves the molecular movement in MWE conditions, cleaving amylopectin and amylose chains [17,18], while A can itself modify starch by the action of the hydrogen ions that react with the oxygen atom of α-(1,4) or α-(1,6) glycosidic bond [38]. Therefore, both factors additively decreased the RS content. High levels produced materials with less concentration of RS, maybe due to severe hydrolysis. Concerning significant interaction effects on WAI (Figure 1(D2,D3)), it can be observed that when moisture content increases from 20 to 40%, the water absorption will always increase, especially when Time = 60 or Power = 6. The non-significant two-way interaction effect, corresponding to M with A (Figure 1(D1)), shows that WAI rises when M increases. This rise is independent of the value of A, probably due to the rupture of starch granules caused by the modification of each factor, thus resulting in the exposure of more hydroxyl groups that can form hydrogen bonds with water similar to that observed by Li et al. [15]. Starches may present a two-step hydrolysis scheme: a rapid initial velocity followed by a later slower velocity. The relatively fast initial rate corresponds to the hydrolysis of the amorphous zones, whereas the slow process is attributed to the simultaneous hydrolysis of the amorphous and crystalline regions. The first hydrolysis stage is influenced by the size of the granules, the pores on the surface, the amylose content, and the amount of lipid-complexed amylose chains. The second hydrolysis step is influenced by the amylopectin content, the distribution of α-(1,6) branches between the amorphous and crystalline lamellae, and the degree of packing of the double helices within the crystallites [10]. The results suggest that with low levels in each factor the hydrolysis of the amorphous parts commence, while with high levels, hydrolysis of the amorphous and crystalline regions may occur. A similar effect was reported for hydrolyzed potato starch using induced electric field and HCl [38].

2.3. Determination of Pasting Properties

The viscosity profiles of NFBS, CFBS, and HFBS are shown in Figure 2 and the pasting parameters in Table 2. Results for NFBS were similar to those previously reported by Ambigaipalan et al. [25] for faba bean starch. The viscosity parameters of CFBS and HFBS with diverse M, HA, and MWE conditions statistically decreased regarding NFBS. The lowest viscosity values were for the starches with the combination of more severe HA and MWE conditions. The use of MWE in acidic conditions allowed the production of starches with different characteristics than other starches treated with MWE [14,15,17,18]. According to Luo et al. [39] the viscosity patterns of waxy and normal starches remained unchanged, while amylomaize V starches changed on microwave treatment. After microwave irradiation, an increase in pasting temperature and a drop in viscosity of the three starches were observed. In a microwave field, polar compounds like water molecules and HCl, vibrate at very high frequency, and then the rapid friction, collision, and vibration between water molecules and starch granules generate heat during a short period, which can cause the physical damage of the starch granules and the degradation of structures in a shorter time [15]. Materials used as encapsulants for food should have some characteristics, including biocompatibility, water solubility, emulsifying and film forming properties, and low viscosity at high concentrations. Additionally, the use of starches with other desirable characteristics includes high amylose and RS [40]. MWE changed the viscosity properties of waxy and non-waxy rice starches containing 20% moisture [41]. In our work, all the starches processed with M, HA, and MWE, decreased in their pasting properties, indicating that the severity of the combined processing parameters resulted in low re-aggregation of starch granules. The breakdown viscosity values also decreased, and HFBS8 showed undetectable parameters of viscosity by RVA. The optimal response variables for potential use as wall material for the encapsulation could be starches with the highest WSI, the lowest final viscosity, and starches with different amylose content. Starch with high amylose content can exhibit a low release behavior, and starch with high amylopectin content can offer high core material stability and protection [42]. Thus, for the subsequent analysis, HFBS2, HFBS5, and HFBS8 treatments can be potentially considered as possible food-grade coating material.

Table 1. Physicochemical characterization of native and hydrolyzed faba bean starches [1].

Sample	Amylose, g/100 g	RS, g/100 g	Water Solubility Index, %	Water Absorption Index, %	Reducing Sugars, g/L
NFBS	38.3 ± 2.63 [b]	12.8 ± 0.57 [a]	1.9 ± 0.36 [f]	2.5 ± 0.22 [d]	0.03 ± 0.020 [e]
CFBS	34.6 ± 0.68 [bc]	10.6 ± 1.21 [ab]	2.1 ± 0.09 [f]	2.7 ± 0.05 [cd]	0.09 ± 0.010 [de]
HFBS1	22.8 ± 3.04 [f]	11.8 ± 0.40 [a]	2.5 ± 0.13 [ef]	2.8 ± 0.16 [cd]	0.06 ± 0.025 [e]
HFBS2	31.1 ± 4.16 [cd]	8.6 ± 0.17 [bc]	8.3 ± 0.29 [b]	3.6 ± 0.48 [b]	0.58 ± 0.066 [b]
HFBS3	30.1 ± 1.99 [cde]	7.5 ± 1.27 [cd]	3.4 ± 0.75 [def]	2.8 ± 0.13 [cd]	0.33 ± 0.144 [cd]
HFBS4	27.3 ± 1.56 [def]	7.4 ± 0.70 [cd]	5.0 ± 0.52 [cd]	3.4 ± 0.15 [bc]	0.11 ± 0.031 [de]
HFBS5	31.4 ± 1.04 [cd]	8.7 ± 0.40 [bc]	4.1 ± 0.82 [cde]	2.7 ± 0.13 [cd]	0.23 ± 0.060 [de]
HFBS6	25.5 ± 0.70 [def]	6.1 ± 0.06 [de]	5.5 ± 0.58 [c]	3.5 ± 0.37 [b]	0.52 ± 0.113 [bc]
HFBS7	24.5 ± 1.66 [ef]	11.0 ± 1.44 [a]	3.0 ± 0.98 [ef]	2.8 ± 0.17 [cd]	0.13 ± 0.020 [de]
HFBS8	82.0 ± 2.05 [a]	4.7 ± 0.31 [e]	49.5 ± 0.93 [a]	4.3 ± 0.18 [a]	4.73 ± 0.147 [a]

[1] RS, resistant starch; NFBS, native faba bean starch; CFBS, conventional hydrolysis faba bean starch; HFBS1, HFBS2, HFBS3, HFBS4, HFBS5, HFBS6, HFBS7, and HFBS8, hydrolyzed faba bean starches. Assays were performed in triplicate. Mean ± SD, values in the same column with different superscript letters are significantly different ($\alpha = 0.05$).

Table 2. Pasting parameters of native and hydrolyzed faba bean starches [1].

Sample	T_{ps} (°C)	η_{pk} (cP)	$\Delta\eta_{bd}$ (cP)	$\Delta\eta_{sb}$ (cP)	η_f (cP)
NFBS	72.5 ± 0.69 [ab]	975.7 ± 24.58 [a]	35.3 ± 5.51 [a]	401.0 ± 15.62 [a]	1341.3 ± 44.24 [a]
CFBS	70.2 ± 2.84 [b]	53.3 ± 40.72 [ef]	8.7 ± 10.01 [b]	7.7 ± 1.53 [de]	52.3 ± 32.01 [fg]
HFBS1	70.8 ± 1.27 [b]	655.3 ± 30.55 [b]	6.3 ± 5.77 [b]	149.3 ± 15.04 [b]	798.3 ± 17.79 [b]
HFBS2	75.1 ± 0.37 [a]	11.3 ± 8.08 [f]	2.0 ± 1.73 [b]	4.7 ± 1.15 [de]	14.0 ± 9.64 [fg]
HFBS3	69.8 ± 2.34 [b]	68.3 ± 11.02 [def]	1.7 ± 1.15 [b]	21.7 ± 7.02 [de]	88.3 ± 17.93 [def]
HFBS4	70.7 ± 1.61 [b]	112.0 ± 27.78 [de]	1.0 ± 0.05 [b]	35.7 ± 3.06 [d]	146.7 ± 38.55 [de]
HFBS5	70.6 ± 1.07 [b]	54.0 ± 13.23 [ef]	1.7 ± 1.15 [b]	15.3 ± 10.78 [de]	67.7 ± 16.07 [efg]
HFBS6	70.2 ± 0.43 [b]	143.3 ± 49.89 [d]	2.3 ± 1.53 [b]	31.0 ± 10.53 [de]	172.0 ± 41.62 [d]
HFBS7	68.9 ± 1.22 [b]	446.3 ± 19.55 [c]	11.7 ± 6.66 [b]	102.7 ± 23.69 [c]	537.3 ± 41.48 [c]
HFBS8	0.0 ± 0.00 [c]	0.0 ± 0.00 [f]	0.0 ± 0.00 [b]	0.0 ± 0.00 [e]	0.0 ± 0.00 [g]

[1] T_{ps}, pasting temperature; η_{pk}, peak viscosity; $\Delta\eta_{bd}$, breakdown viscosity; $\Delta\eta_{sb}$, setback viscosity; η_f, viscosity at end of final holding. NFBS, native faba bean starch; CFBS, conventional hydrolysis faba bean starch; HFBS1, HFBS2, HFBS3, HFBS4, HFBS5, HFBS6, HFBS7, and HFBS8, hydrolyzed faba bean starches. Assays were performed in triplicate. Mean ± SD, values in the same column with different superscript letters are significantly different ($\alpha = 0.05$).

2.4. Thermal Properties by Differential Scanning Calorimetry (DSC)

In Table 3, the parameters of NFBS, CFBS, and HFBS are presented. Results for NFBS for the same legume were similar to those previously reported by Ambigaipalan et al. [25], Li et al. [27], and Piecyk et al. [1]. The gelatinization temperatures increased on the hydrolyzed starches (CFBS, HFBS2, HFBS5, and HFBS8) in comparison to NFBS. The microwave melted off weak crystallites and formed strong ones, increasing the T_o, T_p, and T_f [15]. This shift is considered to reflect an increase in gelatinization temperature due to the increased molecular order in acid hydrolyzed starch. Further interpretations include the preferential hydrolysis of amorphous regions that attenuate the destabilizing effect of swelling in amorphous regions on the melting of the crystallites, or longer amylopectin double helices may be formed as a result of the removal of branch points [10]. In addition, the gelatinization enthalpy (ΔH) of CFBS was not significantly different from NFBS, similar to the one reported by Atichokudomchai et al. [43] in tapioca starch, hydrolyzed by HCl at room temperature for 192 h. It is suggested that the partially hydrolyzed amylose retrograded, leveling the double helix content (which otherwise would decrease with hydrolysis), so that ΔH did not decrease. Even in the presence of a certain hydrolysis of the crystalline domain, the enthalpy of fusion did not diminish. The ΔH of the samples HFBS2 and HFBS8 was decreased by 76.4% and 94.5%, respectively. ΔH reflected mainly the loss of the double-helix and the molecular order in the crystalline region. The lower ΔH suggested the lack of order of the crystalline region and decreased stability of the crystalline and amorphous regions, while the energy required for further destruction of the crystalline regions was diminished [15]. Therefore, the results suggest that the conditions of moisture content (40%) and power level (6) in HFBS2 and HFBS8 were sufficient to lose the molecular order of the starch. The difference in enthalpy between them is due to the acid concentration and time of exposure to MWE. In the HFBS2 treatment, the levels of these factors were 2 mL/100 mL of HA and 30 s, while for HFBS8, they were 4 mL/100 mL and 60 s, respectively. The polar molecules and ions vibrate with the application of MWE, generating thermal energy [44]. Hence, as there were more polar molecules in the medium and the time was longer, the temperature increased as well as the modification of the starch (HFBS8 > HFBS2). Finally, for HFBS5, no statistical changes were observed because the moisture (20%) of the treatment was not enough to modify the molecular order in the crystalline region, similar to that reported by Li et al. [15].

2.5. X-ray Diffraction

The XRD spectra of NFBS, CFBS, and HFBS are shown in Figure 3. NFBS exhibited the characteristic "C-type" XRD patterns, which agreed with previous reports [1,25,26] for the same legume. The acid does not modify the crystalline characteristics of A- and B-type starches and is commonly used to investigate the allomorph distribution of C-type starch [45]. The XRD spectra of acid hydrolysis starches (CFBS, HFBS2, and HFBS5) showed that the peak at 5.6° 2θ was reduced in intensity and the peak at 17°, 18°, and 23° 2θ sharpened (Figure 3), similar to the one reported in faba bean starch hydrolyzed by α-amylase-HCl [9]. Hydrolyzed potato starch using induced electric field-HCl treatment did not cause variation of the crystalline type [38]. Nevertheless, HFBS8 showed that the peak at 5.6° 2θ appeared with weak intensity and the peak at 17°, 18°, 22.5°, and 24° 2θ sharpened (Figure 3). These results showed a pattern of B-type starch, which agrees with the results of Polesi et al. [46] for retrograded chickpea starch. Retrogradation at low temperatures leads to the formation of B-type crystallinity. As mentioned earlier, in HFBS8, the moisture and temperature conditions were suitable for the gelatinization of the starch and its subsequent recrystallization. The transition from C to A or C to B could be predominantly attributed to the preferential hydrolysis of one polymorph followed by the possible rearrangement of decoupled double helices [10]. The relative crystallinity (RC) of NFBS was 27.7% and hydrolyzed starches were in the range of 26.8–28.8% (Figure 3). Ambigaipalan et al. [25] have shown that native faba bean starch has a 20.2–21.9% of RC, lower than our results. However, the values are within the range (17.0–34.0%) reported for other pulse starches [28]. RC in CFBS increases a little (From 27.7 to 28.8%). The difference in RC could be attributed to the size of the crystal and the orientation of the double helix in the crystal domain [47]. There was a significant correlation between the degree of retrogradation and crystallinity, attributed to the development of a more ordered or crystalline state formed during the retrogradation [48]. When starch granules are submitted to acidic hydrolysis, the RC increases with the time of hydrolysis. Several hypotheses have been suggested for the enhanced crystallinity in the initial stages of acid hydrolysis. First, the cleavage of some of the amylose chains running through the amorphous regions may allow reordering of the newly released chain ends into a more crystalline structure. Second, the reordering of the crystalline structure during acid hydrolysis would result in increased crystallinity by partial filling of water channels in the crystallite cavities with double helices. Third, increased crystallinity may also result from the retrogradation of hydrolyzed free amylose into double helices, which rearrange into crystalline regions that are resistant to acid hydrolysis [10]. The lowest RC was 26.8% for HFBS8, probably due to the more intense rupture of the gelatinized starch through the action of acid. Similar results were reported for chickpea starch, retrograded and gelatinized by acid hydrolysis [46]. Progressive loss of starch crystallinity upon increased duration of the microwave treatment was also reported by Kumar et al. [14], caused by the capacity of microwaves to directly damage the lamellar arrangement of amylopectin crystals by resonating water molecules, and in this work also by HCl molecules.

2.6. Fourier Transform-Infrared Spectroscopy

The FTIR spectra of NFBS, CFBS, and HFBS are shown in Figure 4A. Broad peaks appeared approximately at 3325 cm^{-1}, indicating the presence of hydroxyl groups (O–H). Peaks at 2935 cm^{-1} and 1650 cm^{-1} representing the C–H stretching vibration and bound water present in the starch, respectively, were also seen in the FTIR spectra. The fingerprint region of the starch spectrum (Figure 4B) has five characteristic peaks between 800 and 1200 cm^{-1}, attributed to C–O bond stretching. The peak around 996 cm^{-1} is attributed to the C–O of the C–O–C in the polysaccharide; the peaks close to 1078 and 1150 cm^{-1} are characteristic of the anhydroglucose ring C–O stretch; the peak near to 930 cm^{-1} was assigned to the skeletal mode vibration of α-(1–4) glycosidic linkage, while the peak near to 860 cm^{-1} corresponds to C–H and CH$_2$ deformations [49]. There was neither the appearance of a new peak nor loss of a peak, suggesting no new formation or loss of chemical bonds, which indicated that microwave treatment did not change the starch molecules, as reported by Li et al. [15].

2.7. Morphology of Starch Granules

SEM micrographs of NFBS, CFBS, and HFBS appear in Figure 5. Granules of NFBS were round, elliptical, and oval shaped, similar to that reported by [1] for the same legume. There were some damaged granules. This could be due to the extraction method used, which can damage the granules because of the dry and wet grinding. However, damaged granules were rarely observed. The granule size of NFBS was of length, 28.2 ± 6.68 µm and width, 17.7 ± 3.92 µm. These findings were consistent with Zhang et al. [26] for native faba bean starch. The starch granule structure of CFBS and HFBS5 was not completely destroyed, and the original aspect of the granules was maintained but presented some agglomerations, characteristic of starches that have been hydrolyzed. The starch granules with a low degree of hydrolysis remained intact, although the outer surface became roughened and could present some agglomerations [10]. Starch systems with low moisture content (<30%) have poor absorption and conversion abilities, allowing the morphology of starch granules to remain completely intact with controlled microwave treatment [44], and this was observed in HFBS5 with 20% moisture. The rough surface and agglomerations observed could be due to the effect of acid in the medium or be associated with the presence of water in the granules with the combination of MWE [29]. The granule sizes of CFBS and HFBS5 were of lengths, 30.2 ± 10.36 and 27.0 ± 9.35 µm, and widths, 19.4 ± 6.38 and 17.9 ± 8.41 µm, respectively. For HFBS2, the surface became roughened with wide cracks and presented agglomerations to form larger starch clusters. The granule size was of length, 63.5 ± 20.65 µm and width, 43.5 ± 13.75 µm. This damage could be explained by the high internal steam pressure formed by the rapidly heating water, which also suggests that the internal structure of the starch disintegrated [15]. It appears that the starch granules were merged with the adjacent granules after the microwave treatment. For HFBS8, the round, elliptical and oval shapes of the starch disappeared entirely, and the large blocks formed, with a length of 40.5 ± 14.82 µm and width, 24.8 ± 7.53 µm. This indicated that when the moisture content was 40%, the excessive water and high temperature could gelatinize the starch, similar to that reported by Li et al. [15]. This treatment also had a high acid concentration and the longest MWE time, contributing to the rapid hydrolysis of the starch and gelatinization.

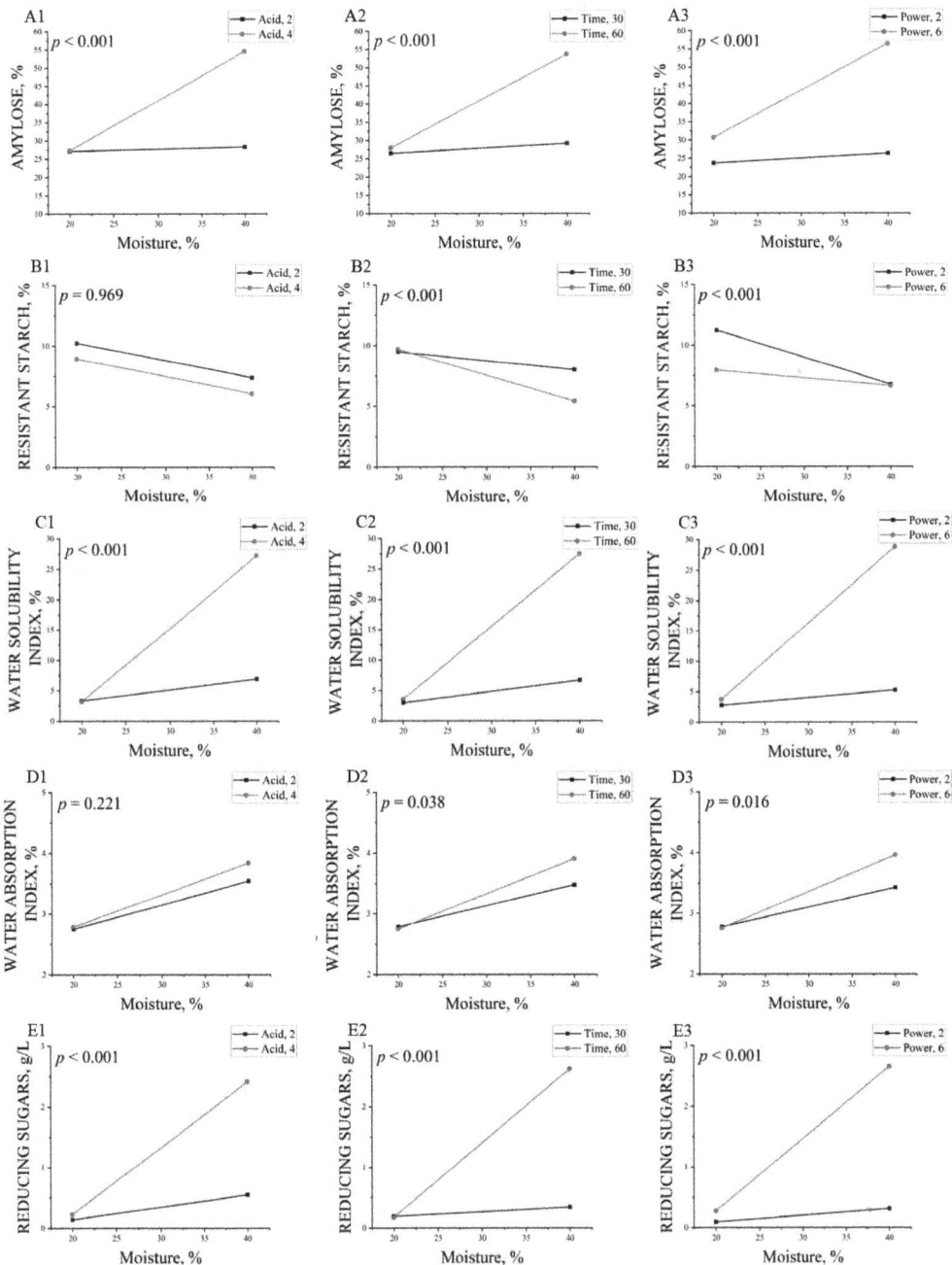

Figure 1. Interaction effects of moisture with pure hydrochloric acid, moisture with time, and moisture with power level. (**A1–A3**), interactions effects for amylose; (**B1–B3**), interactions effects for resistant starch; (**C1–C3**), interactions effects for water solubility index; (**D1–D3**), interactions effects for water absorption index; (**E1–E3**), interactions effects for reducing sugars.

Table 3. Thermal parameters of native and hydrolyzed faba bean starches [1].

Sample	T_o (°C)	T_p (°C)	T_f (°C)	ΔH (J/g)
NFBS	62.5 ± 2.14 [b]	68.0 ± 0.48 [c]	73.8 ± 0.78 [c]	10.9 ± 0.75 [a]
CFBS	64.1 ± 0.17 [b]	71.5 ± 0.45 [b]	80.0 ± 0.19 [b]	12.3 ± 0.77 [a]
HFBS2	74.1 ± 1.12 [a]	79.4 ± 1.86 [a]	86.1 ± 0.95 [a]	2.6 ± 1.28 [b]
HFBS5	63.3 ± 0.07 [b]	71.1 ± 0.11 [bc]	80.4 ± 0.28 [b]	11.0 ± 0.77 [a]
HFBS8	65.4 ± 1.79 [b]	70.7 ± 1.96 [bc]	79.5 ± 0.54 [b]	0.6 ± 0.11 [b]

[1] T_o, onset temperature; T_p, peak temperature; T_f, final temperature; ΔH, gelatinization enthalpy. NFBS, native faba bean starch; CFBS, conventional hydrolysis faba bean starch; HFBS2, HFBS5, and HFBS8, hydrolyzed faba bean starches. Assays were performed in triplicate. Mean ± SD, values in the same column with different superscript letters are significantly different (α = 0.05).

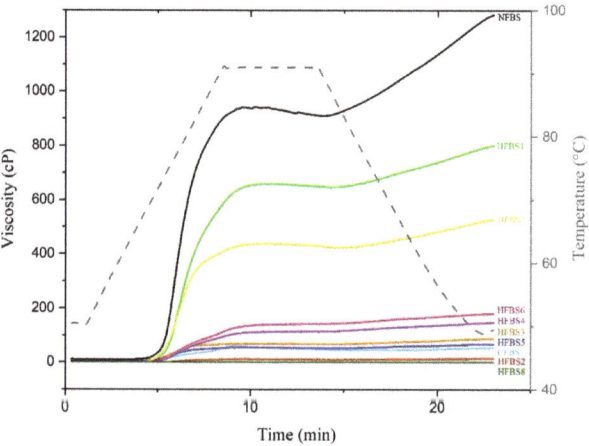

Figure 2. Brabender viscosity curves of native and hydrolyzed faba bean starches. NFBS, native faba bean starch; CFBS, conventional hydrolysis faba bean starch; HFBS1, HFBS2, HFBS3, HFBS4, HFBS5, HFBS6, HFBS7, and HFBS8, hydrolyzed faba bean starches.

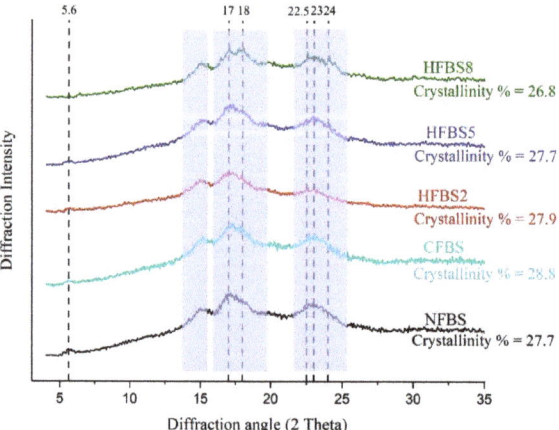

Figure 3. XRD patterns and relative crystallinity of native and hydrolyzed faba bean starches. Regions highlighted in color are the most affected peaks. NFBS, native faba bean starch; CFBS, conventional hydrolysis faba bean starch; HFBS2, HFBS5, and HFBS8, hydrolyzed faba bean starches.

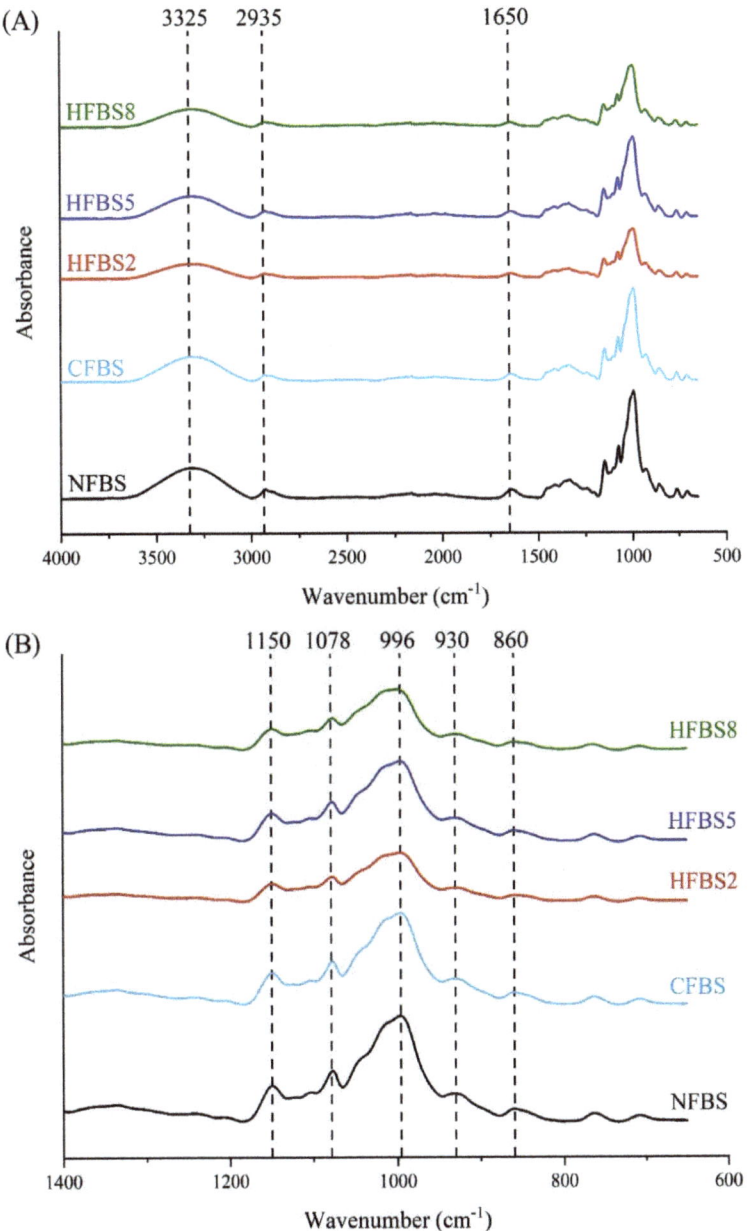

Figure 4. FTIR spectra. (**A**) Patterns of native and hydrolyzed faba bean starches. (**B**) The fingerprint characteristic peaks. NFBS, native faba bean starch; CFBS, conventional hydrolysis faba bean starch; HFBS2, HFBS5, and HFBS8, hydrolyzed faba bean starches.

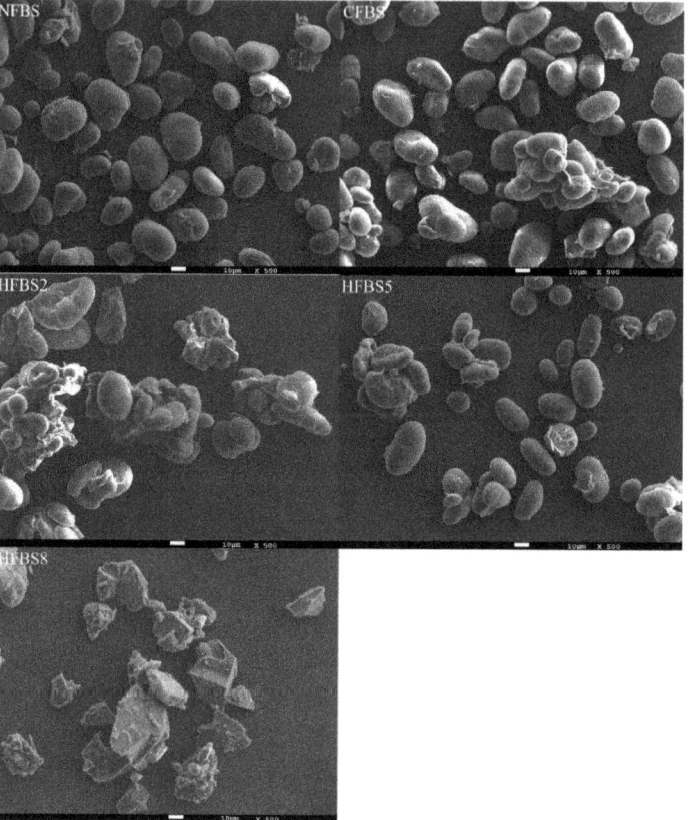

Figure 5. SEM images of native and hydrolyzed faba bean starch. The magnification of images was 500x. NFBS, native faba bean starch; CFBS, conventional hydrolysis faba bean starch; HFBS2, HFBS5, and HFBS8, hydrolyzed faba bean starches.

3. Materials and Methods

3.1. Materials

Faba bean seeds (*Vicia faba* var. major) were acquired in a local market of Queretaro, Queretaro, Mexico (Latitude: 20°35′17″ N; longitude: 100°23′17″ W). All reagents were analytical grade.

3.2. Starch Isolation

Starch was isolated from faba bean seeds according to Li et al. [15] with minor modifications. Seeds were ground in a granulator mill with a blade system and an integrated mesh to obtain a fine powder (flour). The flour (~300 g) was suspended in 750 mL of distilled water and stirred for 1 h. After the supernatant was carefully decanted, the protein layer on top of the precipitate was removed. This step was repeated twice. The pellet was resuspended in 600 mL of distilled water, followed by adding 2 g/100 mL NaOH and stirring for 1 h. The suspension was allowed to stand 12 h at room temperature. Subsequently, a wet grinding was carried out with a stone mill (FUMASA). The suspension obtained was submitted to sieving analysis, with the following set of Tyler sieves: 30, 100, 200 mesh, and a pan (U.S.). The residues were washed with water. The starch was left to stand overnight at 5 °C with 0.2 g/100 mL sodium bisulfite. The supernatant was removed, and the residue was dried at 40 °C in a convection oven for 24 h. The dried starch was ground and sieved in a mesh of 250 µm opening size. The NFBS was placed in polyethylene bags and stored at room temperature. NFBS was analyzed for moisture, protein, fat, and ash contents as described by the AOAC [50] methods 925.10, 920.87, 920.85, and 923.03, respectively.

3.3. Starch Hydrolysis

3.3.1. Conventional Acid Hydrolysis

Starch was hydrolyzed with hydrochloric acid according to the methodology described by Murúa-Pagola et al. [51]. The dried starch was ground and sieved in a mesh of 250 µm opening size. The CFBS was placed in polyethylene bags and stored at room temperature.

3.3.2. Starch Acid Hydrolysis by Microwave Treatment

A conventional microwave oven (Panasonic NN-SB646S, Beijing, China) was used, the working frequency was 2450 MHz, and the maximum electric power was 1100 W. A fractional factorial design (2^{4-1}) for the acid hydrolysis was evaluated. Factors and levels were as follows: moisture, 20% and 40%; pure hydrochloric acid concentration, 2 and 4 mL/100 mL; time, 30 and 60 s; and power level, 2 and 6. The hydrolysis of the starch was carried out as follows. The NFBS was conditioned to moisture content and concentration of HCl according to the design matrix (Table 4). Later, it was stabilized for 12 h at 4 °C. The starch samples were subjected to microwave treatments according to Table 4. After treatment, starch samples were allowed to cool down, and their pH was adjusted to 7.0 with 2 g/100 mL NaOH. Then, they were dried at 40 °C for 24 h, processed in a mill (Krups GX4100, México City, Mexico), and sieved with 60 mesh to obtain the HFBS.

Table 4. The 2^{4-1} fractional factorial design to explore starch acid hydrolysis by microwave energy.

Treatment Identification	Moisture, % (M)	Pure Hydrochloric Acid, mL/100 mL (A)	Time, s (T)	Power Level (P)
HFBS1	20	2	30	2
HFBS2	40	2	30	6
HFBS3	20	4	30	6
HFBS4	40	4	30	2
HFBS5	20	2	60	6
HFBS6	40	2	60	2
HFBS7	20	4	60	2
HFBS8	40	4	60	6

3.4. Characterization of Starches

Samples of NFBS, CFBS, and HFBS were analyzed. Enzyme kits determined amylose and RS according to the specifications (Megazyme K-RSTAR and K-AMYL). WAI and WSI were determined using the method described by Anderson et al. [52]. Reducing sugar content was measured by the 3,5-dinitrosalysilic acid method [53].

3.5. Determination of Viscosity Properties (RVA)

The viscosity profile was evaluated using a Rapid Visco Analyzer (RVA) model Super 4 (Newport Scientific PTY Ltd., Sydney, Australia). Sample amount of 2 g (dry base) was weighed out, and 28 mL of distilled water was added. Measurements were made according to the specifications of the AACC [54], 61-02 method. The results obtained from the equipment were expressed in units of viscosity (cP).

3.6. Thermal Properties by Differential Scanning Calorimetry

Thermal properties were performed using a differential scanning calorimeter (DSC Mettler Toledo, model 821). Sample amount of 3 mg was weighed out, and 7 mg of distilled water was placed in a 40 µL aluminum crucible at room temperature (25 °C), subsequently sealed and allowed to stand for 12 h at room temperature for even distribution of water. The sample was subjected to a heating ramp from 30 to 100 °C, with a 10 °C/min rate. An empty aluminum crucible was used as a reference. The start temperature, the maximum temperature, the completion temperature, and the enthalpy of gelatinization (ΔH) were recorded [15].

3.7. X-ray Diffraction

The X-ray diffraction was recorded using an X-ray diffractometer (Rigaku X-ray diffractometer DMAX-2100). Operating conditions included a CuKα radiation of λ = 1.5405, 30 kV, and an electric current of 16 mA. Approximately 1 g of sample was loaded onto a glass plate and scanned in the range of 5° to 60° Bragg angles in steps of 1°/0.03 s at room temperature. The relative crystallinity was determined from the ratio of the areas of the diffraction peaks to the area of the whole diffraction pattern subtracting amorphous background patterns using OriginPro 2018 Software (OriginLab Corp., Northampton, MA, USA) [15].

3.8. Fourier Transform-Infrared Spectroscopy

FTIR spectra were produced with a Spectrum GX spectrometer (Perkin Elmer, Waltham, MA, USA) with a diffuse reflectance accessory (Pike Technology model). Samples were prepared by finely grinding starch with KBr in a ratio of 1:100 (w/w) and scanned over a wavenumber range from 400 to 4000 cm^{-1} [51].

3.9. Scanning Electron Microscopy

Morphologies of samples were studied and analyzed by a field emission scanning electron microscopy (JXA-8530F, JEOL, Tokyo, Japan). A drop of suspension was mounted on a silicon wafer and allowed to dry on a desiccator. Subsequently, the wafer was mounted on specimen stubs with carbon black tape and sputter-coated (Denton Vacuum Desk V sputter) with gold and an exposure time of 60 s before observation [51].

3.10. Statistical Analysis

All experiments were conducted with three replications. Mean values and standard deviations (SD) were computed. The experimental data were analyzed using analysis of variance (ANOVA). All analysis was performed using R software (version 3.6.1, Vienna, Austria).

4. Conclusions

Microwave energy reduced the time required to obtain HFBS from hours to seconds. Microwave treatments combined with acid decreased the amylose content from 38.3 to 22.8 g/100 g (except for HFBS8), while RS, water solubility, and reducing sugars were increased. Furthermore, MWE modified the viscosity profile of starch granules, compared with NFBS and CFBS; the viscosity values statistically decreased. The most severe conditions resulted in the lowest viscosity values, the highest amylose content, solubility, absorption indexes, and reducing sugars. The starch granule structure of samples CFBS and HFBS5 was not totally destroyed, and the original appearance of the granules was maintained but presented some agglomerations. In HFBS8, the original shape of the starch disappeared completely. In summary, we report a broadly applicable and rapid microwave treatment protocol for the development of HFBS with high amylose and low viscosity contents that can be applied as an alternative encapsulant material.

Author Contributions: M.E.G.-M. devised the main conceptual ideas, designed the study, performed the experiments, analyzed the data obtained and wrote the original draft. F.M.-B. and S.L.A.-L. contributed equally in this manuscript in designing and supervising the experiments and analyzing the results. E.C.-T. provided support in the statistical design, analysis, and in interpreting the results. All authors have read and agreed to the published version of the manuscript.

Funding: This research received no external funding.

Institutional Review Board Statement: Not applicable.

Informed Consent Statement: Not applicable.

Data Availability Statement: Not applicable.

Acknowledgments: Special acknowledgements are given to Martínez-Bustos (CINVESTAV-IPN), for the support in the development of methodology. González-Mendoza acknowledges CONACYT for the PhD scholarship.

Conflicts of Interest: The authors declare no conflict of interest.

Sample Availability: Samples of the compounds are available from the authors.

References

1. Piecyk, M.; Domian, K. Effects of heat–moisture treatment conditions on the physicochemical properties and digestibility of field bean starch (*Vicia faba* var. minor). *Int. J. Biol. Macromol.* **2021**, *182*, 425–433. [CrossRef] [PubMed]
2. Dona, A.C.; Pages, G.; Gilbert, R.G.; Kuchel, P.W. Digestion of starch: In vivo and in vitro kinetic models used to characterize oligosaccharide or glucose release. *Carbohydr. Polym.* **2010**, *80*, 599–617. [CrossRef]
3. Ambigaipalan, P.; Hoover, R.; Donner, E.; Liu, Q. Retrogradation characteristics of pulse starches. *Int. Food Res. J.* **2013**, *54*, 203–212. [CrossRef]
4. Tharanathan, R.; Mahadevamma, S. Grain legumes—A boon to human nutrition. *Trends Food Sci Technol.* **2003**, *14*, 507–518. [CrossRef]
5. Englyst, H.N.; Kingman, S.M.; Cummings, J.H. Classification and measurement of nutritionally important starch fractions. *Eur. J. Clin. Nutr.* **1992**, *46*, S33–S50.
6. Lunn, J.; Buttriss, J.L. Carbohydrates and dietary fibre. *Nutr. Bull.* **2007**, *32*, 21–64. [CrossRef]
7. Sajilata, M.G.; Singhal, R.S.; Kulkarni, P.R. Resistant starch—A review. *Compr. Rev. Food Sci. Food Saf.* **2006**, *5*, 1–17. [CrossRef]
8. Suárez-Diéguez, T.; Pérez-Moreno, F.; Ariza-Ortega, J.A.; López-Rodríguez, G.; Nieto, J.A. Obtention and characterization of resistant starch from creole faba bean (Vicia faba L. creole) as a promising functional ingredient. *LWT-Food Sci. Technol.* **2021**, *145*, 111247. [CrossRef]
9. Cai, J.; Cai, C.; Man, J.; Zhou, W.; Wei, C. Structural and functional properties of C-type starches. *Carbohydr. Polym.* **2014**, *101*, 289–300. [CrossRef]
10. Wang, S.; Copeland, L. Effect of acid hydrolysis on starch structure and functionality: A review. *Crit. Rev. Food Sci. Nutr.* **2015**, *55*, 1081–1097. [CrossRef]
11. Muhammad, Z.; Ramzan, R.; Zhang, R.; Zhang, M. Resistant starch-based edible coating composites for spray-dried microencapsulation of Lactobacillus acidophilus, comparative assessment of thermal protection, in vitro digestion and physicochemical characteristics. *Coatings* **2021**, *11*, 587. [CrossRef]
12. Ashwar, B.A.; Gani, A.; Gani, A.; Shah, A.; Masoodi, F.A. Production of RS4 from rice starch and its utilization as an encapsulating agent for targeted delivery of probiotics. *Food Chem.* **2018**, *239*, 287–294. [CrossRef] [PubMed]
13. Lu, K.; Zhu, J.; Bao, X.; Liu, H.; Yu, L.; Chen, L. Effect of starch microstructure on microwave-assisted esterification. *Int. J. Biol. Macromol.* **2020**, *164*, 2550–2557. [CrossRef] [PubMed]
14. Kumar, Y.; Singh, L.; Sharanagat, V.S.; Patel, A.; Kumar, K. Effect of microwave treatment (low power and varying time) on potato starch: Microstructure, thermo-functional, pasting and rheological properties. *Int. J. Biol. Macromol.* **2020**, *155*, 27–35. [CrossRef] [PubMed]
15. Li, Y.; Hu, A.; Wang, X.; Zheng, J. Physicochemical and in vitro digestion of millet starch: Effect of moisture content in microwave. *Int. J. Biol. Macromol.* **2019**, *134*, 308–315. [CrossRef] [PubMed]
16. Lin, D.; Zhou, W.; He, Q.; Xing, B.; Wu, Z.; Chen, H.; Wu, D.; Zhang, Q.; Qin, W. Study on preparation and physicochemical properties of hydroxypropylated starch with different degree of substitution under microwave assistance. *Int. J. Biol. Macromol.* **2018**, *125*, 290–299. [CrossRef] [PubMed]
17. Yang, Q.; Qi, L.; Luo, Z.; Kong, X.; Xiao, Z.; Wang, P.; Peng, X. Effect of microwave irradiation on internal molecular structure and physical properties of waxy maize starch. *Food Hydrocoll.* **2017**, *69*, 473–482. [CrossRef]
18. Zhong, Y.; Liang, W.; Pu, H.; Blennow, A.; Liu, X.; Guo, D. Short-time microwave treatment affects the multi-scale structure and digestive properties of high-amylose maize starch. *Int. J. Biol. Macromol.* **2019**, *137*, 870–877. [CrossRef]
19. Xie, Y.; Yan, M.; Yuan, S.; Sun, S.; Huo, Q. Effect of microwave treatment on the physicochemical properties of potato starch granules. *Chem. Cent. J.* **2013**, *7*, 113. [CrossRef]
20. Zhu, J.; Li, L.; Zhang, S.; Li, X.; Zhang, B. Multi-scale structural changes of starch-based material during microwave and conventional heating. *Int. J. Biol. Macromol.* **2016**, *92*, 270–277. [CrossRef]
21. Zeng, S.; Chen, B.; Zeng, H.; Guo, Z.; Lu, X.; Zhang, Y.; Zheng, B. Effect of microwave irradiation on the physicochemical and digestive properties of Lotus seed starch. *J. Agric. Food Chem.* **2016**, *64*, 2442–2449. [CrossRef] [PubMed]
22. Yu, H.; Chen, S.; Suree, P.; Nuansri, R.; Wang, K. Effect of microwave irradiation on acid-catalyzed hydrolysis of starch. *J. Org. Chem.* **1996**, *61*, 9608–9609. [CrossRef]
23. Warrand, J.; Janssen, H.-G. Controlled production of oligosaccharides from amylose by acid-hydrolysis under microwave treatment: Comparison with conventional heating. *Carbohydr. Polym.* **2007**, *69*, 353–362. [CrossRef]

24. Sunarti, T.C.; Dwiko, M.; Derosya, V.; Meryandini, A. Effect of Microwave Treatment on Acid and Enzymes Susceptibilities of Sago Pith. *Procedia Chem.* **2012**, *4*, 301–307. [CrossRef]
25. Ambigaipalan, P.; Hoover, R.; Donner, E.; Liu, Q.; Jaiswal, S.; Chibbar, R.; Nantanga, K.K.M.; Seetharaman, K. Structure of faba bean, black bean and pinto bean starches at different levels of granule organization and their physicochemical properties. *Food Res. Int.* **2011**, *44*, 2962–2974. [CrossRef]
26. Zhang, Z.; Tian, X.; Wang, P.; Jiang, H.; Li, W. Compositional, morphological, and physicochemical properties of starches from red adzuki bean, chickpea, faba bean, and baiyue bean grown in China. *Food Sci. Nutr.* **2019**, *7*, 2485–2494. [CrossRef]
27. Li, L.; Yuan, T.Z.; Setia, R.; Raja, R.B.; Zhang, B.; Ai, Y. Characteristics of pea, lentil and faba bean starches isolated from air-classified flours in comparison with commercial starches. *Food Chem.* **2019**, *276*, 599–607. [CrossRef]
28. Hoover, R.; Hughes, T.; Chung, H.J.; Liu, Q. Composition, molecular structure, properties, and modification of pulse starches: A review. *Int. Food Res.* **2010**, *43*, 399–413. [CrossRef]
29. Zailani, M.A.; Kamilah, H.; Husaini, A.; Awang Seruji, A.Z.R.; Sarbini, S.R. Functional and digestibility properties of sago (*Metroxylon sagu*) starch modified by microwave heat treatment. *Food Hydrocoll.* **2022**, *122*, 107042. [CrossRef]
30. Atichokudomchai, N.; Shobsngob, S.; Varavinit, S. Morphological properties of acid-modified tapioca starch. *Starch/Stärke* **2000**, *52*, 283–289. [CrossRef]
31. Amaya-Llano, S.L.; Martínez-Bustos, F.; Martínez Alegría, A.L.; de Jesús Zazueta-Morales, J. Comparative studies on some physico-chemical, thermal, morphological, and pasting properties of acid-thinned jicama and maize starches. *Food Bioprocess Technol.* **2008**, *4*, 48–60. [CrossRef]
32. Aaliya, B.; Sunooj, K.V.; John, N.E.; Navaf, M.; Akhila, P.P.; Sudheesh, C.; Sabu, S.; Sasidharan, A.; Mir, S.A.; George, J. Impact of microwave irradiation on chemically modified talipot starches: A characterization study on heterogeneous dual modifications. *Int. J. Biol. Macromol.* **2022**, *209*, 1943–1955. [CrossRef] [PubMed]
33. Cruz-Benítez, M.M.; Gómez-Aldapa, C.A.; Castro-Rosas, J.; Hernández-Hernández, E.; Gómez-Hernández, E.; Fonseca-Florido, H.A. Effect of amylose content and chemical modification of cassava starch on the microencapsulation of *Lactobacillus pentosus*. *LWT-Food Sci. Technol.* **2019**, *105*, 110–117. [CrossRef]
34. González, Z.; Pérez, E. Evaluation of lentil starches modified by microwave irradiation and extrusion cooking. *Food Res. Int.* **2002**, *35*, 415–420. [CrossRef]
35. Deka, D.; Sit, N. Dual modification of taro starch by microwave and other heat moisture treatments. *Int. J. Biol. Macromol.* **2016**, *92*, 416–422. [CrossRef]
36. Espinosa-Solis, V.; Sanchez-Ambriz, S.L.; Hamaker, B.R.; Bello-Pérez, L.A. Fine structural characteristics related to digestion properties of acid-treated fruit starches. *Starch/Stärke* **2011**, *63*, 717–727. [CrossRef]
37. Wang, M.; Sun, M.; Zhang, Y.; Chen, Y.; Wu, Y.; Ouyang, J. Effect of microwave irradiation-retrogradation treatment on the digestive and physicochemical properties of starches with different crystallinity. *Food Chem.* **2019**, *298*, 125015. [CrossRef]
38. Li, D.; Yang, N.; Zhou, X.; Jin, Y.; Guo, L.; Xie, Z.; Jin, Z.; Xu, X. Characterization of acid hydrolysis of granular potato starch under induced electric field. *Food Hydrocoll.* **2017**, *71*, 198–206. [CrossRef]
39. Luo, Z.; He, X.; Fu, X.; Luo, F.; Gao, Q. Effect of microwave radiation on the physicochemical properties of normal maize, waxy maize and amylomaize V starches. *Starch/Stärke* **2006**, *58*, 468–474. [CrossRef]
40. Márquez-Gómez, M.; Galicia-García, T.; Márquez-Meléndez, R.; Ruiz-Gutiérrez, M.; Quintero-Ramos, A. Spray-dried microencapsulation of orange essential oil using modified rice starch as wall material. *J. Food Process. Preserv.* **2017**, *42*, e13428. [CrossRef]
41. Anderson, A.K.; Guraya, H.S. Effects of microwave heat-moisture treatment on properties of waxy and non-waxy rice starches. *Food Chem.* **2006**, *97*, 318–323. [CrossRef]
42. Hoyos-Leyva, J.D.; Bello-Pérez, L.A.; Alvarez-Ramirez, J.; Garcia, H.S. Microencapsulation using starch as wall material: A review. *Food Rev. Int.* **2017**, *34*, 148–161. [CrossRef]
43. Atichokudomchai, N.; Varavinit, S. Characterization and utilization of acid-modified cross-linked tapioca starch in pharmaceutical tablets. *Carbohydr. Polym.* **2003**, *53*, 263–270. [CrossRef]
44. Tao, Y.; Yan, B.; Fan, D.; Zhang, N.; Ma, S.; Wang, L.; Wu, Y.; Wang, M.; Zhao, J.; Zhang, H. Structural changes of starch subjected to microwave heating: A review from the perspective of dielectric properties. *Trends Food Sci Technol.* **2020**, *99*, 593–607. [CrossRef]
45. Wang, S.; Yu, J.; Zhu, Q.; Yu, J.; Jin, F. Granular structure and allomorph position in C-type Chinese yam starch granule revealed by SEM, 13C CP/MAS NMR and XRD. *Food Hydrocoll.* **2019**, *23*, 426–433. [CrossRef]
46. Polesi, L.F.; Sarmento, S.B.S. Structural and physicochemical characterization of RS prepared using hydrolysis and heat treatments of chickpea starch. *Starch/Stärke* **2011**, *63*, 226–235. [CrossRef]
47. Hoover, R.; Ratnayake, W.S. Starch characteristics of black bean, chick pea, lentil, navy bean and pinto bean cultivars grown in Canada. *Food Chem.* **2002**, *78*, 489–498. [CrossRef]
48. Xie, Y.-Y.; Hu, X.-P.; Jin, Z.-Y.; Xu, X.-M.; Chen, H.-Q. Effect of temperature-cycled retrogradation on in vitro digestibility and structural characteristics of waxy potato starch. *Int. J. Biol. Macromol.* **2014**, *67*, 79–84. [CrossRef]
49. Simsek, S.; Ovando-Martinez, M.; Marefati, A.; Sjöö, M.; Rayner, M. Chemical composition, digestibility and emulsification properties of octenyl succinic esters of various starches. *Food Res. Int.* **2015**, *75*, 41–49. [CrossRef]
50. AOAC. *Official Methods of Analysis of the Association of Official Analytical Chemists*, 15th ed.; AOAC: Rockville, MD, USA, 1990.

51. Murúa-Pagola, B.; Beristain-Guevara, C.I.; Martínez-Bustos, F. Preparation of starch derivatives using reactive extrusion and evaluation of modified starches as shell materials for encapsulation of flavoring agents by spray drying. *J. Food Eng.* **2009**, *91*, 380–386. [CrossRef]
52. Anderson, R.; Conway, H.; Pfeifer, V.; Griffin, E. Gelatinization of corn grits by roll and extrusion cooking. *J. Cereal Sci.* **1970**, *14*, 4–7. [CrossRef]
53. Miller, G.L. Use of dinitrosalicylic acid reagent for determination of reducing sugar. *Anal. Chem.* **1959**, *31*, 426–428. [CrossRef]
54. AACC. *Methods of the AACC*, 3rd ed.; American Association of Cereal Chemists, Academic Press Inc.: Cambridge, MA, USA, 1999.

Article

Beneficial Effect of Kidney Bean Resistant Starch on Hyperlipidemia—Induced Acute Pancreatitis and Related Intestinal Barrier Damage in Rats

Zhaohang Zuo [1,†], Shuting Liu [1,†], Weiqiao Pang [1], Baoxin Lu [1], Wei Sun [1], Naidan Zhang [1], Xinyu Zhou [1], Dongjie Zhang [1,2,*] and Ying Wang [1,2,*]

1 College of Food Science, Heilongjiang Bayi Agricultural University, Daqing 163319, China; byndzzh1994@163.com (Z.Z.); xkljcg@163.com (S.L.); hljbypwq@163.com (W.P.); bynd2020ztgzy@163.com (B.L.); swsw0102@163.com (W.S.); zndan1025@163.com (N.Z.); zxy199810271015@163.com (X.Z.)
2 National Coarse Cereals Engineering Research Center, Daqing 163319, China
* Correspondence: zhangdongjie@byau.edu.cn (D.Z.); wangying@byau.edu.cn (Y.W.)
† These authors contributed equally to this work.

Abstract: Accumulating attention has been focused on resistant starch (RS) due to its blood-lipid-lowering activities. However, reports on the potential bioactivities of RS for preventing hyperlipidemia acute pancreatitis (HLAP) are limited. Therefore, in this study, an acute pancreatitis model was set up by feeding a hyperlipidemia diet to rats, and subsequently evaluating the anti-HLAP effect of RS in kidney beans. The results show that the IL-6, IL-1β, and TNF-α of serum in each RS group were decreased by 18.67–50.00%, 7.92–22.89%, and 8.06–34.04%, respectively, compared with the model group (MOD). In addition, the mRNA expression of tight junction protein ZO-1, occludin, and antibacterial peptides CRAMP and DEFB1 of rats in each RS group increased by 26.43–60.07%, 229.98–279.90%, 75.80–111.20%, and 77.86–109.07%, respectively. The height of the villi in the small intestine and the thickness of the muscle layer of rats were also increased, while the depth of the crypt decreased. The present study indicates that RS relieves intestinal inflammation, inhibits oxidative stress, and prevents related intestinal barrier damage. These results support the supplementation of RS as an effective nutritional intervention for HLAP and associated intestinal injury.

Keywords: kidney bean resistant starch; hyperlipidemia; acute pancreatitis; intestinal barrier damage

Citation: Zuo, Z.; Liu, S.; Pang, W.; Lu, B.; Sun, W.; Zhang, N.; Zhou, X.; Zhang, D.; Wang, Y. Beneficial Effect of Kidney Bean Resistant Starch on Hyperlipidemia—Induced Acute Pancreatitis and Related Intestinal Barrier Damage in Rats. *Molecules* 2022, 27, 2783. https://doi.org/10.3390/molecules27092783

Academic Editors: Litao Tong and Lili Wang

Received: 1 April 2022
Accepted: 26 April 2022
Published: 27 April 2022

Publisher's Note: MDPI stays neutral with regard to jurisdictional claims in published maps and institutional affiliations.

Copyright: © 2022 by the authors. Licensee MDPI, Basel, Switzerland. This article is an open access article distributed under the terms and conditions of the Creative Commons Attribution (CC BY) license (https://creativecommons.org/licenses/by/4.0/).

1. Introduction

Acute pancreatitis (AP) is a serious comprehensive digestive system disease caused by a variety of exogenous or endogenous factors that cause the activation and release of pancreatic acinar trypsin, which induces pancreatic autophagy and digestion, thereby triggering local pancreatic inflammation and systemic organ failure [1]. Clinical investigations have found that about 34% of patients with acute pancreatitis also suffer from dyslipidemia, indicating that hyperlipidemia has gradually become a common etiology of hyperlipidemia acute pancreatitis (HLAP) [2,3]. During the early phase of HLAP, trypsinogen produced in situ in the pancreas is abnormally activated, leading to pancreatic tissue self-destruction. This injury induces a strong systemic inflammatory response released into circulation by pro-inflammatory molecules (such as interleukins, tumor necrosis factor, and platelet activation factor) [4]. The above inflammatory factors can damage the intestinal mucosal barrier and cause intestinal dysfunction, which is the inevitable result of the development of acute pancreatitis. Therefore, effective control of intestinal tissue structure and function damage can delay the development of HLAP [5,6]. On another side, the triglyceride (TG), total cholesterol (TC), and low-density lipoprotein cholesterol (LDL-C) in the lipid profile of HLAP were significantly increased ($p < 0.05$). HLAP is related to lipid metabolism [7].

In line with the side effects of statins and the intense pain patients suffer, there is accumulating interest in using dietary botanical supplements for prevention and adjuvant therapy. Studies on dietary factors and blood lipid levels have shown that different dietary patterns impact blood lipid levels to a certain extent, and a plant-based diet is a protective factor for dyslipidemia [8]. The compact molecular structure of RS can hinder the digestion of various enzymes [9]. RS improves gastrointestinal health, affects the composition of the intestinal flora, potentially affects blood lipid metabolism, and inhibits inflammatory cytokines by producing gas (methane, hydrogen, carbon dioxide) and short-chain fatty acids (formic acid, acetic acid, propionic acid, etc.) during colonic fermentation [10]. In addition, RS can regulate glucose metabolism by down-regulating the expression level of key enzymes, and also restore normal levels of insulin by repairing damaged pancreatic β cells [11]. After ingesting RS, the concentration of HDL-C in a hamster study increased, and the concentration of TG, TC, and LDL-C reduced, repairing the damage to the cecum and colon tissue caused by a high-fat diet [12]. RS supplementation has obvious effects on lowering TC and LDL-C, and a longer time (>4 weeks) of supplementation can provide a more robust effect [13]. The metabolic effects of RS3 on T2DM have been studied, indicating that RS3 can down-regulate the levels of blood glucose, improve dyslipidemia, reduce insulin resistance, and enhance insulin sensitivity [14]. GB (rich in resistant starch) in the diet of high-fat-fed mice increased SCFA production, down-regulated the expression of genes involved in lipogenesis, and enhanced the expression of transport proteins involved in lipid excretion [15]. The rice starch-FA complex (resistant starch V) can reduce the bodyweight of rats under a high-fat diet, and improve serum lipid profiles, oxidative stress, and liver function [16]. In summary, RS may be a promising food in diet therapy for obesity, hyperlipidemia, and T2DM. However, the effect of RS on HLAP is still unknown.

In view of these facts, we explored whether kidney bean RS and its metabolites can alleviate HLAP in rats and restore the intestinal barrier damage caused by inflammatory cytokines.

2. Results

2.1. Molecular Structure and Digestibility of RS

Comparing purified RS with starch, it was found that kidney bean RS granules have an irregular and angular polygonal structure with a compact texture, rough surface, and a lamellar structure on the cross-section (Figure 1A). X-ray diffraction results show that RS based on a renal bean starch crystal structure had new diffraction peaks at 6°, 22°, and 24°, and the new diffraction peak was the characteristic diffraction peak of B-type crystals, so RS exhibited a combination of A and B to form a C-type crystal structure (Figure 1B). Compared with the original starch, the volume average particle size of the RS granules increased significantly, and the specific surface reduced significantly. The reduction in a specific surface area can avoid excessive contact between RS and enzymes, thereby enhancing its resistance to enzymatic hydrolysis (Table 1, Figure 1C). The above results show that the internal structure of RS is stable from the perspective of micro-morphology, crystal structure, molecular mass, and particle size.

Subsequently, we applied RS and starch to obtain the vitro digestion ratio (Figure 1D,E). The digestion process of the sample involved three steps to simulate oral chewing in the mouth, digestion by the stomach, and the small intestine in vitro digestion [17]. We were intuitively exploring the final digestion rate of carbohydrates through releasing sugars in the digestive juice of the small intestine. The results show that the glucose concentration in the RS digestive fluid was significantly lower than that of starch at the same digestion time. At the same time, the digestibility of resistant starch was much lower than that of native starch in the same period. Combined with the previous structural properties of resistant starch, it is known that resistant starch has good resistance to enzymatic hydrolysis in the digestive system, which may be beneficial to its physiological function in the intestine [18].

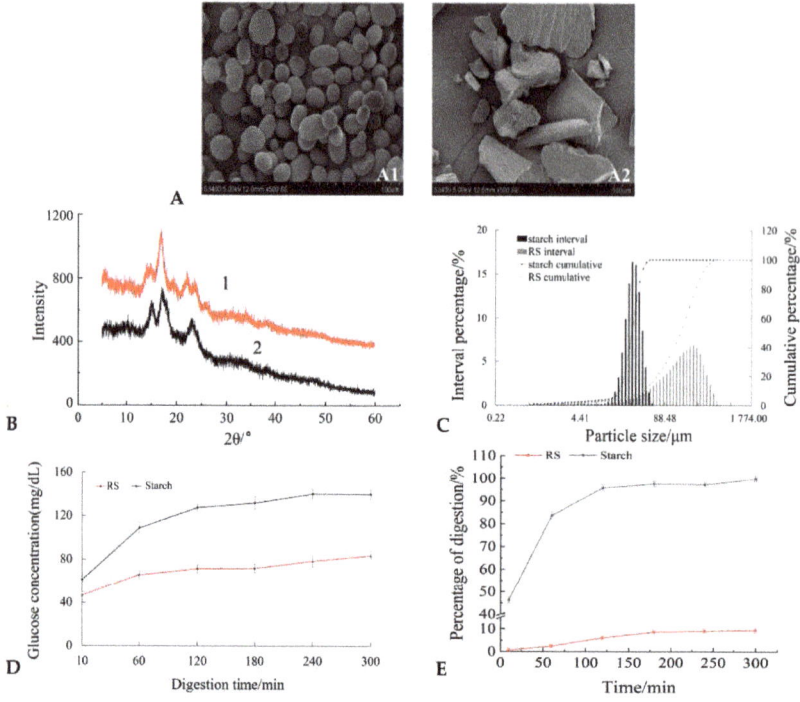

Figure 1. Molecular structure and digestibility of RS. (**A**) Scanning diagram of the morphology of kidney bean starch (A1) and RS (A2); (**B**) X-ray diffraction spectra of kidney bean RS (1) and starch (2); (**C**) size distribution of kidney bean RS and starch; (**D**) digestive characteristics of kidney bean RS and starch; (**E**) digestion ratio of kidney bean RS and starch at different times.

Table 1. Mean particle size of kidney bean starch and resistant starch.

Sample	Low Particle Size D10/μm	Median Particle Size D50/μm	High Particle Size D90/μm	Volume Average Particle Size/μm	Specific Surface Area/m²·g
Starch	18.65 ± 0.18 [a]	28.77 ± 0.75 [a]	40.85 ± 3.54 [a]	28.88 ± 1.46 [a]	0.119 ± 0.011 [a]
RS	41.63 ± 0.22 [b]	166.30 ± 2.87 [b]	355.2 ± 2.12 [b]	184.60 ± 3.39 [b]	0.044 ± 0.013 [b]

Values followed by different lower-case letters in the same column are significantly different from each other ($p < 0.05$).

2.2. Dietary Intake Affected the Body Weight and Pancreas Weight of HLAP Rats

The body weight and pancreas mass of the rats were measured, and the trend of body weight and related organ index were monitored. The results show that the rats in the CON had a slower weight increase trend than those in the other groups because they were fed with a common diet. After oral administration of RS and simvastatin, compared with the MOD, the rats in the SV and the RS groups grew slowly or even slightly, indicating that the RS can control the weight to a normal extent (Table 2). The pancreas coefficient of the MOD was significantly different from that of the CON ($p < 0.05$), indicating that intraperitoneal injection caused damage to the rat pancreas, and the pancreas index decreased, indicating that the pancreas may suffer from atrophy or degenerative disease. After treatment with positive drugs and RS, the pancreas index of each RS group of rats was 12.46%, 34.26%, and 44.64% higher than that in the MOD.

Table 2. Dietary intake of RS affected the body weight and pancreas weight of HLAP rats.

Parameter	Model Group (MOD)	Control Group (CON)	Simvastatin Group (SV)	Low-Dose RS Group (L-RS)	Medium-Dose RS Group (M-RS)	High-Dose RS Group (H-RS)
Body mass/g	471.16 ± 8.08 [f]	378.73 ± 10.15 [a]	405.88 ± 7.19 [b]	451.67 ± 6.26 [e]	429.78 ± 8.98 [d]	418.35 ± 5.19 [c]
Pancreas mass/g	1.295 ± 0.054 [a]	1.789 ± 0.019 [f]	1.520 ± 0.023 [c]	1.468 ± 0.024 [b]	1.666 ± 0.035 [d]	1.748 ± 0.041 [e]
Pancreas index/%	0.289 ± 0.042 [a]	0.455 ± 0.041 [e]	0.375 ± 0.012 [c]	0.325 ± 0.010 [b]	0.388 ± 0.016 [cd]	0.418 ± 0.014 [d]

Values followed by different lower-case letters in the same line are significantly different from each other ($p < 0.05$).

2.3. The Effect of RS on Serum Pathological Parameters in Rats

Serum triglyceride, amylase, and lipase concentrations are common indicators of clinical diagnosis of acute pancreatitis. The blood lipid and serum levels of AMLY and LIPA in the rats are shown in Table 3. After being fed a high-fat diet and given intraperitoneal injection of meninges, the contents of TC, TG, and AMLY and LIPA in the serum of the MOD increased significantly ($p < 0.05$) compared with the CON, indicating that the HLAP model was successfully induced. After RS and positive drug intervention, compared with the MOD, the blood lipid and serum levels of AMLY and LIPA in the rats in each RS group increased by 16.68–31.05% and 21.97–48.55%, respectively. Moreover, the TC and TG content in each RS group increased significantly by 43.14–69.36% and 12.61–43.69% compared to that in the MOD ($p < 0.05$). All of the improvements in the H-RS were significantly better than those in the L-RS.

Table 3. Blood-serum-related parameter levels in each group of rats.

Parameter	MOD	CON	SV	L-RS	M-RS	H-RS
TC (mmol/L)	4.08 ± 0.13 [e]	0.81 ± 0.14 [a]	2.44 ± 0.11 [d]	2.32 ± 0.21 [d]	1.55 ± 0.17 [c]	1.25 ± 0.14 [b]
TG (mmol/L)	2.22 ± 0.12 [f]	0.67 ± 0.05 [a]	1.48 ± 0.07 [c]	1.94 ± 0.13 [e]	1.74 ± 0.05 [d]	1.25 ± 0.08 [b]
AMLY (U/L)	4366.16 ± 117.84 [f]	1280.27 ± 71.93 [a]	2660.19 ± 105.55 [b]	3637.9 ± 61.05 [e]	3218.61 ± 60.65 [d]	3010.32 ± 65.86 [c]
LIPA (U/L)	1016.58 ± 35.73 [e]	107.97 ± 19.91 [a]	780.51 ± 20.57 [d]	793.22 ± 15.39 [d]	674.44 ± 32.76 [c]	523.05 ± 27.66 [b]
TNF-α (pg/mL)	197.60 ± 2.72 [f]	60.13 ± 4.10 [a]	110.60 ± 2.37 [b]	181.67 ± 3.25 [e]	140.67 ± 4.39 [d]	130.33 ± 4.29 [c]
IL-6 (pg/mL)	110.67 ± 4.59 [d]	12.33 ± 1.53 [a]	50.67 ± 3.61 [bc]	90.00 ± 2.00 [c]	75.33 ± 4.18 [b]	55.33 ± 2.07 [b]
IL-1β (pg/mL)	86.88 ± 2.46 [f]	15.00 ± 1.43 [a]	53.80 ± 0.82 [b]	80.00 ± 1.26 [e]	72.82 ± 1.81 [d]	67.00 ± 2.03 [c]
DAO (ng/mL)	95.07 ± 4.21 [e]	57.58 ± 3.51 [a]	70.51 ± 1.41 [b]	85.98 ± 3.32 [d]	72.13 ± 4.96 [bc]	75.31 ± 2.31 [c]
DLA (μmol/L)	38.76 ± 2.27 [f]	23.57 ± 0.62 [a]	26.72 ± 1.34 [b]	36.76 ± 1.05 [e]	35.08 ± 1.14 [d]	28.41 ± 0.92 [c]
ET (EU/mL)	71.84 ± 1.87 [e]	53.16 ± 2.58 [a]	57.26 ± 1.82 [c]	67.74 ± 1.12 [d]	56.37 ± 1.70 [bc]	54.84 ± 2.27 [ab]
sIgA (μg/mL)	19.01 ± 0.73 [a]	25.04 ± 1.16 [c]	26.75 ± 1.75 [d]	19.49 ± 0.78 [a]	21.28 ± 1.39 [b]	22.67 ± 1.38 [b]

Values followed by different lower-case letters in the same line are significantly different from each other ($p < 0.05$).

2.4. Intaking of RS Affects the Degree of Edema and Myeloperoxidase Activity in Pancreatic Tissue of Rats in Each Group

The degree of pancreatic organ edema could directly reflect the severity of pancreatic injury and HLAP in rats. As shown in Figure 2A, the wet/dry mass ratio of the pancreas in the MOD was significantly higher than that in the CON, indicating that the MOD had pancreatic damage and obvious edema. Compared with the MOD, the degree of pancreatic edema of H-RS and M-RS decreased to varying degrees (25.24% and 19.46%), after the intervention of RS. However, they are less effective than SV (33.84%). Myeloperoxidase (MPO), as a heme protease, mainly exists in neutrophils and monocytes, and the activity change is the key symbol of neutrophil function and activation. As shown in Figure 2B, the MPO activity of pancreatic tissue in the MOD was significantly higher than that in the CON ($p < 0.001$), indicating that the inflammatory cells in the MOD were severely aggregated and infiltrated. After gavage treatment and positive drug intervention, compared with the MOD, the MPO activity of each RS group decreased by 11.53–44.81%. The inhibitory effect of the H-RS and M-RS on MPO activity was significantly better than that of L-RS, and even the H-RS had a better effect than SV.

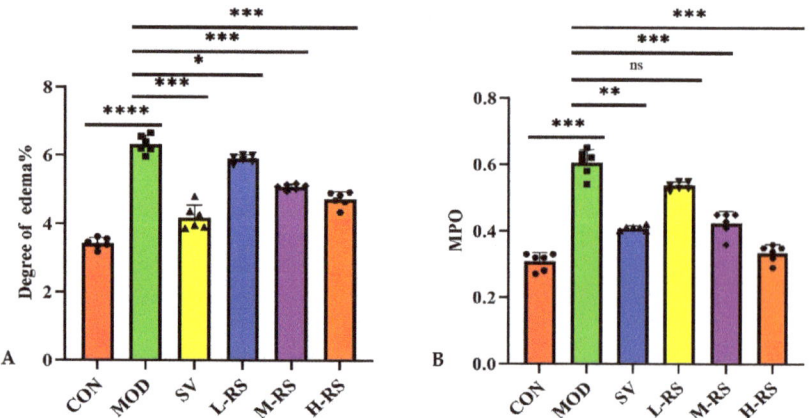

Figure 2. The degree of pancreatic edema on rats (**A**) and MPO activity in pancreatic tissues of rats (**B**) in each group (ns $p > 0.05$; * $p < 0.05$; ** $p < 0.01$; *** $p < 0.001$; **** $p < 0.0001$ versus MOD, and the points of different shapes represent the actual value of each sample, $n = 6$).

2.5. The Results of the Pancreas Pathology Section

The pancreas tissue of rats in the CON was in the normal range (Figure 3). The acetic acid original was intact and uniformly distributed. The lobular veins were clear, and there was no cell necrosis or inflammatory cell infiltration. Conversely, the pancreatic lobular space of the MOD was significantly wider than that of the CON, and a large number of monocytes and neutrophils infiltrated around the necrotic focus, which was consistent with the symptoms of edema-type acute pancreatitis. In the SV, after oral administration of simvastatin and intravenous injection of simvastatin, the pancreatic tissue structure was relatively clear, acinar vacuoles and swelling and necrosis were reduced, and inflammatory cell infiltration was relieved. After gavage of RS, compared with MOD rats, pancreatic tissue damage reduced to varying degrees in rats in the RS groups; pancreas tissue structure was more complete in M-RS and H-RS, with dense lobular stroma, as well as acinar cell necrosis and congestion. In addition, inflammatory cell infiltration reduced significantly; in the L-RS, the pancreas tissue of the rats still showed lobular interstitial looseness and focal acinar cell necrosis, which revealed that high-dose RS could effectively alleviate acute edema-induced pancreatic damage in rats induced by the peritoneum.

2.6. Effect of Kidney Bean RS on Secretion of Inflammatory Cells in Rats

The onset of acute pancreatitis and hyperlipidemia induces pancreatic acinar cells to release pro-inflammatory cytokines, such as IL-6, IL-1β, TNF-α, etc., through activating the nuclear transcription factor NF-κB to produce a large number of inflammatory mediators, which trigger multiple organ dysfunction or systemic inflammatory response syndrome to aggravate the condition. Therefore, the course of HLAP and the degree of inflammation development can be analyzed according to the concentration of inflammatory factors. The inflammatory cell secretion level of the rats in each group was measured, and the result is shown in Table 3. The serum level inflammatory factors of rats in the MOD were significantly higher than those in the CON ($p < 0.05$). After the intervention of positive drugs and RS, compared with the MOD, the serum levels of IL-6, IL-1β, and TNF-α in each RS group reduced by 18.67–50.00%, 7.92–22.89%, and 8.06–34.04%, respectively. The regulatory effect of RS on the inflammatory factors was positively correlated with the dose; especially, H-RS could effectively inhibit the release of inflammatory factors.

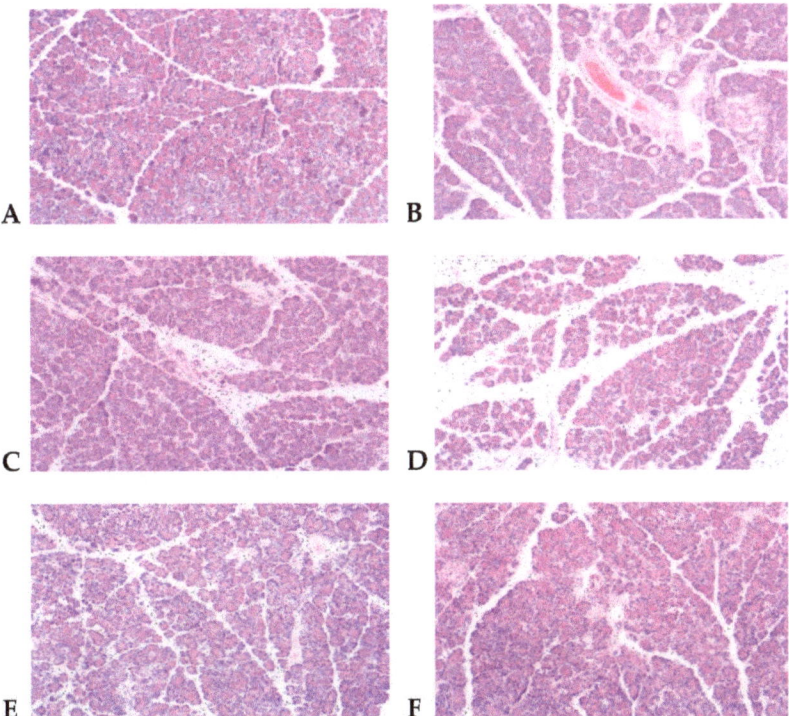

Figure 3. HE staining observation of rat pancreas pathology (×100). (**A**) CON; (**B**) MOD; (**C**) SV; (**D**) L-RS; (**E**) M-RS; (**F**) H-RS.

2.7. Effects of Kidney Bean RS on Serum DAO, DLA, ET, and sIgA Levels in Rats

The levels of serum DAO, DLA, and ET in the MOD were higher than those in the CON, while the levels of sIgA decreased ($p < 0.05$) (Table 3). After gavage treatment, the serum DAO, DLA, and ET levels of rats in each RS group, respectively, decreased to varying degrees (9.56–24.13%, 5.16–26.70%, and 5.71–23.66%) compared with the MOD. The levels of sIgA increased by 2.52–19.25%. Except for the DLA in L-RS, the levels of DAO, DLA, ET, and sIgA in the serum were significantly different from the MOD ($p < 0.05$).

2.8. Effect of Kidney Bean RS on Histopathological Sections of Rat Small Intestinal Mucosa

In the MOD, the intestinal mucosa was damaged, and the intestinal villi were lodging and falling off. The height of the intestinal villi was reduced and uneven. The thickness of the intestinal mucosa muscle layer reduced, the gap was wide, and the crypt depth increased. After oral administration, the intestinal morphology of the rats in each group was relatively complete and hierarchical. Compared with the MOD, the intestinal mucosa infiltration was seen in the L-RS, and the inflammatory cell infiltration in the M-RS and H-RS was relatively less. The height of the villi in the small intestine and the thickness of the muscle layer of rats increased, and the depth of the crypt decreased, in each RS group. The intestinal mucosa was tightly connected, and the columnar epithelial cells were relatively complete. It was shown that kidney bean RS could prevent small intestinal mucosal mechanical barrier damage in HLAP rats to a certain extent (Figure 4).

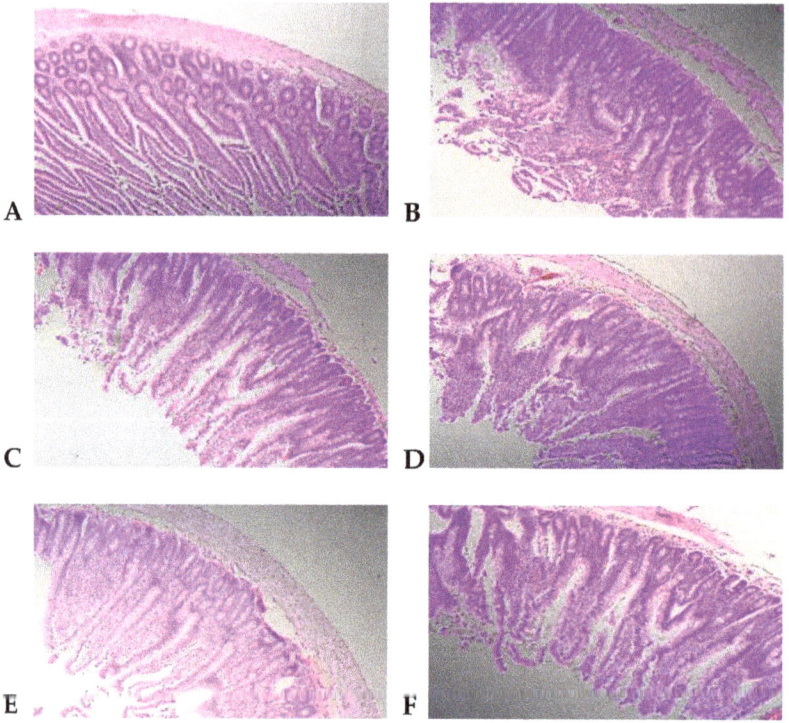

Figure 4. HE staining of the small intestine in each group of rats (×100). (**A**) CON; (**B**) MOD; (**C**) SV; (**D**) L-RS; (**E**) M-RS; (**F**) H-RS.

2.9. Intestinal Functional Protein mRNA Level and Protein Expression

Gene expression levels of rats in the CON were set as controls, and mRNA expression levels of ZO-1, occludin, CRAMP, and DEFB1 in the MOD rats reduced significantly ($p < 0.0001$). The RS was gavaged before caerulein induced HLAP. Compared with the MOD, the mRNA expression levels of tight junction protein ZO-1, occludin, and antibacterial peptides CRAMP and DEFB1 of rats in each RS group increased by 26.43–60.07%, 229.98–279.90%, 75.80–111.20%, and 77.86–109.07%, respectively (Figure 5).

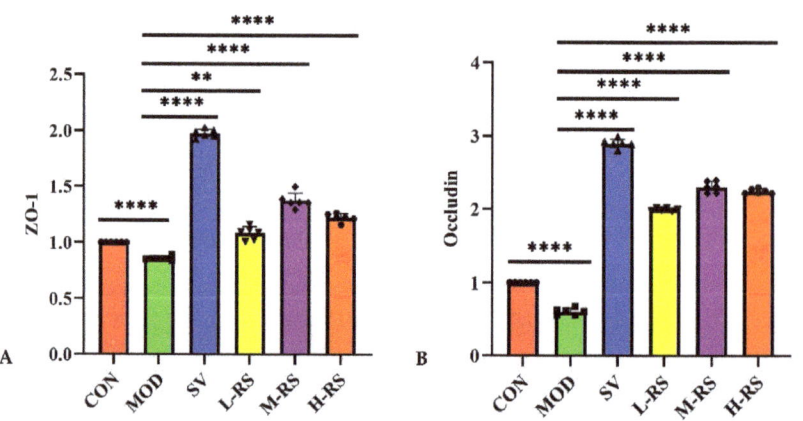

Figure 5. *Cont.*

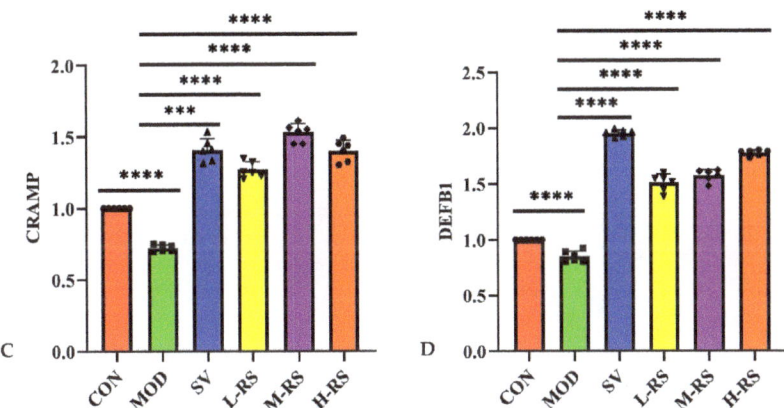

Figure 5. Effect of kidney bean RS on ZO-1 (**A**), occludin (**B**), CRAMP (**C**), and DEFB1 (**D**) gene expression in rat intestine (** $p < 0.01$; *** $p < 0.001$; **** $p < 0.0001$ versus MOD, and the points of different shapes represent the actual value of each sample, $n = 6$).

3. Discussion

This study explored the protective effect of RS on acute pancreatitis and related intestinal barrier damage in rats with hyperlipidemia. The core index results of acute pancreatitis in each group of rats confirmed that HLAP raised the serum AMLY and LIPA levels to 3 times higher than normal levels. Meanwhile, it damaged the pancreatic tissues and led to edema of the pancreatic organs. However, dietary supplementation of RS effectively reduced the levels of AMLY and LIPA in the serum of rats among groups, reduced the degree of pancreatic edema, and inhibited the activity of MPO in pancreatic tissue. Combined with pathological sections, the pancreas of the rats in the RS groups had no obvious lipid vacuoles, and the inflammatory infiltration was reduced, which further proved that the intake of RS can effectively protect the pancreas from damage. Researchers have found that dietary fiber intake correlates with pancreatic enzyme activity. Dietary fiber can inhibit trypsinogen digestive activity through a certain pathway, preventing pancreatic tissue damage [19,20]. At present, some relevant studies have proved that HLAP is a fatal disease that can lead to organ failure [21,22]. Its premature activation of pancreatic proteases in pancreatic acinar cells causes the process of self-digestion of the pancreas parallel to the immune response, and hyperlipidemia aggravates the immune response, which promotes the strength of the immune response to determine systemic complications and disease severity [23]. Therefore, we explored the protective effect of RS on HLAP rats and intestinal barrier damage from the following two aspects: (1) its regulation of blood lipid levels; (2) its secretion of inflammatory cytokines during the course of HLAP inhibition.

As the acinar body is damaged during its own digestion process, it excessively stimulates pancreatic parenchymal inflammation, which implies that the pathogenesis of HLAP-induced neutrophils and macrophages is infiltrated by inflammatory mediators to release inflammatory markers such as IL-6, IL-1β, TNF-α, etc. These factors activate NF-κB and then produce more inflammatory mediators, causing more serious MODS and SIRS [24,25]. Among them, TNF-α is the first pro-inflammatory cytokine released in the course of HLAP, which could stimulate the secretion of IL-6 and IL-1β. IL-6 and IL-1β are genera of ILs, which promote the development and differentiation of specific lymphocytes and activate inflammatory mediators, which in turn stimulate other immune cells to secrete TNF-α, and the three interact with each other in a vicious cycle, causing severe damage to the pancreas and related organ tissues [26,27]. The results show that RS reduces the inflammatory cytokine cycle and has a certain immune regulating effect. Nilsson et al. found that consumption of RS by the body increased glucagon-like peptide (GLP-2), affected the activity of intestinal hormones, and thus suppressed the increase in inflammatory markers [28]. In this experiment, the levels of serum TNF-α, IL-6, and IL-1β in rats in each RS group after gavage were lower than those in the MOD, especially in the M-RS and H-RS, which were significantly reduced. It was proved that the intake of RS could effectively inhibit the inflammatory response of HLAP rats, and that the medium and high doses of RS had a stronger inhibitory effect on the secretion of inflammatory cytokines.

From a pathological point of view, researchers generally believe that HLAP is caused by too many free fatty acids decomposed by triglycerides, which disrupt many acinar functions like exocytosis enzyme activation [29,30]. It causes premature activation of trypsinogen and self-digestion, thereby damaging pancreatic tissue. Zhang found that TG levels are directly related to the severity of AP disease. In the onset of AP, rapidly reducing TG levels can block the progress of SIRS in time [31]. The results of this experiment confirm that the body weight and blood lipid concentration of rats in the RS groups were significantly lower than in the MOD. Related studies have demonstrated that dietary intake of RS reduces the levels of serum TC and TG by stimulating the secretion of bile acids and promoting cholesterol and lipid excretion, and stimulates the synthesis of gastrointestinal peptide hormones to avoid diet-induced obesity and hyperlipidemia [32,33].

In addition, the intestinal barrier serves as an effective defense system for the body, separating the internal and external environments to prevent the passage of potentially harmful substances [34]. Researchers have found that the closure of the intestinal mucosa requires a complete epithelial structure and an efficient apical cell connection complex [35,36]. However, under the dual influence of HL changing intestinal flora and AP releasing pro-inflammatory factors, intestinal mucosal permeability improves to break the closed state. This phenomenon further stimulates the body to release inflammatory cytokines, which exacerbate the HLAP condition to the level of damage to the intestinal functional barrier. In this study, different doses of RS reduced the levels of serum D-LA, DAO, and ET; increased the content of sIgA; and significantly increased the expression level of mRNA of ZO-1, occludin, CRAMP, and DEFB1 in the intestinal mucosa of rats in each RS group. Related pathological section results also characterize the protective effect of RS on the microstructure of the small intestinal mucosa. In summary, it was proved that RS could up-regulate the mRNA expression levels of the main tight junction proteins and antimicrobial peptides in the intestine, and promote the secretion of immune antibodies, thereby improving the permeability of the intestinal mucosa and maintaining the integrity of the intestinal barrier.

4. Materials and Methods

4.1. Materials

Purple speckled kidney beans (*Phaseolus vulgaris*) were provided by the reclamation area of Heihe (China). Male Wistar rats (SPF, Permit number: SCXK (Liao) 2020-0001) were purchased from Changsheng Biotechnology Co., Ltd. (Benxi, China). Caerulein (≥99%) was obtained from Med Chem Express (Monmouth Junction, NJ, USA). Octreotide acetate injection was from Novartis Pharmaceutical Co., Ltd. (Beijing, China). Assay kits for tumor necrosis factor-α (TNF-α), interleukine-1 beta (IL-1β), interleukine-6 (IL-6), diamine oxidase (DAO), endotoxin (ET), and D-lactic acid (DLA) were provided by Calvin Biotechnology Co., Ltd. (Suzhou, China). Assay kits for triglyceride (TG), total cholesterol (TC), and myeloperoxidase (MPO) were purchased from the Institute of Bioengineering (Nanjing, China). The NCBI primer designing tool designed RT-PCR amplification primers, and the primer sequence was synthesized by Shenggong Biotechnology Co., Ltd. (Shanghai, China).

4.2. Preparation and Purification of Kidney Bean RS

The kidney bean starch was weighed and mixed in an Erlenmeyer flask to a starch suspension with a mass fraction of 20%. The temperature of the ultrasonic-microwave collaborative reactor was set at 40 °C, the microwave power at 300 W, and coordinated processing took place for 20 min. After autoclaving at 121 °C for 30 min, we then added pullulanase at a ratio of 9 ASPU/g dry basis, which was shaken at 55 °C for 10 h, and we deactivated the enzyme by boiling in water for 10 min. It was then refrigerated at 4 °C for 24 h. The aged gelatin starch was dried in an oven at 50 °C for 12 h. It was crushed at high speed through an 80-mesh sieve to obtain kidney bean RS. Excessive thermostable α-amylase and saccharification enzymes were utilized to purify RS.

4.3. Structure of RS

The structural characteristics of kidney bean RS were determined using a scanning electron microscope, X-ray diffractometer (Empyrean, PANalytical B.V., Almelo, Overijssel, Netherlands), and laser particle size analyzer (Bettersize2000, Better, China) [37–39]. Meanwhile, a GI20 automatic in vitro simulated digestion system (NutriScan GI20, Next Instruments, Condell Park, NSW, Australia) was applied to determine the glycemic index of RS.

4.4. Animal Feeding and Administration

Normal male Wistar rats were housed in rat cages for animal experiments and maintained under a simulated ambient environment that was kept at a cycle of illumination for 12 h, a constant temperature (24 ± 2) °C, and a relative humidity (50 ± 5)%. The experimental rats were randomly divided into six groups based on body weight after seven days of adaptive feeding ($n = 6$), including control group (CON), model group (MOD), simvastatin group (SV), and RS groups (H-RS, M-RS, and L-RS). In the initial stage of the experiment, except for the CON rats the other groups of rats were fed a hyperlipidemia diet (Table 4) for 4 weeks to induce a hyperlipidemia model. During this period, it was judged whether the model was successful based on the serum TG and TC concentration of rats (the model was established when the contents of serum TG and TC in each group increased to 2~3 times those of the CON). The acute pancreatitis model was induced by intraperitoneal injection of caerulein (50 µg·kg^{-1}) for 8 consecutive times [40–42]. The SIM was pre-fed with simvastatin (10 mg·kg^{-1}) and injected with octreotide acetate in the tail vein after intraperitoneal injection (10 mg·kg^{-1}). RS groups were pre-fed with different doses of RS (5.4 g·kg^{-1}, 2.7 g·kg^{-1}, 1.35 g·kg^{-1}) for 6 weeks. Rats were humanely sacrificed to obtain blood and related issues. All experiments were carried out according to the P.R. China legislation and we strictly followed the international guidelines of the Institutional Animal Care and Use Committee. The rat handling and the experimental protocol were also performed according to the Directive 2010/63/EU on the protection of animals used for scientific purposes (European Parliament and Council, Directive 2010/63/EU of 22 September 2010 on the protection of animals used for scientific purposes. Off. J. Eur. Union 2010, L276, 33–79).

Table 4. The feed composition of rats in each group.

Ingredients (%)	CON	MOD	SV	L-RS	M-RS	H-RS
Corn starch	73.5	53.51	53.51	53.51	53.51	53.51
Wheat bran	20	14.6	14.6	14.6	14.6	14.6
Fish meal	5	3.6	3.6	3.6	3.6	3.6
Farina	1	0.73	0.73	0.73	0.73	0.73
Sodium salt	0.5	0.56	0.56	0.56	0.56	0.56
Cholesterol	/	1.2	1.2	1.2	1.2	1.2
Egg yolk powder	/	5.8	5.8	5.8	5.8	5.8
Sucrose	/	10	10	10	10	10
Lard	/	10	10	10	10	10

4.5. Serum Analysis

Blood was collected from the portal vein and centrifuged at 4000× g (4 °C, 20 min). The serum was collected and dispensed by using sterile pipets; one part was used to determine serum lipase and amylase, and the others were stored at −80 °C until analysis. Serum samples were equilibrated at 4 °C for 10 min and then amylase and lipase levels were analyzed using a Vet Test 8008 automatic biochemical analyzer (IDEXX Bioresearch, Westbrook, ME, USA, USA). Results are expressed in international units per liter (U/L).

Serum was removed from −80 °C and thawed to analyze related indicators. The expression and activity of TNF-α, IL-1β, IL-6, ET, sIgA, DAO, and DLA were determined according to kit instructions.

4.6. Histology Analysis

Portions of fresh, independent pancreatic tissue in each group of rats were randomly collected and weighed. The degree of pancreatic edema was assessed by the ratio of the mass lost from a completely dried pancreas to the wet weight of the pancreas [43]. Pancreatic tissue was formulated into a homogenized medium according to experimental methods and evaluated based on the mechanism by which myeloperoxidase reduces hydrogen peroxide.

Then, the fresh pancreatic and colon tissue were fixed in a volume fraction of 10% formaldehyde solution and dehydrated with gradient ethanol, soaked to transparent xylene, waxed, and embedded to prepare hematoxylin-eosin-stained sections. The pathological changes in pancreatic tissue were observed under a ICC50 light microscope (LEICA, Wetzlar, Germany).

4.7. RNA Extraction and Quantitative PCR

RNA was separated using a UNIQ-10 column Trizol total RNA extraction kit (SK1321). The frozen colon tissue was homogenized by a homogenizer at a ratio of 0.5 mL Trizol per 15 to 25 mg of tissue. The homogenate was allowed to stand at room temperature for 5~10 min to separate the nucleoprotein from the nucleic acid completely, and then the total RNA was obtained according to the instructions. The cDNA was synthesized by reverse transcription kit manipulation instructions for reverse transcription. Specific primers were designed based on the gene sequence (Table 5) [43]. β-actin was used as the reference gene for quantitative detection by real-time fluorescent quantitative PCR, and the cycling conditions refer to Piorkowski, with appropriate adjustments as follows: the thermocycler profile was 3 min at 95 °C, followed by 5 s at 95 °C and 30 s at 60 °C for 45 cycles [44]. Finally, the data were analyzed using the $2^{-\Delta\Delta Ct}$ method.

Table 5. The gene sequence of each primer.

Gene Name	Forward Primer	Temperature/°C	Length/bp
ZO-1	GAGATGAGCGGGCTACCTTA	57.2	210
	GCTGTGGAGACTGTGTGGAAT	57.0	
Occludin	TGGGACAGAGCCTATGGAAC	57.2	197
	ACCAAGGAAGCGATGAAGC	57.5	
CRAMP	TCACTGTCACTGCTATTGCTCCT	59.3	208
	CCTTCACTCGGAACCTCACAT	58.9	
DEFB1	CTGCCCATCTCATACCAAACTAC	58.4	112
	TTTACAATCCCTTGCTGTCCTT	58.5	

4.8. Statistical Analysis

All data are expressed as the mean ± standard error of the mean. Differences between data were analyzed using an independent sample T-test and one-way ANOVA. Statistical analyses were conducted using SPSS 20.0 statistical software with $p < 0.05$ considered statistically significant.

5. Conclusions

In conclusion, the internal composition and stable internal structure of RS reduced the area of contact between RS and digestive enzymes, and increased its residence time in the intestine. These factors contributed to it performing its physiological functions. In addition, the RS reduced serum lipid content, improved pancreatic tissue edema and inflammatory damage, increased the expression of tight junction proteins, and reduced intestinal mucosal permeability. This study demonstrates that RS has the potential to prevent HLAP, and it will hopefully provide a foundation for the design and development of kidney-bean-based functional foods.

Author Contributions: Conceptualization, Y.W. and D.Z.; data curation, W.P. and X.Z.; formal analysis, Y.W.; methodology, Z.Z. and S.L.; resources, B.L. and D.Z.; software, W.S. and N.Z.; validation, S.L.; writing—original draft, Z.Z.; writing—review and editing, Z.Z. All authors have read and agreed to the published version of the manuscript.

Funding: This work was supported by the National Key Research and Development Project of China (2018YFE0206300 and 2020YFD1001400), the Research Team Project of the Natural Science Foundation of Heilongjiang Province, China (TD2020C003), and the Basic Scientific Research Operating Expenses Projects of the Provincial University of Heilongjiang Province (ZDZX202106).

Institutional Review Board Statement: Not applicable.

Informed Consent Statement: Not applicable.

Data Availability Statement: The data presented in this study are available on request from the corresponding author.

Conflicts of Interest: The authors declare no conflict of interest.

Sample Availability: Samples of the compounds are not available from the first author.

References

1. Lankisch, P.G.; Apte, M.; Banks, P.A. Acute pancreatitis. *Lancet* **2015**, *386*, 85–96. [CrossRef]
2. Yang, A.L.; Mcnabb, B.J. Hypertriglyceridemia and acute pancreatitis. *Pancreatology* **2020**, *20*, 795–800. [CrossRef] [PubMed]
3. Reed, J.M.; Hogan, B.M.; Nasser-Ghodsi, N.; Loftus, C.G. Management of Hypertriglyceridemia-Induced Acute Pancreatitis in a Nondiabetic Patient. *Mayo Clin. Proc. Innov. Qual. Outcomes* **2021**, *5*, 520–524. [CrossRef] [PubMed]
4. Prasada, R.; Muktesh, G.; Samanta, J.; Sarma, P.; Singh, S.; Arora, S.K.; Dhaka, N.; Ramachandran, R.; Gupta, V.; Sinha, S.K.; et al. Natural history and profile of selective cytokines in patients of acute pancreatitis with acute kidney injury. *Cytokine* **2020**, *133*, 155177. [CrossRef]
5. Rychter, J.W.; Minnen, L.; Verheem, A.; Timmerman, H.M.; Rijkers, G.T.; Schipper, M.E.I.; Gooszen, H.G.; Akkermans, L.M.A.; Kroese, A.B.A. Pretreatment but not treatment with probiotics abolishes mouse intestinal barrier dysfunction in acute pancreatitis. *Surgery* **2009**, *145*, 157–167. [CrossRef]
6. Tan, C.C.; Ling, Z.X.; Huang, Y.; Cao, Y.D.; Liu, Q.; Cai, T.; Yuan, H.; Liu, C.J.; Li, Y.F.; Xu, K.Q. Dysbiosis of Intestinal Microbiota Associated with Inflammation Involved in the Progression of Acute Pancreatitis. *Pancreas* **2014**, *44*, 868–875. [CrossRef]
7. Hong, W.D.; Zimmer, V.; Basharat, Z.; Zippi, M.; Stock, S.; Geng, W.J.; Bao, X.Q.; Dong, J.F.; Pan, J.Y.; Zhou, M.T. Association of total cholesterol with severe acute pancreatitis: A U-shaped relationship. *Clin. Nutr.* **2019**, *39*, 250–257. [CrossRef]
8. Pang, L.; Yang, Z.H.; Wu, Y.; Yin, R.X.; Liao, Y.H.; Wang, J.W.; Gao, B.X.; Zhang, L.X. The prevalence, awareness, treatment and control of dyslipidemia among adults in China. *Atherosclerosis* **2016**, *248*, 2–9. [CrossRef]
9. Zou, J.; Xu, M.J.; Wen, L.R.; Yang, B. Structure and physicochemical properties of native starch and resistant starch in Chinese yam (*Dioscorea opposita* Thunb.). *Carbohydr. Polym.* **2020**, *237*, 116188. [CrossRef]
10. Grooms, K.N.; Ommerborn, M.J.; Pham, D.Q.; Djousse, L.; Clark, C.R. Dietary Fiber Intake and Cardiometabolic Risks among US Adults, NHANES 1999–2010. *Am. J. Med.* **2013**, *126*, 1059–1067. [CrossRef]
11. Sun, H.; Ma, X.H.; Zhang, S.Q.; Zhao, D.; Liu, X. Resistant starch produces antidiabetic effects by enhancing glucose metabolism and ameliorating pancreatic dysfunction in type 2 diabetic rats. *Int. J. Biol. Macromol.* **2017**, *110*, 276–284. [CrossRef] [PubMed]
12. Li, T.; Teng, H.; An, F.P.; Huang, Q.; Chen, L.; Song, H.B. The beneficial effects of purple yam (*Dioscorea alata* L.) resistant starch on hyperlipidemia in high-fat-fed hamsters. *Food Funct.* **2019**, *10*, 2642–2650. [CrossRef] [PubMed]
13. Yuan, H.C.; Meng, Y.; Bai, H.; Shen, D.Q.; Wan, B.C.; Chen, L.Y. Meta-analysis indicates that resistant starch lowers serum total cholesterol and low-density cholesterol. *Nutr. Res.* **2018**, *54*, 1–11. [CrossRef] [PubMed]
14. Zhang, C.J.; Dong, L.; Wu, J.H.; Qiao, S.Y.; Xu, W.J.; Ma, S.S.; Zhao, B.S.; Wang, X.Y. Intervention of resistant starch 3 on type 2 diabetes mellitus and its mechanism based on urine metabonomics by liquid chromatography-tandem mass spectrometry. *Biomed. Pharmacother.* **2020**, *128*, 110350. [CrossRef]
15. Rosado, C.P.; Rosa, V.H.C.; Martins, B.C.; Soares, A.C.; Santos, I.B.; Monteiro, E.B.; Moura-Nunes, N.; Costa, C.A.D.; Mulder, A.D.R.P.; Daleprane, J.B. Resistant starch from green banana (*Musa* sp.) attenuates non-alcoholic fat liver accumulation and increases short-chain fatty acids production in high-fat diet-induced obesity in mice. *Int. J. Biol. Macromol.* **2020**, *145*, 1066–1072. [CrossRef]
16. Zheng, B.; Wang, T.T.; Wang, H.W.; Chen, L.; Zhou, Z.K. Studies on nutritional intervention of rice starch- oleic acid complex (resistant starch type V) in rats fed by high-fat diet. *Carbohydr. Polym.* **2020**, *246*, 116637. [CrossRef]

17. Kung, B.; Turgeon, S.L.; Rioux, L.E.; Anderson, G.H.; Wright, A.J.; Goff, H.D. Correlating in vitro digestion viscosities and bioaccessible nutrients of milks containing enhanced protein concentration and normal or modified protein ratio to human trials. *Food Funct.* **2019**, *10*, 7687–7696. [CrossRef]
18. Reddy, C.K.; Haripriya, S.; Mohamed, A.N.; Suriya, M. Preparation and characterization of resistant starch III from elephant foot yam (*Amorphophallus paeoniifolius*) starch. *Food Chem.* **2014**, *155*, 38–44. [CrossRef]
19. Dutta, S.K.; Hlasko, J. Dietary fiber in pancreatic disease: Effect of high fiber diet on fat malabsorption in pancreatic insufficiency and in vitro study of the interaction of dietary fiber with pancreatic enzymes. *Am. J. Clin. Nutr.* **1985**, *41*, 517–525. [CrossRef]
20. Dhital, S.; Gidley, M.J.; Warren, F.J. Inhibition of α-amylase activity by cellulose: Kinetic analysis and nutritional implications. *Carbohydr. Polym.* **2015**, *123*, 305–312. [CrossRef]
21. Wang, W.J.; He, B.; Shi, W.; Liang, X.L.; Ma, J.C.; Shan, Z.X.; Hu, Z.Y.; Danesh, F.R. Deletion of scavenger receptor A protects mice from progressive nephropathy independent of lipid control during diet-induced hyperlipidemia. *Kidney Int.* **2012**, *81*, 1002–1014. [CrossRef] [PubMed]
22. Khlifi, R.; Lahmar, A.; Dhaouefi, Z.; Kalboussi, Z.; Maatouk, M.; Kilani-Jaziri, S.; Ghedira, K.; Chekir-Ghedira, L. Assessment of hypolipidemic, anti-inflammatory and antioxidant properties of medicinal plant Erica multiflora in triton WR-1339-induced hyperlipidemia and liver function repair in rats: A comparison with fenofibrate-ScienceDirect. *Regul. Toxicol. Pharmacol.* **2019**, *107*, 104404. [CrossRef] [PubMed]
23. Yuan, L.; Tang, M.D.; Huang, L.; Gao, Y.; Li, X.L. Risk Factors of Hyperglycemia in Patients after a First Episode of Acute Pancreatitis: A Retrospective Cohort. *Pancreas* **2016**, *46*, 209–218. [CrossRef] [PubMed]
24. Chen, C.R.; Wang, J.J.; Chen, J.F.; Zhou, L.L.; Wang, H.; Chen, J.N.; Xu, Z.H.; Zhu, S.J.; Liu, W.; Yu, R.J.; et al. Morusin alleviates mycoplasma pneumonia via the inhibition of Wnt/β-catenin and NF-κB signaling. *Biosci. Rep.* **2019**, *39*, BSR20190190. [CrossRef] [PubMed]
25. Yang, B.; Yan, P.; Yang, G.Z.; Cao, H.L.; Wang, F.; Li, B. Triptolide reduces ischemia/reperfusion injury in rats and H9C2 cells via inhibition of NF-κB, ROS and the ERK1/2 pathway. *Int. J. Mol. Med.* **2018**, *41*, 3127–3136. [CrossRef] [PubMed]
26. Landahl, P.; Ansari, D.; Andersson, R. Severe Acute Pancreatitis: Gut Barrier Failure, Systemic Inflammatory Response, Acute Lung Injury, and the Role of the Mesenteric Lymph. *Surg. Infect.* **2015**, *16*, 651–656. [CrossRef]
27. Perriot, S.; Mathias, A.; Perriard, G.; Canales, M.; Jonkmans, N.; Merienne, N.; Meunier, C.; Kassar, L.E.; Perrier, A.L.; Laplaud, D.A.; et al. Human Induced Pluripotent Stem Cell-Derived Astrocytes Are Differentially Activated by Multiple Sclerosis-Associated Cytokines. *Stem Cell Rep.* **2018**, *11*, 1199–1210. [CrossRef]
28. Nilsson, A.C.; Johansson-Boll, E.V.; Björck, I.M.E. Increased gut hormones and insulin sensitivity index following a 3-d intervention with a barley kernel-based product: A randomised cross-over study in healthy middle-aged subjects. *Br. J. Nutr.* **2015**, *114*, 899–907. [CrossRef]
29. Dias, G.; Queiroz, L.R.; Máfia, R.; Stephane, B.; Medeiros, G.; Martins, V. Understanding the International Consensus for Acute Pancreatitis: Classification of Atlanta 2012. *ABCD Arq. Bras. Cir. Dig.* **2016**, *29*, 206–210. [CrossRef]
30. Campos, T.D.; Parreira, J.G.; Assef, J.C.; Rizoli, S.; Nascimento, B.; Fraga, G.P. Classification of acute pancreatitis. *Rev. Do Colégio Bras. De Cir.* **2013**, *40*, 164–168. [CrossRef]
31. Zhang, Q.Y.; Qin, M.B.; Liang, Z.H.; Huang, H.L.; Tang, Y.F.; Qin, L.Y.; Wei, Z.P.; Xu, M.T.; Tang, G.D. The relationship between serum triglyceride levels and acute pancreatitis in an animal model and a 14-year retrospective clinical study. *Lipids Health Dis.* **2019**, *18*, 183–192. [CrossRef] [PubMed]
32. Shang, W.T.; Xu, S.; Zhou, Z.K.; Wang, J.X.; Strappe, P.; Blanchard, C. Studies on the unique properties of resistant starch and chito-oligosaccharide complexes for reducing high-fat diet-induced obesity and dyslipidemia in rats. *J. Funct. Foods* **2017**, *38*, 20–27. [CrossRef]
33. Keenan, M.J.; Zhou, J.; McCutcheon, K.L.; Raggio, A.M.; Bateman, H.G.; Todd, E.; Jones, C.K.; Tulley, R.T.; Melton, S.; Martin, R.J.; et al. Effects of resistant starch, a non-digestible fermentable fiber, on reducing body fat. *Obesity* **2006**, *14*, 1523–1534. [CrossRef] [PubMed]
34. Lin, S.B.; Han, Y.R.; Jenkin, K.; Lee, S.J.; Sasaki, M.; Klapproth, J.M.; He, P.J.; Yun, C.C. Lysophosphatidic Acid Receptor 1 Is Important for Intestinal Epithelial Barrier Function and Susceptibility to Colitis. *Am. J. Pathol.* **2018**, *188*, 353–366. [CrossRef]
35. Ge, P.; Luo, Y.L.; Okoye, C.S.; Chen, H.Y.; Liu, J.Y.; Zhang, G.X.; Xu, C.M.; Chen, H.L. Intestinal barrier damage, systemic inflammatory response syndrome, and acute lung injury: A troublesome trio for acute pancreatitis. *Biomed. Pharmacother.* **2020**, *132*, 110770. [CrossRef]
36. Yang, Y.T.; Chen, L.; Tian, Y.; Ye, J.; Liu, Y.; Song, L.L.; Pan, Q.; He, Y.H.; Chen, W.S.; Peng, Z.H.; et al. Numb modulates the paracellular permeability of intestinal epithelial cells through regulating apical junctional complex assembly and myosin light chain phosphorylation. *Exp. Cell Res.* **2013**, *319*, 3214–3225. [CrossRef]
37. Ma, Z.; Yin, X.X.; Hu, X.Z.; Li, X.P.; Liu, L.; Boye, J.I. Structural characterization of resistant starch isolated from Laird lentils (*Lens culinaris*) seeds subjected to different processing treatments. *Food Chem.* **2018**, *263*, 163–170. [CrossRef]
38. Li, S.L.; Ward, R.; Gao, Q.Y. Effect of heat-moisture treatment on the formation and physicochemical properties of resistant starch from mung bean (*Phaseolus radiatus*) starch. *Food Hydrocoll.* **2011**, *25*, 1702–1709. [CrossRef]
39. Reddy, C.K.; Suriya, M.; Haripriya, S. Physico-chemical and functional properties of Resistant starch prepared from red kidney beans (*Phaseolus vulgaris* L.) starch by enzymatic method. *Carbohydr. Polym.* **2013**, *95*, 220–226. [CrossRef]

40. Barreto, S.G.; Carati, C.J.; Schloithe, A.C.; Mathison, R.; Davison, J.S.; Toouli, J.; Saccone, G.T.P. The efficacy of combining feG and galantide in mild caerulein-induced acute pancreatitis in mice. *Peptides* **2010**, *31*, 1076–1082. [CrossRef]
41. Xue, D.B.; Zhang, W.H.; Zhang, Y.M.; Wang, H.Y.; Zheng, B.; Shi, X.Y. Adjusting effects of baicalin for nuclear factor-κB and tumor necrosis factor-α on rats with caerulein-induced acute pancreatitis. *Mediat. Inflamm.* **2006**, *5*, e26295. [CrossRef] [PubMed]
42. Lerch, M.M.; Gorelick, F.S. Models of Acute and Chronic Pancreatitis. *Gastroenterology* **2013**, *144*, 1180–1193. [CrossRef] [PubMed]
43. Sun, Y.J.; He, Y.; Wang, F.; Zhang, H.; Vos, P.D.; Sun, J. Low-methoxyl lemon pectin attenuates inflammatory responses and improves intestinal barrier integrity in caerulein-induced experimental acute pancreatitis. *Mol. Nutr. Food Res.* **2017**, *61*, 1600885. [CrossRef] [PubMed]
44. Piorkowski, G.; Baronti, C.; Lamballerie, X.D.; Fabritus, L.D.; Bichaud, L.; Pastorino, B.A.; Bessaud, M. Development of generic Taqman PCR and RT-PCR assays for the detection of DNA and mRNA of β-actin-encoding sequences in a wide range of animal species. *J. Virol. Methods* **2014**, *202*, 101–105. [CrossRef] [PubMed]

Review

The Iodine/Iodide/Starch Supramolecular Complex

Szilard Pesek and Radu Silaghi-Dumitrescu *

Department of Chemistry, Faculty of Chemistry and Chemical Engineering, Babeș-Bolyai University, 11 Arany Janos Street, 400028 Cluj-Napoca, Romania; szilard.pesek@ubbcluj.ro
* Correspondence: radu.silaghi@ubbcluj.ro; Tel.: +40-264-593833

Abstract: The nature of the blue color in the iodine–starch reaction (or, in most cases, iodine–iodide-starch reaction, i.e., I_2 as well as I^- are typically present) has for decades elicited debate. The intensity of the color suggests a clear charge-transfer nature of the band at ~600 nm, and there is consensus regarding the fact that the hydrophobic interior of the amylose helix is the location where iodine binds. Three types of possible sources of charge transfer have been proposed: (1) chains of neutral I_2 molecules, (2) chains of poly-iodine anions (complicated by the complex speciation of the I_2-I^- mixture), or (3) mixtures of I_2 molecules and iodide or polyiodide anions. An extended literature review of the topic is provided here. According to the most recent data, the best candidate for the "blue complex" is an I_2-I_5^--I_2 unit, which is expected to occur in a repetitive manner inside the amylose helix.

Keywords: amylose; iodine; iodide; starch; UV-vis; DFT

Citation: Pesek, S.; Silaghi-Dumitrescu, R. The Iodine/Iodide/Starch Supramolecular Complex. *Molecules* 2024, 29, 641. https://doi.org/10.3390/molecules29030641

Academic Editors: Litao Tong and Lili Wang

Received: 15 August 2023
Revised: 21 January 2024
Accepted: 25 January 2024
Published: 30 January 2024

Copyright: © 2024 by the authors. Licensee MDPI, Basel, Switzerland. This article is an open access article distributed under the terms and conditions of the Creative Commons Attribution (CC BY) license (https://creativecommons.org/licenses/by/4.0/).

1. Introduction

Starch is a mixture of amylose and amylopectin in a ratio of ~20–30% to 70–80%, both of which are polymers of glucopyranose. Amylose is a linear polymer typically containing 300–3000 (or sometimes much more) monomeric units that are interconnected via α(1→4) glycosidic bonds (cf. Figure 1) [1]. Amylopectin has a similar structure, additionally featuring ramifications via α(1→6) bonds. A small number of ramifications (of the same type as amylose) are in fact also found in amylose. The average number of branches in amylose molecules is size-dependent, ranging from zero to 2–4 branches per molecule. The variations have been generally attributed to differences in chain length, with the observation that most branching occurs at the early stage of amylose synthesis [2]. Both polymers are sparsely soluble in water, with amylose more so than amylopectin. Inside living organisms, the enzyme amylase is mainly responsible for hydrolyzing starch. Dextrins, which are polysaccharides with a low degree of polymerization, are produced by partial hydrolysis of starch; with complete hydrolysis, glucose is formed [3].

Amylose can exist in a disordered amorphous conformation or in two types of helical forms (Figure 2). The first type is a double helix with itself (including forms A and B). The second type is the V form, which consists of a single helix; this is the structure typically discussed in biochemistry textbooks. The V form features an internal cavity that is large enough to accommodate hydrophobic guest molecules such as iodine, fatty acids, or small aromatic compounds. This property has in fact been recently exploited in extensive attempts to design smart drug delivery systems, with amylose/starch serving as the host for inclusion complexes with various molecules of therapeutic potential/use [4–6].

In 1930, Katz studied the aging and cooking of bread, and with the help of X-ray diffraction, in the analyzed powder he found, in addition to the type A and B models known until then from native starch, he found another crystalline form of starch, which he called model V. To describe the contribution of this form, he used the German word "Verkleitsterung," which means gelatinization [7]. He found the same V-type pattern when he prepared pasta, which was precipitated with alcohols [8]. Bear identified other different V-type patterns depending on the precipitation agent used [9]. Native starch was fractionated by Meyer et al., separating amylose from amylopectin using hot water [10].

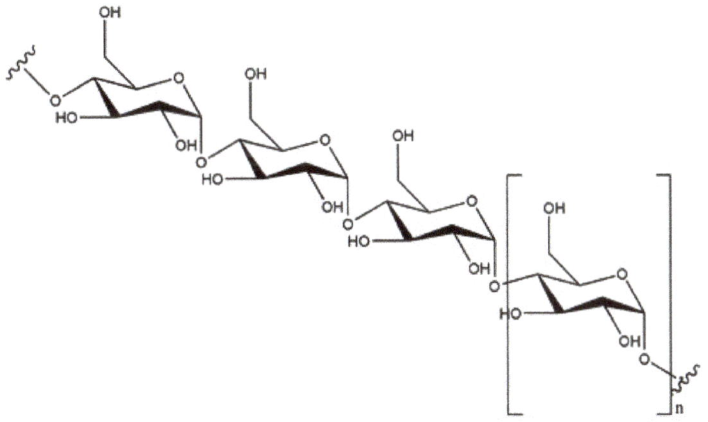

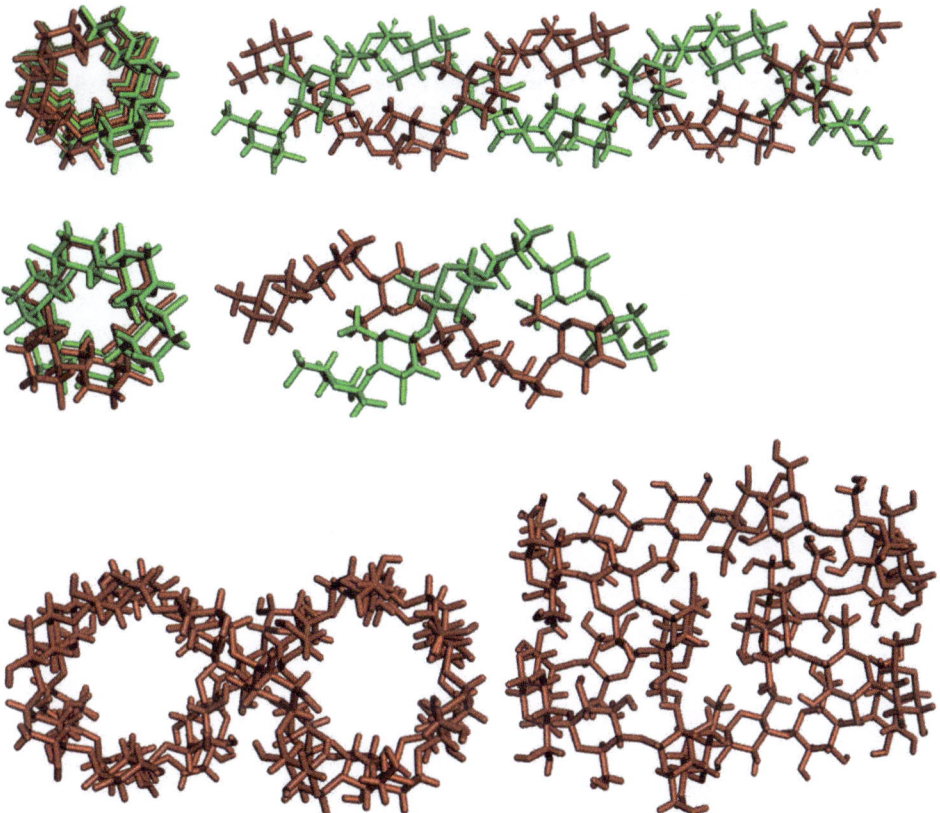

Figure 1. Molecular structure of amylose.

Figure 2. Molecular models of helices of A-type, B-type, and V-type cycloamylose.

A- and B-amylose both form parallel-stranded double helices of 6 × 2 glucoses per turn and right-handed [11–16] or left-handed [17] turns. Structurally, these two structures differ from each other only in their packing arrangements and water contents. They also differ in biological locations, with A preferentially in grains and B in tubers [11]. A polymorph C of amylose has also been described, consisting of a mixture of A- and B- unit cells [18].

The V form is also found naturally and is structured as a single left-handed helix with six glucose units per turn and a step height of 7.91 to 8.17 Å [11–13]. The V form can be isolated through its precipitation from aqueous solution using alcohols, ketones, fatty acids, iodine, or salts that form inclusion complexes (thus behaving similarly to cyclodextrins, α-cyclodextrin, or cyclohexaamylose) [19,20]. The glucose units in V-amylose (as in cyclodextrins) are all in the *syn* orientation. This entails hydrogen bonds between the secondary hydroxyl groups $O(3)_n \cdots O(2)_{n+1}$, as well as $O(6)_n \cdots O(2)_{n+6}$ hydrogen bonds between turns [11–13].

Variants of the V-type structure of amylose have been described, each marked with a subscript typically indicating the number of glucose units per shift. The most common variant is the V_6 form [14], though V_8 and V_7 have also been described. The latter would provide even more space for the guest molecule to bind [21]. Depending on the solvent, slightly different crystallographic structures are formed. The crystals grown with water as the solvent, dubbed amylose V_h, feature six glucose units in the unit cell, aligned as a single strand. Within the helix, there are three water molecules, statistically positioned on six positions. In the spaces between helices, there is one water molecule each, on one of three possible positions. The helices point upwards and downwards in a random fashion [14].

Amylose V is the allomorph known for its deep blue complexes with iodine. The iodine molecules are trapped in the channels within the helices, where molecules of the solvent can also be present. As detailed in the following sections, extensive research has been devoted to understanding the structural features of the amylose–iodine complexes, especially as the blue complexes generally form only when iodide anions are present alongside iodine. This implies the presence of polyiodide species, of which tri-iodide is the most often invoked. Figure 3 illustrates the proposed structures of amylose–I_2 and amylose–tri-iodide species based on electronic structure calculations. Figure 4 illustrates the representative structures proposed for the iodine/iodide arrangement inside the amylose helix [22].

Cyclodextrins are glucose oligomers featuring the same type of glycosidic bonds as starch, and they can be obtained enzymatically. Depending on their size, some cyclodextrins can arrange themselves in a circular fashion, appearing similar to single turns of the amylose helix, and then forming dimers which generate large cylinders, similar to the amylose helix in its overall structure.

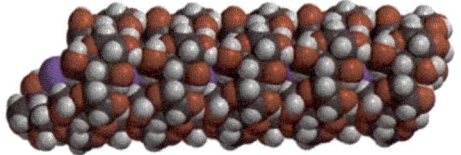  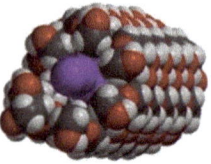

Figure 3. *Cont.*

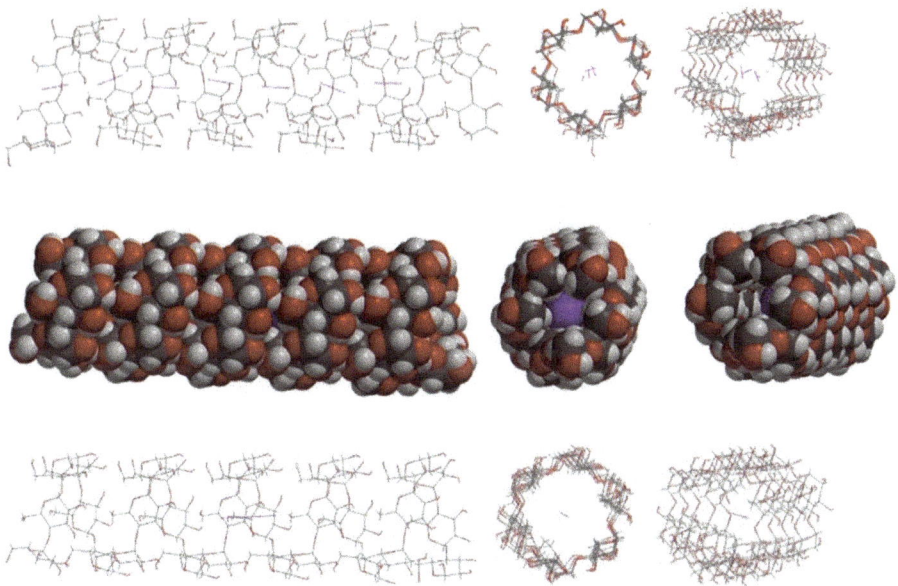

Figure 3. Structures of AM1-optimized amylose models: A-I$_2$ and A-I$_3^-$ [22].

$$\text{I-I}---\text{I-I}---\text{I-I}---\text{I-I}---\text{I-I}---\text{I-I}---\text{I-I}$$
$$\text{[I-I-I]}^- ---\text{[I-I-I]}^- ---\text{[I-I-I]}^- ---\text{[I-I-I]}^- ---\text{[I-I-I]}^-$$
$$\text{[I-I-I-I-I]}^- ---\text{[I-I-I-I-I]}^- ---\text{[I-I-I-I-I]}^- ---\text{[I-I-I-I-I]}^- ---\text{[I-I-I-I-I]}^-$$
$$\text{[I-I-I]}^- ---\text{I-I}---\text{[I-I-I]}^- ---\text{I-I}---\text{[I-I-I]}^- ---\text{I-I}---\text{[I-I-I]}^- ---\text{I-I}---\text{[I-I-I]}^-$$
$$\text{[I-I-I-I-I]}^- ---\text{I-I}---\text{[I-I-I-I-I]}^- ---\text{I-I}---\text{[I-I-I-I-I]}^- ---\text{I-I}---\text{[I-I-I-I-I]}^- ---\text{I-I}---\text{[I-I-I-I-I]}^-$$

Figure 4. Structures of I$_2$, I$_3^-$, I$_5^-$, I$_3^-$-I$_2$-I$_3^-$, I$_5^-$-I$_2$-I$_5^-$, proposed as possible sources of the blue color in the iodine–iodide–amylose complexes.

2. General Considerations on the Reaction of Starch with Iodine

Colin and Claubry discovered in 1814 the reaction between starch and iodine (or, more specifically, with iodide–iodine solutions, as molecular I$_2$ is otherwise very insoluble in water in the absence of I$^-$ ions). Since then, the reaction has been found both in organic chemistry classes in school and in qualitative and quantitative analysis courses. Over time, many experiments have shown that the starch–iodine complex shows absorption at ~600 nm: a strong dark blue color [23]. A more detailed description of the reaction was provided by H. H. Landolt in 1886 [24]. The color is mainly due to the complex of iodine with the amylose complex that absorbs at ~620 nm. The affinity of iodine for amylopectin is distinctly smaller (~20 times), and the resulting complex is reddish-violet, with a maximum at ~540 nm. In 1948, Gilbert and Marriott [25] showed that at higher concentrations of iodide, the ratio of iodide ions to iodine molecules increases to at least one, leading to a purple tint in the blue complex.

The nature of the amylose–iodine complex has been debated for many decades, especially in terms of the stoichiometry and charge of the poly-iodine substructures within the helix (see, e.g., Figure 4). It is now generally accepted that iodide ions are also required in this process. Thus, the iodine atoms would align inside the amylose helix as a mixture/combination of I$_2$ and I$^-$ [21,22,24–27] with an unusual metallic-like structure [23]. This mixture would not entail isolated I$^-$ ions, but rather I$_n^-$ polyiodides (n = 3, 5, 9...) [28–32]. The wavelength of the absorbance maximum in the amylose–iodine complex is known to depend on the chain length. Thus, glucose chains of 4 to 6 units yield no color, while those 8 to 12 yield a red color with a peak at 520 nm reminiscent of amylopectin. Longer chains progressively show a bathochromic shift until a length of 30 to 35 units, when the blue color is reached with a peak at 600–620 nm. Very similar spectra are also obtained through the synthetic action of potato phosphorylase on starch when chains of 50–150 units are obtained.

This relationship between chain length and iodine color has also been applied to branched polysaccharides. Comparing the spectra of the iodine/iodide complexes with dextrin, amylopectin, glycogen, or various synthetic oligosaccharides, estimations of the extent of helical portions available for iodine binding in amylopectin and glycogen were formulated, at 8–18 glucose residues. In line with these observations, hydrolysis of amylose by α-amylase or under acid catalysis gradually changes the absorption maximum from blue to red. On the other hand, hydrolysis with β-amylase leads to a hypochromic but no hypsochromic shift. This can be explained by the fact that β-amylase remains bound to its substrate until it completely degrades it, rather than gradually degrading all the polymer chains at the same time. In this way, little or no red-colored intermediate dextrins remain in the mixture [33]. Variations in color for iodine–amylopectin complexes can be observed due to the difference in structures [34], branched chain length [35], and branching points in amylopectin [36]. If the degree of polymerization in amylopectin entails 15 glucose units, complexation of iodine is not observed at any temperature. If it exceeds 30, then iodine binds to amylopectin at either 1.5 °C or 20 °C [37]. Mould and Synge [38] used potentiometric and spectrophotometric titrations for complexation of iodine/iodide with the products of enzymatic hydrolysis of amylose, i.e., dextrins with different molecular masses. Dextrins of less than 10 glucose units were essentially unreactive, those of 10–25 units were orange, those 25–40 were red, and those of 40–130 were blue. Ono [39] showed that the λ_{max} of the amylose–iodine complex shifted to shorter wavelengths with increasing iodide concentrations; this was interpreted as evidence of the breaking the polyiodine chains by the permeated iodide ions.

3. Dependence on the Nature of the Organic (Bio)Polymer

In addition to starch, there is a long list of natural polymers that afford colored complexes with iodine. These include chitosan, glycogen, silk, wool, albumin, cellulose, xylan, and natural rubber. A large number of synthetic polymers have also been described to react with iodine. Examples include poly(vinyl alcohol) (PVA), poly(vinyl pyrrolidone) (PVP), nylons, poly(Schiff base)s, polyaniline, and unsaturated polyhydrocarbons (carbon nanotubes, fullerenes C_{60}/C_{70}, and polyacetylene) [40]. It is important to note that most of these polymers do not feature helical structures of the type seen in amylose; binding of the iodine on the outside of the organic polymer, or between polymer chains, is likely occurring in such cases.

Differences in the blue color were reported depending on the size (and, implicitly, biological source) of the polymer, be it amylose or related poly and oligo saccharides or other organic polymers [34,36,40–54]. Yu et al. [55] showed that when complexation occurs between I_2/KI and potato amylose, the speciation of iodide varies. Data from Raman and UV-visible spectroscopy were interpreted as evidence that the primary forms of iodine were the monoanions I_3^- and I_5^-, but larger units were also present such as I_{93}^-, I_{113}^-, I_{133}^-, and I_{153}^- (with bands at 460–480, 560–590, 660–700, and 710–740 nm, respectively), with higher iodide concentrations expectedly favoring shorter polyiodide chains. By adding an iodine–potassium iodide solution to a solution of cellulose, Abe [56] obtained an intense blue solution. At 80 °C, the color disappeared, but upon cooling, it reappeared. Takahashi [57] reported a dark purple adduct upon treating chitin with I_2/KI solution for 24 h at room temperature. Depending on the amount reactant ratios, the iodine content of the adduct was in the range of 9–20%, with one molecular iodine per 6.4 chitin residues. Yajima et al. [58] prepared a purple complex (maximum at 550 nm) by freezing a mixture of chitosan and I_2/KI solution at −20 °C and then thawing it at 4 °C.

Glycogen, as also discussed above, yields a reddish-brown complex with iodine. The largest maximum is at 395 nm, [59] but as shown by Kumari et al. [60] and Lecker et al. [61], a series of UV absorption bands is also present at 408, 453, 496, 560, 650, and 698 nm. The bands at 408, 453, 560, and 650 nm were comparable to those of amylopectin (412, 458, 550, and 640 nm), and the bands at 496 and 698 nm were assigned to an I_4 species. The bands at 453 and 560 nm were more visible at higher iodine concentrations.

Stromeyer, in 1815, mentions that wool and silk (i.e., the protein structure therein) give yellow colors on exposure to iodine [37]. Amyloid peptides also yield a blue color with molecular iodine. In the presence of sulfuric acid, this shifts to blue–violet, cf. Aterman [62]. Dzwolak [63] described the formation of an insulin–amyloid complex in the presence of I_2/KI that is stable up to 90 °C, interpreted as an inclusion complex between the amyloid fibrils and iodine.

Pritchard and Serra [64] reported that reaction of poly(vinyl acetate) (PVAc) with molecular iodine in methanol in the presence of aqueous KI yields a deep-red precipitate (darker at higher iodine concentrations), interpreted as an I_2-PVAc adduct. Two UV-vis absorbance bands were reported for the complex at 520 and 510 nm. Hughes et al. [65] noted that the red color of the I_2-PVAc complex is independent of the method through which the polymer is prepared. Solutions of the PVA–iodine complex (I_2-PVA) feature intense 600–620 nm, 650–680 nm, 480–500 nm, and 350 nm [66] bands depending on the conditions. The complex of I_2 with poly(vinylpyrrolidone) (povidone or PVP) has a maximum at 361 nm, with the iodine molecule reported to be attached to the PVP matrix through non-covalent interactions with the carbonyl groups. A poly(N-methyl-4-vinyl pyridinium) triiodide was reported to produce a dark-brown powder with molecular iodine in a water–alcohol solution, with bands at 295 nm, 367 nm, and 460 nm [67].

Charge transfer complexes of iodine with ferrocenyl-bearing Schiff bases have also been described [68]. Liu et al. [69] showed that the brown poly(ferrocenyl-Schiff) bases turn black after treatment with iodine in acetone. In the infrared (IR) spectra, iodine binding resulted in a decrease of the ~1610 cm^{-1} signal of the Schiff base, accompanied by the appearance of a new wide absorption band at 450–480 nm with a long tail up to ~900 nm in the UV-vis spectra.

Shirakawa [70] examined the reaction of polyacetylene (PAC) with iodine. Strong IR bands were noted at 870 cm^{-1} and 1390 cm^{-1}. In the UV-vis spectrum analysis, bands at 365 nm and 502 nm were noted to be suggestive of the presence of I_3^- and I_2, respectively. A band at 280 nm characteristic of unreacted PAC was noted to decrease in intensity upon doping.

Treatment of poly(p-phenylene vinylene) or phen(p-phenyvinylene) (PPV) with iodine vapor showed a significant effect on its luminescence [71].

In principle, due to the differences in reactivity between amylose and amylopectin, the iodine reaction can be used for assessing the amylose content of starch. However, a number of other methods may be more appropriate and accurate, such as polarimetry, anthrone, FTIR, penetrometry, Luff school, gravimetry, X-ray diffraction (XRD), and size exclusion chromatography (SEC) techniques including multi-angle laser light scattering–differential refractive index detection (SEC-MALS-DRI) [72–77].

Three bands in the UV-visible–NIR spectra, at 688 nm, 1724 nm, and 2292 nm, were interpreted as evidence for a PPV–iodine complex. Polybutadiene, poly(cis-isoprene), and their copolymers were reported to display an intense absorption band at 305 nm in the UV spectrum (with some variations depending on the solvent) upon treatment with iodine [78]. Sreeja et al. [79] found that acrylonitrile butadiene rubber developed two new broad bands appearing between 300 nm and 500 nm, while the 262 nm band attributed to the C=C bonds decreased. Poly(β-pinene) treated with iodine vapor was shown by Vippa et al. [80] to produce bands at 310 nm and 400 nm in the UV-vis spectrum, with the latter interpreted as evidence for a charge transfer between the double bond and I_2.

4. Geometrical Data

Crystalline units in starch and in its partial degradation products have long been explored [81–83]. Transmission electron microscopy [84], light microscopy [85], the use of atomic force microscopy (AFM) [86–92], and X-ray and neutron small angle scattering [93] have allowed for some partial visualization of repeating units in amylose. Microstructures were observed, which were called nodules, initially with diameter variations between 150 and 300 nm [84], and also smaller particles of 20–50 nm [85,87,88,93] (e.g., 130–250 nm for pea starch [92], 20–50 nm for potato starch [90], or 10–30 nm for corn granules [84]). Using combined methods of X-ray diffraction and stereochemical packing analysis of the amylose–iodine complex, Bluhm and Zugenmaier [36] reported that the iodine atoms aligned almost linearly in the center of the amylose chain. Eight hydration water molecules were found per unit cell, located in the amylose helices. The amylose left-handed helix was reported to have an outer diameter of ~13 Å and a pitch of 8 Å (six 1,4-glucose units per pitch), hosting an internal cavity of ~5 Å [28,30–32,34,36,94–97]. Some statistical disorder was noted within the polyiodide chain, with an average iodine–iodine distance of ~3.1 Å. This value is larger than the 2.67 Å in molecular I_2 as well as than the 2.90 Å in ionic I_3^-, but it is distinctly shorter than the 4.3 Å sum of van der Waals radii for two iodine atoms. With the help of the electron–gas theory, the 620 nm maximum in the UV-vis spectrum was suggested to be due to 14 iodine atoms when assuming an equidistant distance of 3.1 Å. [98,99] However, the X-ray diffraction data so far cannot distinguish between a structure where the iodine atoms are placed equidistantly at 3.1 Å vs. a structure consisting of I_2 and/or I_n^- units (with internal I-I distances lower than 3.1 Å) placed in non-covalent contact with each other. For instance, I_n^- units with n = 5–7 (and assuming the same internal I-I distance as in I_3^-) placed at 4.3 Å from each other would allow an average distance of 3.1 Å across the crystal structure, in line with experiments. If assuming the intermolecular distance to be shorter than the sum of van der Waals radii (as expected in complexes displaying charge-transfer bands), the experimentally observed iodine–iodine distance of 3.1 Å could be reached with, e.g., I_3^- units placed at 3.7 Å from each other or I_5^- units placed at 4.1 Å. The same average distance could be obtained by a chain of I_2 molecules placed within 3.6 Å of each other.

5. The I_2-Only Hypothesis

Many textbooks still list the amylose structure as featuring I_2 units aligned inside the helical channel. The driving force for this arrangement would be the hydrophobic character of the I_2 molecules, which thus escape the solvent to align in a more hydrophobic environment inside the helix [100]. The non-covalent interactions between I_2 would then facilitate charge-transfer bands to appear, resulting in the intense blue color. NMR and UV-vis studies have shown that I^- ions are not involved in the iodine–amylose helix, but they help to dissolve iodine in water [55]. Simulations of the UV-vis spectra using semiempirical INDO configuration interactions have been reported to support a $(C_6H_{10}O_5)_{16.5}\cdots(I_2)_3$ stoichiometry for the amylose–iodine complex, refuting instead I_n^- (n = 3, 5, or 7) as possible candidates [101].

Water molecules were reported to modulate the structure of the iodine–amylose complex [102]. It was also noted that neither the dimensions of the amylose helix nor the rigidity or the helical vs. random coil secondary structure within the amylose polymer were affected by iodine binding—all of which suggest no strong specific inter-molecular bonding between amylose and iodine. This is also consistent with a simple inclusion complex involving neutral molecules. In contrast, the charge in I_n^- may have been expected to induce local changes in the neighboring polysaccharide units [103–110].

According to semiempirical calculations, the binding of water or of interspersed water and iodine molecules inside the amylose helix results in slight steric distortions of the polymer [22]. The water molecules were found to adhere closely to the walls of the internal channel of amylose and to exclude I_2 from interactions, suggesting that I_2 molecules alone would be unable to dislocate water molecules from inside the amylose helix. In fact, I_2 solutions (in alcohol, so as to not require iodide for solubilization) were only found to be effective in binding to starch at high temperatures, while at lower temperatures, vapor I_2 easily adsorbs/binds to solid/dry amylose in the absence of water [111].

6. Poly-Iodine Anions as Candidates

It is in fact now generally accepted that the formation of the intense blue color in the amylose–iodine reaction requires iodide ions not only simply as an accessory that allows for the solubilization of I_2 in water, but also because a combination of I_2 and I^- chains is present inside the amylose helix, most likely involving I_n^- units [30–36]. The (poly) anionic character of these guest ligands may be taken, as discussed above, to be at odds with the hydrophobic nature of the interior cavity of the helix, especially as no counter ions have been discussed or are presumed to be present throughout the cavity. This issue has been addressed by proposing a structure consisting of alternating tri-iodide units and I_2 molecules, where the more hydrophilic ends/entrances of the helix would remain unoccupied by iodine [112]. To explain the fact that at room temperature, molecular iodine can bind to amylose in the solid state but not in solution, it was noted that the secondary structure of amylose, including the internal diameter of the helix, varies depending on the environment (e.g., solvent, ionic strength, pH, temperature, surfactants) [49,50,54,113–117]. Moreover, the solid-state amylose–I_2 complex is not stable in water [118]. To complicate matters, it has also been reported that in the iodine–amylose complex (somewhat similarly to the above-discussed cases of other organic polymers), large amounts of iodine can associate outside the helical cavity, with inter- and intra-chain associations also important [29]. Theoretical studies of the complex have been interpreted as evidence that at low temperatures, an I_6 structure dominates, while at higher temperatures, nonlinear geometries also appear. However, the experimental data have led to descriptions of the amylose-bound iodine chains as featuring 3–4 to 14–15 atoms, and as high as 160 atoms [31–36]. Potentiometric titrations at low iodide concentrations have been interpreted as evidence for a $3/2\ I_2/I^-$ ratio (hence, formally I_8^{2-}). However, as the structure appears to be consistently affected by the concentrations of the reactants (especially the iodine–iodide ratios, as followed, e.g., by UV-vis and circular dichroism titrations), I_4^-, I_7^-, I_9^-, I_6^{2-}, I_8^{2-}, I_{10}^{2-}, I_4^{2-}, I_6^-, and I_{24}^{2-} structures have also been proposed [29,30,32,50,52,53,119–122]. Rawlings and Schneider [123], using the statistics of binding isotherms of I_2/I_3^- and the large variation of the value of the term of the ratio (R), determined the intrinsic binding constant of I_3^- to amylose, whose value was much higher than that of I_2. Likewise, the mixed binding energy between I_2-I_3^- exceeded those between I_2-I_2 and I_3^--I_3^- species [29]. Statements [97] which assume that I^- is not required sometimes do not take into account that it is formed through the hydrolysis of I_2 itself. Also, it is known that by adding an acid to an aqueous solution of iodine, the blue color is suppressed [26,124].

$$I_2 + H_2O = I^- + H^+ + HOI$$

$$3I_2 + 3H_2O = IO_3^- + 5^- + 6H^+$$

$$I_2 + H_2O = H_2OI+ + I^-$$

In addition to classical short units such as I_3^- ions and iodine molecules, the presence of I_5^- [125] and I_7^- [126] ions was also suggested. These proposals can be based on the fact that the composition of the unit is variable and influenced by the degree of polymerization of the amylose and the concentration of the iodide ion. At the end of the amylose chain, there are 7–8 glucose residues, which do not participate in iodine binding [118].

Stopped-flow UV-vis and circular dichroism (CD) kinetics have revealed that shorter chains of iodine enter the amylose helix very fast (less than 1 millisecond) and then rearrange rapidly inside the helix without further contributions from the excess iodine/iodide in the solution [28,127]. The optical rotatory dispersion (ORD) spectrum of the amylose–iodine complex was noted to change even when no changes in the UV-vis spectrum were observed; this was interpreted as evidence for complex dynamics of the helix during which the length of the poly-iodine chains remained unaffected [128]. In the structure of the synthetic complex (benzamide)$_z$ H$^+$I$_3^-$, a poly-I$_3$ structure was reported, and a range of kinetic, spectroscopic (UV-vis, CD, Raman, X-ray absorption), and thermodynamic data have also been interpreted to support such a structure in the amylose–iodine complex [122,129–131]. On the other hand, the X-ray diffraction data on the amylose–iodine complex have been interpreted to be inconsistent with arrangements consisting of only single I$_2$ or only I$_3^-$ [35,132,133]. An I$_5^-$ structure was proposed for the amylose–iodine complex based on the similarities between the Raman and Mössbauer spectra and those of polycrystalline (trimesic acid···H$_2$O)$_{10}$H$^+$I$_5^-$ [129,134]. Of the main bands in the Raman spectra of starch–iodine complexes, at 27, 55, 109, and 160 cm^{-1}, three were assigned to I$_5^-$ as the dominant species, while the 109 cm^{-1} band was assigned to I$_3^-$ being present as a minor species or impurity [135–139]. Further resonance Raman, ^{129}I Mössbauer spectroscopy [129,140] and X-ray diffraction [132,141,142] data have supported I$_5^-$ or I$_2$·I$_3^-$ as the dominant species in the amylose–iodine complex. Equilibrium studies in solution have confirmed that I$_5^-$ is present as a free species, with I$_4^{2-}$ and I$_6^{2-}$ also being present at higher I$_2$ or I$^-$ concentrations [143–145]. Teitelbaum et al [125] presented evidence that the I$_5^-$ ion is present in the helix. Part of this explanation may be that the formation of the I$_5^-$ ion in the solid complex is produced by the hydrolysis or alcoholysis of iodine, and the amylose studied in this case was freed from water and alcohols [118,125]. Analyzing the absorption spectra of an aqueous solution of iodine at pH = 4.8 in acetate buffer, three peaks were found at 286, 350, and 460 nm [146]. The peaks at 286 and 350 nm were attributed to the presence of I$_3^-$ ions. If amylose was added to this solution, the peak at 286 nm disappeared, the peak at 350 nm had a slight enhancement, and the peak at 460 nm shifted to 620–650 nm. By adding iodic acid to amylose, the blue color does not appear; this indicates the necessity of the presence of anions for the formation of the colored complex [26,124,147].

7. Structural Role of the Solvent

Benesi and Hildebrand [148] in 1949 showed that iodine is pink–red in benzene, purple in CCl$_4$ (similarly to vapor iodine), and reddish-brown in alcohol—all of which were interpreted as evidence for charge transfer complexes. Bernal-Uruchurtu et al. [149] provided a microscopic explanation of the interaction between iodine and water. Kereev and Shnyrev studied iodine, iodide [150], and triiodide in water and noted a band at 203 nm, which they attributed to iodine, and a band at 461 nm, which was then assigned to a H$_2$O-I$_2$ charge–transfer complex. To complicate such studies, I$_3^-$ is formed by the oxidation of water with iodine. Analyzing iodine solution in water, the absorption bands of I$_3^-$ are at 284 nm and 351 nm. Density functional theory (DFT) calculations showed that two I$_2$ molecules can bind to two O lone pairs in H$_2$O in a nearly tetrahedral geometry with a stabilization energy of 8.3 kcal/mol. This complex was described as involving a charge-transfer character, with a band in the UV-vis spectrum due to the excitation of electrons from the lone pair of oxygen of the molecular complex at 202 nm, where the molecular orbitals of iodine are destabilized and those of water are stabilized. When iodine is bound to water, the lone pairs on oxygen become equivalent due to hybridization [151]. A small effect of water molecules on the structure of the iodine–amylose complex was described, as previously discussed [102].

8. The I$_5^-$-I$_2$ Hypothesis

Recent DFT calculations on iodine/iodide chains have been interpreted as evidence that without iodide, the blue color cannot be formed in the starch–iodine system [22]. These simulations propose that the nature of the complex consists of alternating sets of I$_2$ and I$_x^-$ units, where the nature of the charge transfer bands responsible for the blue color involves transfer from the I$_x^-$ σ* orbitals (HOMO) to the I$_2$ σ* LUMO orbitals (cf. Figure 5). By analyzing the TD-DFT-computed (time-dependent density functional) UV-vis spectra of various candidates (I_2 chains vs. mixtures of I_2 and I_x^- with various values of x) and cross-checking with DFT geometry optimizations, a unit of I_2-I_5^--I_2, in a repetitive manner within the amylose helix was the only structure that would fit the experimental data. [22] Poly-I_2 structures were shown to be responsible for the enhanced blue color under certain conditions (e.g., consistent with the experimental observations on dry/solid amylose). Based on semiempirical calculations, poly-I_n^- structures were found to be unlikely to exist inside the amylose helix (as no distinct local energy minima were identified for such arrangements). Moreover, TD-DFT simulations of the UV-vis spectra of such chains were found to be less consistent with the experiments compared to I_2/I_n^- pairs. Charge transfer bands from the occupied I_n^- ($n > 3$) σ* to the empty I_2 σ* orbital were instead found to be reasonably responsible for the blue color. Of these, the I_2-I_5^--I_2 trimeric assemblies (i.e., n = 5) were the smallest units that represented the local minima in DFT geometry optimizations. These DFT-optimized units remarkably showed average iodine–iodine distances essentially identical to the 3.1 Å value seen experimentally in the iodine-amylose complex. The distinct charge-transfer character of the UV-vis bands (cf. Figure 5) also brings about a strong dependence on the dielectric constant in the region ε~1–30, which in turn was proposed to explain at least part of the dependence of the UV-vis properties of the amylose–iodine/iodine complexes on various external factors that may subtly affect amylose architecture and hence exposure of the interior cavity to solvent (e.g., temperature, other solutes, solvents, chain length) [22].

HOMO-4   LUMO

Figure 5. Molecular orbitals in the I_2-I_5^--I_2 that are proposed to be responsible for the ~600 nm band in the iodine/iodide–amylose complex. Iodine atoms are shown in violet, HOMO/LUMO lobes are shown in red/green [22].

9. Conclusions

The iodine–starch (or, more specifically, iodine–amylose) reaction has a two-century history and a wide range of practical applications. Similar reactions occur with other organic polymers. The nature of the reaction is generally accepted to entail the alignment of iodine atoms inside the amylose helix. However, the structural details are still a source of confusion in many current reference sources, with alternative explanations given such as poly-I_2 (chain of neutral iodine molecules), poly-I_3^- (chain of I_3 anions), poly-I_x (chains of anionic structures of various lengths), or mixtures of I_2 and I_x^-. The most recent data suggest that the best explanation is a (probably repetitive) I_2-I_5^--I_2 unit.

Author Contributions: Conceptualization, R.S.-D.; formal analysis, S.P. and R.S.-D.; resources, R.S.-D.; writing—original draft preparation, S.P.; writing—review and editing, R.S.-D. and S.P. All authors have read and agreed to the published version of the manuscript.

Funding: Funding from the Romanian Ministry of Education and Research (projects PN-III-P4-ID-PCCF-2016-0142 and PN-III-P1-1.1-PD-2021-0279) is gratefully acknowledged.

Institutional Review Board Statement: Not applicable.

Informed Consent Statement: Not applicable.

Data Availability Statement: Not applicable.

Conflicts of Interest: The authors declare no conflicts of interest.

References

1. Nelson, D.L.; Cox, M.M. Principles of Biochemistry. In *Principles of Biochemistry*, 5th ed.; W.H. Freeman & Company: New York, NY, USA, 2008.
2. Wang, K.; Vilaplana, F.; Wu, A.; Hasjim, J.; Gilbert, R.G. The Size Dependence of the Average Number of Branches in Amylose. *Carbohydr. Polym.* **2019**, *223*, 115134. [CrossRef]
3. Green, M.M.; Blankenhorn, G.; Hart, H. Which Starch Fraction Is Water-Soluble, Amylose or Amylopectin? *J. Chem. Educ.* **1975**, *52*, 729. [CrossRef]
4. Nouri, A.; Khoee, S. Preparation of Amylose-Poly(Methyl Methacrylate) Inclusion Complex as a Smart Nanocarrier with Switchable Surface Hydrophilicity. *Carbohydr. Polym.* **2020**, *246*, 116662. [CrossRef] [PubMed]
5. Wang, Y.; Zhang, Y.; Guan, L.; Wang, S.; Zhang, J.; Tan, L.; Kong, L.; Zhang, H. Lipophilization and Amylose Inclusion Complexation Enhance the Stability and Release of Catechin. *Carbohydr. Polym.* **2021**, *269*, 118251. [CrossRef]
6. Prasher, P.; Fatima, R.; Sharma, M. Therapeutic Delivery with V-Amylose. *Drug Dev. Res.* **2021**, *82*, 727–729. [CrossRef]
7. Katz, J.R. Abhandlungen Zur Physikalischen Chemie Der Stärke Und Der Brotbereitung. *Z. Phys. Chem.* **1930**, *150A*, 37–59. [CrossRef]
8. Katz, J.R.; Derksen, J.C. Abhandlungen Zur Physikalischen Chemie Der Stärke Und Der Brotbereitung. *Z. Phys. Chem.* **1933**, *167A*, 129–136. [CrossRef]
9. Bear, R.S. The Significance of the "V" X-Ray Diffraction Patterns of Starches. *J. Am. Chem. Soc.* **1942**, *64*, 1388–1392. [CrossRef]
10. Meyer, K.H.; Bernfeld, P.; Wolf, E. Recherches Sur l'amidon III. Fractionnement et Purification de l'amylose de Maïs Naturel. *Helv. Chim. Acta* **1940**, *23*, 854–864. [CrossRef]
11. Sarko, A.; Zugenmaier, P. Fiber Diffraction Methods. *Am. Chem. Soc.* **1980**, *141*, 459–482.
12. Rappenecker, G.; Zugenmaier, P. Detailed refinement of the crystal structure of Vh-amylose. *Carbohydr Res.* **1981**, *89*, 11–19. [CrossRef]
13. Murphy, V.G.; Zaslow, B.; French, A.D. The structure of V amylose dehydrate: A combined X-ray and stereochemical approach. *Biopolymers* **1975**, *14*, 1487–1501. [CrossRef]
14. Brisson, J.; Chanzy, H.; Winter, W.T. The crystal and molecular structure of VH amylose by electron diffraction analysis. *Int. J. Biol. Macromol.* **1991**, *13*, 31–39. [CrossRef]
15. Veregin, R.P.; Fyfe, C.A.; Marchessault, R.H. Investigation of the crystalline "V" amylose complexes by high-resolution carbon-13 CP/MAS NMR spectroscopy. *Macromolecules* **1987**, *20*, 3007–3012. [CrossRef]
16. Gidley, M.J.; Bodek, S.M. Carbon-13 CP/MAS NMR studies of amylose inclusion complexes, cyclodextrins, and the amorphous phase of starch granules: Relationships between glycosidic linkage conformation and solid-state carbon-13 chemical shifts. *J. Am. Chem. Soc.* **1988**, *110*, 3820–3829. [CrossRef]
17. Imberty, A.; Chanzy, H.; Pérez, S.; Buléon, A.; Tran, V. The Double-Helical Nature of the Crystalline Part of A-Starch. *J. Mol. Biol.* **1988**, *201*, 365–378. [CrossRef] [PubMed]
18. Sarko, A.; Wu, H.-C.H. The Crystal Structures of A-, B- and C-Polymorphs of Amylose and Starch. *Starch-Starke* **1978**, *30*, 73–78. [CrossRef]
19. Saenger, W. *Inclusion Compounds*; Atwood, J.L., Davies, J.E.D., MacNicol, D.D., Eds.; Academic Press: London, UK, 1984; Volume 2.
20. Harata, K. *Comprehensive Supramolecular Chemistry*; Atwood, J.L., Davies, J.E.D., MacNicol, D.D., Eds.; Pergamon: Oxford, UK, 1996; Volume 3.
21. Cohen, R.; Orlova, Y.; Kovalev, V.; Ungar, Y.; Shimoni, E. Structural and Functional Properties of Amylose Complexes with Genistein. *J. Agric. Food Chem.* **2008**, *56*, 4212–4218. [CrossRef]
22. Pesek, S.; Lehene, M.; Brânzanic, A.M.V.; Silaghi-Dumitrescu, R. On the Origin of the Blue Color in The Iodine/Iodide/Starch Supramolecular Complex. *Molecules* **2022**, *27*, 8974. [CrossRef] [PubMed]
23. Rani, A.; Ali, U. Degree-Based Topological Indices of Polysaccharides: Amylose and Blue Starch-Iodine Complex. *J. Chem.* **2021**, *2021*, 6652014. [CrossRef]
24. Landolt, H. Uber Die Zeitdauer Der Reaction Zwischen Jodsaure Und Schwefliger Saure. *Ber. Dtsch. Chem. Ges.* **1886**, *19*, 1317–1365. [CrossRef]
25. Gilbert, G.A.; Marriott, J.V.R. Starch-Iodine Complexes. Part I. *Trans. Faraday Soc.* **1948**, *44*, 84–93. [CrossRef]
26. Thoma, J.A.; French, D. The Starch-Iodine-Iodide Interaction. Part I. Spectrophotometric Investigations 1. *J. Am. Chem. Soc.* **1960**, *82*, 4144–4147. [CrossRef]
27. Stein, R.S.; Rundle, R.E. On the Nature of the Interaction between Starch and Iodine. *J. Chem. Phys.* **1948**, *16*, 195–207. [CrossRef]
28. Hiromi, K.; Shibaoka, T.; Ono, S. Kinetic Studies of Amylose-Iodine-Iodide Reaction by Stopped-Flow Method. *J. Biochem.* **1970**, *68*, 205–214. [CrossRef] [PubMed]
29. Yajima, H.; Nishimura, T.; Ishii, T.; Handa, T. Effect of Concentration of Iodide on the Bound Species of I_2/I^-_3 in the Amylose-Iodine Complex. *Carbohydr. Res.* **1987**, *163*, 155–167. [CrossRef]
30. Cronan, C.L.; Schneider, F.W. Cooperativity and Composition of the Linear Amylose-Iodine-Iodide Complex. *J. Phys. Chem.* **1969**, *73*, 3990–4004. [CrossRef]
31. Cramer, F.; Herbst, W. Die Lichtabsorption von Jodkettenmolekeln. *Naturwissenschaften* **1952**, *39*, 256. [CrossRef]
32. Bersohn, R.; Isenberg, I. Metallic Nature of the Starch-Iodine Complex. *J. Chem. Phys.* **1961**, *35*, 1640–1643. [CrossRef]

33. Rundle, R.E.; Baldwin, R.R. The Configuration of Starch and Starch-Iodine Complex. I. The Dichroism of Flow of Starch-Iodine Solutions. *J. Am. Chem. Soc.* **1943**, *65*, 554–558. [CrossRef]
34. Rundle, R.E. The Configuration of Starch in the Starch-Iodine Complex. V. Fourier Projections from X-ray Diagrams. *J. Am. Chem. Soc.* **1947**, *69*, 1769–1772. [CrossRef]
35. Noltemeyer, M.; Saenger, W. X-Ray Studies of Linear Polyiodide Chains in α-Cyclodextrin Channels and a Model for the Starch-Iodine Complex. *Nature* **1976**, *259*, 629–632. [CrossRef]
36. Bluhm, T.L.; Zugenmaier, P. Detailed Structure of the Vh-Amylose-Iodine Complex: A Liner Polyiodine Chain. *Carbohydr. Res.* **1981**, *89*, 1–10. [CrossRef]
37. Handa, T.; Yajima, H. Conformation of Amylose-Iodine-Iodide Complex in Aqueous Solution. *Biopolymers* **1981**, *20*, 2051–2072. [CrossRef]
38. Mould, D.L.; Synge, R.M. Separations of Polysaccharides Related to Starch by Electrokinetic Ultrafiltration in Collodion Membranes. *Biochem. J.* **1954**, *58*, 571–600. [CrossRef] [PubMed]
39. Ono, S.; Tsuchihashi, S.; Kuge, T. On the Starch-Iodine Complex. *J. Am. Chem. Soc.* **1953**, *75*, 3601–3602. [CrossRef]
40. Moulay, S. Molecular Iodine/Polymer Complexes. *J. Polym. Eng.* **2013**, *33*, 389–443. [CrossRef]
41. Séne, M.; Thévenot, C.; Prioul, J.L. Simultaneous Spectrophotometric Determination of Amylose and Amylopectin in Starch from Maize Kernel by Multi-Wavelength Analysis. *J. Cereal Sci.* **1997**, *26*, 211–221. [CrossRef]
42. Sashio, M.; Tanaka, M. Thermal Reaction of Poly(Vinyl Alcohol)-Iodine Complex Membranes. *J. Polym. Sci. Polym. Chem. Ed.* **1985**, *23*, 905–909. [CrossRef]
43. Dintzis, F.R. Instability of Solutions of Amylose-Iodine Complex in Concentrated Calcium Chloride. *Starch-Stärke* **1974**, *26*, 56–58. [CrossRef]
44. Tashiro, K.; Gakhutishvili, M. Crystal Structure of Cellulose-Iodine Complex. *Polymer* **2019**, *171*, 140–148. [CrossRef]
45. Konishi, T.; Tanaka, W.; Kawai, T.; Fujikawa, T. Iodine L-Edge XAFS Study of Linear Polyiodide Chains in Amylose and α-Cyclodextrin. *J. Synchrotron Radiat.* **2001**, *8*, 737–739. [CrossRef]
46. Knutson, C.A. Evaluation of Variations in Amylose–Iodine Absorbance Spectra. *Carbohydr. Polym.* **2000**, *42*, 65–72. [CrossRef]
47. Nishimura, T.; Yajima, H.; Ishii, T.; Endo, R. Effect of Molecular Weight of Amylose on the Iodine Coloring Species Responsible for the Optical Properties of Amylose-Iodine Complexes. *Kobunshi Ronbunshu* **1989**, *46*, 537–544. [CrossRef]
48. SenGupta, U.K.; MuKherjee, A.K.; SenGupta, K.K. Spectrophotometric Studies on Amylose-Iodine and Amylopectin-Iodine Complexes. *Kolloid-Z. Z. Polym.* **1966**, *208*, 32–34. [CrossRef]
49. Sakajiri, T.; Kikuchi, T.; Simon, I.; Uchida, K.; Yamamura, T.; Ishii, T.; Yajima, H. Molecular Dynamics Approach to Study the Discrepancies in the Thermal Behavior of Amylose and Chitosan Conformations. *J. Mol. Struct. THEOCHEM* **2006**, *764*, 133–140. [CrossRef]
50. Szejtli, J.; Augustat, S.; Richter, M. Molecular Configuration of Amylose and Its Complexes in Aqueous Solutions. Part III. Investigation of the DP Distribution of Helical Segments in Amylose-Iodine Complexes. *Biopolymers* **1967**, *5*, 17–26. [CrossRef]
51. McMullan, R.K.; Saenger, W.; Fayos, J.; Mootz, D. Topography of Cyclodextrin Inclusion Complexes. *Carbohydr. Res.* **1973**, *31*, 211–227. [CrossRef]
52. Baldwin, R.R.; Bear, R.S.; Rundle, R.E. The Relation of Starch—Iodine Absorption Spectra to the Structure of Starch and Starch Components 1. *J. Am. Chem. Soc.* **1944**, *66*, 111–115. [CrossRef]
53. Davis, H.; Khan, A. Determining the Chromophore in the Amylopectin-Iodine Complex by Theoretical and Experimental Studies. *J. Polym. Sci. A Polym. Chem.* **1994**, *32*, 2257–2265. [CrossRef]
54. Rendleman, J.A. The Reaction of Starch with Iodine Vapor. Determination of Iodide-Ion Content of Starch–Iodine Complexes. *Carbohydr. Polym.* **2003**, *51*, 191–202. [CrossRef]
55. Yu, X.; Houtman, C.; Atalla, R.H. The Complex of Amylose and Iodine. *Carbohydr. Res.* **1996**, *292*, 129–141. [CrossRef]
56. Abe, T. The Visible and Ultraviolet Absorption Spectra of Cellulose- and Amylose-Iodine Complexes. *Bull. Chem. Soc. Jpn.* **1958**, *31*, 661–662. [CrossRef]
57. Takahashi, Y.J. Binding Properties of Alginic Acid and Chitin. *Inclus. Phenom.* **1978**, *5*, 525–534. [CrossRef]
58. Yajima, H.; Morita, M.; Hashimoto, M.; Sashiwa, H.; Kikuchi, T.; Ishii, T. Complex Formation of Chitosan with Iodine and Its Strucutre and Spectroscopic Properties—Molecular Assembly and Thermal Hysteresis Behavior. *Int. J. Thermophys.* **2001**, *22*, 1265–1283. [CrossRef]
59. Gunasekaran, M. Physiological Studies on Phymatotrichum Omnivorum II. Physicochemical Properties of Glycogen. *Arch. Mikrobiol.* **1972**, *84*, 69–76. [CrossRef] [PubMed]
60. Kumari, S.; Roman, A.; Khan, A. Chromophore and Spectrum of the Glycogen-Iodine Complex. *J. Polym. Sci. Part A Polym. Chem.* **1996**, *34*, 2975–2980. [CrossRef]
61. Lecker, D.N.; Kumari, S.; Khan, A. Iodine Binding Capacity and Iodine Binding Energy of Glycogen. *J. Polym. Sci. Part A Polym. Chem.* **1997**, *35*, 1409–1412. [CrossRef]
62. Aterman, K. A Historical Note on the Iodine-Sulphuric Acid Reaction of Amyloid. *Histochemistry* **1976**, *49*, 131–143. [CrossRef]
63. Dzwolak, W. Insulin Amyloid Fibrils Form an Inclusion Complex with Molecular Iodine: A Misfolded Protein as a Nanoscale Scaffold. *Biochemistry* **2007**, *46*, 1568–1572. [CrossRef]
64. Pritchard, J.G.; Serra, F.T. Complexation of Polyvinyl Acetate with Iodine. *Talanta* **1973**, *20*, 541–546. [CrossRef]

65. Hughes, J. Analytical Behaviour of Poly(Vinyl Acetate) and Its Hydrolysis Products with Iodine. *Talanta* **1979**, *26*, 1161–1163. [CrossRef]
66. Schulz, R.C.; Fleischer, D.; Henglein, A.; Bössler, H.M.; Trisnadi, J.; Tanaka, H. Addition Compounds and Complexes with Polymers and Models. *Pure Appl. Chem.* **1974**, *38*, 227–247. [CrossRef]
67. Chernov'yants, M.S.; Burykin, I.V.; Pisanov, R.V.; Shalu, O.A. Synthesis and Antimicrobial Activity of Poly(N-Methyl-4-Vinylpyridinium Triiodide). *Pharm. Chem. J.* **2010**, *44*, 61–63. [CrossRef]
68. Pal, S.K.; Krishnan, A.; Das, P.K.; Samuelson, A.G. Schiff Base Linked Ferrocenyl Complexes for Second-Order Nonlinear Optics. *J. Organomet. Chem.* **2000**, *604*, 248–259. [CrossRef]
69. Liu, W.-J.; Xiong, G.-X.; Zeng, D.-H. Synthesis and Electrical Properties of Three Novel Poly (Ferrocenyl-Schiff Bases) and Their Charge Transfer Complexes with Iodine. *J. Inorg. Organomet. Polym.* **2010**, *20*, 97–103. [CrossRef]
70. Shirakawa, H.; Louis, E.J.; MacDiarmid, A.G.; Chiang, C.F.; Heeger, A.J. Synthesis of Electrically Conducting Organic Polymers: Halogen Derivatives of Polyacetylene, $(CH)_x$. *J. Chem. Soc. Commun.* **1977**, 578–580. [CrossRef]
71. Bakueva, L.; Matheson, D.; Musikhin, S.; Sargent, E.H. Luminescence of Pure and Iodine Doped PPV: Internal Energetic Structure Revealed through Spectral Signatures. *Synth. Met.* **2002**, *126*, 207–211. [CrossRef]
72. Fasahat, P.; Rahman, S.; Ratnam, W. Genetic Controls on Starch Amylose Content in Wheat and Rice Grains. *J. Genet.* **2014**, *93*, 279–292. [CrossRef] [PubMed]
73. Ashogbon, A.O.; Akintayo, E.T.; Oladebeye, A.O.; Oluwafemi, A.D.; Akinsola, A.F.; Imanah, O.E. Developments in the Isolation, Composition, and Physicochemical Properties of Legume Starches. *Crit. Rev. Food Sci. Nutr.* **2021**, *61*, 2938–2959. [CrossRef] [PubMed]
74. Wang, L.; Liu, L.; Zhao, J.; Li, C.; Wu, H.; Zhao, H.; Wu, Q. Granule-Bound Starch Synthase in Plants: Towards an Understanding of Their Evolution, Regulatory Mechanisms, Applications, and Perspectives. *Plant Sci.* **2023**, *336*, 111843. [CrossRef]
75. Guo, K.; Liang, W.; Wang, S.; Guo, D.; Liu, F.; Persson, S.; Herburger, K.; Petersen, B.L.; Liu, X.; Blennow, A.; et al. Strategies for Starch Customization: Agricultural Modification. *Carbohydr. Polym.* **2023**, *321*, 121336. [CrossRef]
76. Chiaramonte, E.; Rhazi, L.; Aussenac, T.; White, D.R. Amylose and Amylopectin in Starch by Asymmetric Flow Field-Flow Fractionation with Multi-Angle Light Scattering and Refractive Index Detection (AF4–MALS–RI). *J. Cereal Sci.* **2012**, *56*, 457–463. [CrossRef]
77. Ulbrich, M.; Scholz, F.; Flöter, E. Chromatographic Study of High Amylose Corn Starch Genotypes—Investigation of Molecular Properties after Specific Enzymatic Digestion. *Starch-Stärke* **2022**, *74*, 2100303. [CrossRef]
78. Tutorskii, I.A.; Sokolova, L.V. Mechanism of the Reaction of Polybutadiene with Molecular Iodine. *Polym. Sci. U.S.S.R.* **1977**, *19*, 176–183. [CrossRef]
79. Sreeja, R.; Najidha, S.; Remya Jayan, S.; Predeep, P.; Mazur, M.; Sharma, P.D. Electro-Optic Materials from Co-Polymeric Elastomer–Acrylonitrile Butadiene Rubber (NBR). *Polymer* **2006**, *47*, 617–623. [CrossRef]
80. Vippa, P.; Rajagopalan, H.; Thakur, M. Electrical and Optical Properties of a Novel Nonconjugated Conductive Polymer, Poly (β-pinene). *J. Polym. Sci. Part B Polym. Phys.* **2005**, *43*, 3695–3698. [CrossRef]
81. Nagelli, W. Beitrage Zur Naheren Kenntniss Der Starkegruppe. *Ann. Chem.* **1874**, *173*, 218–227. [CrossRef]
82. Helbert, W.; Chanzy, H. The Ultrastructure of Starch from Ultrathin Sectioning in Melamine Resin. *Starch-Starke* **1996**, *48*, 185–188. [CrossRef]
83. Atkin, N.J.; Abeysekera, R.M.; Cheng, S.L.; Robards, A.W. An Experimentally-Based Predictive Model for the Separation of Amylopectin Subunits during Starch Gelatinization. *Carbohydr. Polym.* **1998**, *36*, 173–192. [CrossRef]
84. Baker, A.A.; Miles, M.J.; Helbert, W. Internal Structure of the Starch Resistant Granule Revealed by AFM. *Carbohydr. Res.* **2001**, *330*, 249–256. [CrossRef]
85. Baldwin, P.M.; Adler, J.; Davies, M.; Melia, D. High Resolution Imaging of Starch Granule Surfaces by Atomic Force Microscopy. *J. Cereal Sci.* **1998**, *27*, 255–265. [CrossRef]
86. Dang, J.M.C.; Copeland, L. Imaging Rice Grains Using Atomic Force Microscopy. *J. Cereal Sci.* **2003**, *37*, 165–170. [CrossRef]
87. Ohtani, T.; Yoshimo, T.; Hagiwara, S.; Maekawa, T. High-Resolution Imaging of Starch Granule Structure Using Atomic Force Microscopy. *Starch-Starke* **2000**, *52*, 150–153. [CrossRef]
88. Park, H.; Xu, S.; Seetharaman, S. A Novel in Situ Atomic Force Microscopy Imaging Technique to Probe Surface Morphological Features of Starch Granules. *Carbohydr. Res.* **2011**, *346*, 847–853. [CrossRef] [PubMed]
89. Ridout, M.J.; Gunning, A.P.; Parker, M.L.; Wilson, R.H.; Morris, V.J. Using AFM to Image the Internal Structure of Starch Grsnules. *Carbohydr. Polym.* **2002**, *50*, 123–132. [CrossRef]
90. Szymonska, J.; Krok, F. Potato Starch Granule Nanostructure Studied by Highresolution Non-Contact AFM. *Int. J. Biol. Macromol.* **2003**, *33*, 1–7. [CrossRef]
91. Waduge, R.N.; Xu, S.; Seetharaman, S. Iodine Absorption Properties and Its Effect on the Crystallinity of Developing Wheat Starch Granules. *Carbohydr. Polym.* **2010**, *82*, 786–794. [CrossRef]
92. Gallant, D.J.; Bouchet, B.; Baldwin, P.M. Microscopy of Starch: Evidence of a New Level of Granule Organization. *Carbohydr. Polym.* **1997**, *32*, 177–191. [CrossRef]
93. Doutch, J.; Gilbert, E.P. Characterisation of Large Scale Structures in Starch Granules via Small-Angle Neutron and X-ray Scattering Techniques. *Carbohydr. Polym.* **2013**, *91*, 444–451. [CrossRef]

94. Barrett, A.J.; Barrett, K.L.; Khan, A. Effects of Acetone, Ethanol, Isopropanol, and Dimethyl Sulfoxide on Amylose-Iodine Complex. *J. Macromol. Sci. Part A* **1998**, *35*, 711–722. [CrossRef]
95. Fonslick, J.; Khan, A. Thermal Stability and Composition of the Amylose–Iodine Complex. *J. Polym. Sci. A Polym. Chem.* **1989**, *27*, 4161–4167. [CrossRef]
96. Rundle, R.E.; French, D. The Configuration of Starch and the Starch—Iodine Complex. II. Optical Properties of Crystalline Starch Fractions 1. *J. Am. Chem. Soc.* **1943**, *65*, 558–561. [CrossRef]
97. Rundle, R.E.; French, D. The Configuration of Starch in the Starch—Iodine Complex. III. X-ray Diffraction Studies of the Starch—Iodine Complex 1. *J. Am. Chem. Soc.* **1943**, *65*, 1707–1710. [CrossRef]
98. Cramer, F.; Windel, H. Über Einschlußverbindungen, X. Mitteil.: Die Blauen Jodverbindungen Der Cumarine Und Anderer Verwandter Verbindungen. *Chem. Ber.* **1956**, *89*, 354–365. [CrossRef]
99. Saenger, W. The Structure of the Blue Starch-Iodine Complex. *Naturwissenschaften* **1984**, *71*, 31–36. [CrossRef]
100. Immel, S.; Lichtenthaler, F.W. The Hydrophobic Topographies of Amylose and Its Blue Iodine Complex. *Starch-Stärke* **2000**, *52*, 1–8. [CrossRef]
101. Minick, M.; Fotta, K.; Khan, A. Polyiodine Units in Starch-Iodine Complex: INDO CI Study of Spectra and Comparison with Experiments. *Biopolymers* **1991**, *31*, 57–63. [CrossRef]
102. Zaslow, B.; Miller, R.L. Hydration of the "V" Amylose Helix 1. *J. Am. Chem. Soc.* **1961**, *83*, 4378–4381. [CrossRef]
103. Hirai, M.; Hirai, T.; Ueki, T. Effect of Branching of Amylopectin on Complexation with Iodine as Steric Hindrance. *Polymer* **1994**, *35*, 2222–2225. [CrossRef]
104. Dintzis, F.R.; Beckwith, A.C.; Babcock, G.E.; Tobin, R. Amylose-Iodine Complex. I. Sedimentation Behavior. *Macromolecules* **1976**, *9*, 471–478. [CrossRef]
105. Moulik, S.P.; Gupta, S. Effects of Solvents on the Spectrophotometric and Hydrodynamic Behavior of Amylose and Its Iodine Complex. *Carbohydr. Res.* **1980**, *81*, 131–143. [CrossRef]
106. Senior, M.B.; Hamori, E. Investigation of the Effect of Amylose/Iodine Complexation on the Conformation of Amylose in Aqueous Solution. *Biopolymers* **1973**, *12*, 65–78. [CrossRef]
107. Vladimirov, A.V.; Volkova, T.V.; Agafonov, A.V. Temperature Dependence of Stability Constants of the Iodine-Iodide-Amylose Complexes. *Russ. J. Phys. Chem. A* **2003**, *77*, 612–615.
108. Zhang, Q.; Lu, Z.; Hu, H.; Yang, W.; Marszalek, P.E. Direct Detection of the Formation of V-Amylose Helix by Single Molecule Force Spectroscopy. *J. Am. Chem. Soc.* **2006**, *128*, 9387–9393. [CrossRef] [PubMed]
109. Dintzis, F.R.; Tobin, R.; Beckwith, A.C. Amylose-Iodine Complex. II. Molecular Weight Estimates. *Macromolecules* **1976**, *9*, 478–482. [CrossRef] [PubMed]
110. Mikus, F.F.; Hixon, R.M.; Rundle, R.E. The Complexes of Fatty Acids with Amylose 1. *J. Am. Chem. Soc.* **1946**, *68*, 1115–1123. [CrossRef]
111. Calabrese, V.T.; Khan, A. Amylose-Iodine Complex Formation without KI: Evidence for Absence of Iodide Ions within the Complex. *J. Polym. Sci. A Polym. Chem.* **1999**, *37*, 2711–2717. [CrossRef]
112. Cesaro, A.; Benegas, J.C.; Ripoll, D.R. Molecular Model of the Cooperative Amylose-Iodine-Triiodide Complex. *J. Phys. Chem.* **1986**, *90*, 2787–2791. [CrossRef]
113. Kuge, T.; Ono, S. Amylose-Iodine Complex. III. Potentiometric and Spectrophotometric Studies. *Bull. Chem. Soc. Jpn.* **1960**, *33*, 1273–1278. [CrossRef]
114. Schulz, W.; Sklenar, H.; Hinrichs, W.; Saenger, W. The Structure of the Left-Handed Antiparallel Amylose Double Helix: Theoretical Studies. *Biopolymers* **1993**, *33*, 363–375. [CrossRef]
115. Moulik, S.P.; Gupta, S. Environment-Induced, Physicochemical Behavior of Amylose-Iodine Complexes. *Carbohydr. Res.* **1979**, *71*, 251–264. [CrossRef]
116. Szejtli, J.; Richter, M.; Augustat, S. Molecular Configuration of Amylose and Its Complexes in Aqueous Solutions. Part IV. Determination OfDP of Amylose by Measuring the Concentration of Free Iodine in Solution of Amylose-Iodine Complex. *Biopolymers* **1968**, *6*, 27–41. [CrossRef]
117. Peng, Q.-J.; Perlin, A.S. Observations on N.M.R. Spectra of Starches in Dimethyl Sulfoxide, Iodine-Complexing, and Solvation in Water-Di-Methyl Sulfoxide. *Carbohydr. Res.* **1987**, *160*, 57–72. [CrossRef]
118. Murdoch, K.A. The Amylose-Iodine Complex. *Carbohydr. Res.* **1992**, *233*, 161–174. [CrossRef]
119. Knutson, C.A.; Cluskey, J.E.; Dintzis, F.R. Properties of Amylose-Iodine Complexes Prepared in the Presence of Excess Iodine. *Carbohydr. Res.* **1982**, *101*, 117–128. [CrossRef]
120. Murakami, H. Electronic Structure of the Amylose-Iodine Complex. *J. Chem. Phys.* **1954**, *22*, 367–374. [CrossRef]
121. Nishimura, T.; Yajima, H.; Kubota, S.; Ishii, T.; Endo, R. Polymer Effect on the Iodine Coloring Species Responsible for the Spectroscopic Properties of Amylose-Iodine Complexes. *Kobunshi Ronbunshu* **1990**, *47*, 717–725. [CrossRef]
122. Nishimura, T.; Yajima, H.; Kubota, S.; Ishii, T.; Endo, R. Effect of I$^-$ Concentration on the Optical Properties of Amylose-Iodine Complexes. *Kobunshi Ronbunshu* **1988**, *45*, 945–952. [CrossRef]
123. Rawlings, P.K.; Schneider, F.W. Models for Competitive Cooperative Linear Adsorption. The Amylose–Iodine-Iodide Complex. *J. Chem. Phys.* **1970**, *52*, 946–952. [CrossRef]
124. Kuge, T.; Ono, S. Advances in Carbohydrate Chemistry and Biochemistry. *Bull. Chem. Sot. Jpn.* **1960**, *33*, 1269–1272. [CrossRef]

125. Teitelbaum, R.C.; Ruby, S.L.; Marks, T.J. A Resonance Raman/Iodine Moessbauer Investigation of the Starch-Iodine Structure. Aqueous Solution and Iodine Vapor Preparations. *J. Am. Chem. Soc.* **1980**, *102*, 3322–3328. [CrossRef]
126. Cesàro, A.; Jerian, E.; Saule, S. Physicochemical Studies of Amylose and Its Derivatives in Aqueous Solutions: Thermodynamics of the Iodine-Triiodide Complex. *Biopolymers* **1980**, *19*, 1491–1506. [CrossRef]
127. Nishimura, T.; Yajima, H.; Ishii, T.; Endo, R. Study of the Bluing Mechanism of Amylose-Iodine Complexes by CD Stopped-Flow Method. *Kobunshi Ronbunshu* **1991**, *48*, 525–528. [CrossRef]
128. Wolf, R.; Schulz, R.C. Optical Rotatory Dispersion of the Starch Iodine Complex. Part 2. *J. Macromol. Sci. Part A Chem.* **1968**, *2*, 821–832. [CrossRef]
129. Agafonov, A.V.; Vladimirov, A.V.; Volkova, T.V. The Concentration Dependences of the Stability Constants of Iodine-Iodide-Amylose Complexes in Aqueous Solutions of Electrolytes. *Russ. J. Phys. Chem. A* **2004**, *78*, 1584–1587.
130. Yamamoto, M.; Sano, T.; Harada, S.; Yasunaga, T. Interaction of Amylose with Iodine. II. Kinetic Studies of the Complex Formation by the Temperature-Jump Method. *Bull. Chem. Soc. Jpn.* **1982**, *55*, 3702–3706. [CrossRef]
131. Foster, J.F.; Zucker, D. Length of the Amylose–Iodine Complex as Determined by Streaming Dichroism. *J. Phys. Chem.* **1952**, *56*, 170–173. [CrossRef]
132. Noltemeyer, M.; Saenger, W. Topography of Cyclodextrin Inclusion Complexes. 12. Structural Chemistry of Linear.Alpha.-Cyclodextrin-Polyiodide Complexes. X-ray Crystal Structures of (.Alpha.-Cyclodextrin)$_2$.LiI$_3$.I$_2$.8H$_2$O and (.Alpha.-Cyclodextrin)$_2$.Cd$_{0.5}$).I$_5$.27H$_2$O. Models for the Blue. *J. Am. Chem. Soc.* **1980**, *102*, 2710–2722. [CrossRef]
133. Betzel, C.; Hingerty, B.; Noltemeyer, M.; Weber, G.; Saenger, W.; Hamilton, J.A. (β-Cyclodextrin)$_2$ KI$_7$ 9 H$_2$O. Spatial Fitting of a Polyiodide Chain to a given Matrix. *J. Incl. Phenom.* **1983**, *1*, 181–191. [CrossRef]
134. Bowmaker, G. Bonding and Nuclear Quadrupole Coupling in Linear Pentaiodide Ions. *Aust. J. Chem.* **1978**, *31*, 2713. [CrossRef]
135. Nimz, O.; Geßler, K.; Usón, I.; Laettig, S.; Welfle, H.; Sheldrick, G.M.; Saenger, W. X-Ray Structure of the Cyclomaltohexaicosaose Triiodide Inclusion Complex Provides a Model for Amylose–Iodine at Atomic Resolution. *Carbohydr. Res.* **2003**, *338*, 977–986. [CrossRef] [PubMed]
136. Ziegast, G.; Pfannemüller, B. Resonance Raman Studies of Amaylose—Iodine Complexes. *Int. J. Biol. Macromol.* **1982**, *4*, 419–424. [CrossRef]
137. Heyde, M.E.; Rimai, L.; Kilponen, R.G.; Gill, D. Resonance-Enhanced Raman Spectra of Iodine Complexes with Amylose and Poly(Vinyl Alcohol), and of Some Iodine-Containing Trihalides. *J. Am. Chem. Soc.* **1972**, *94*, 5222–5227. [CrossRef]
138. Okuda, M.; Hiramatsu, T.; Yasuda, M.; Ishigaki, M.; Ozaki, Y.; Hayashi, M.; Tominaga, K.; Chatani, E. Theoretical Modeling of Electronic Structures of Polyiodide Species Included in α-Cyclodextrin. *J. Phys. Chem. B* **2020**, *124*, 1089–1096. [CrossRef]
139. Mizuno, M.; Tanaka, J.; Harada, I. Electronic Spectra and Structures of Polyiodide Chain Complexes. *J. Phys. Chem.* **1981**, *85*, 1789–1794. [CrossRef]
140. Teitelbaum, R.C.; Ruby, S.L.; Marks, T.J. On the Structure of Starch-Iodine. *J. Am. Chem. Soc.* **1978**, *100*, 3215–3217. [CrossRef]
141. Hach, R.J.; Rundle, R.E. The Structure of Tetramethylammonium Pentaiodide [1,1a]. *J. Am. Chem. Soc.* **1951**, *73*, 4321–4324. [CrossRef]
142. Herbstein, F.H.; Kapon, M. Zigzag Chains of Alternating Molecules and Triiodide Ions in Crystalline (Phenacetin)$_2$.HI$_5$. *Nat. Phys. Sci.* **1972**, *239*, 153–154. [CrossRef]
143. Haddock, A.; Steidemann, M.; Readnour, M. Polyiodide Equilibria in Aqueous Solutions of Iodine and Iodide. *Synth. React. Inorg. Met. Org. Chem.* **1979**, *9*, 39–56. [CrossRef]
144. Ramette, R.W.; Sandford, R.W. Thermodynamics of Iodine Solubility and Triiodide Ion Formation in Water and in Deuterium Oxide. *J. Am. Chem. Soc.* **1965**, *87*, 5001–5005. [CrossRef]
145. Sekine, T. Abstracts. *Nippon Kagaku Zassi* **1969**, *90*, 951–983. [CrossRef]
146. Mould, D.L. Potentiometric and Spectrophotometric Studies of Complexes of Hydrolysis Products of Amylose with Iodine and Potassium Iodide. *Biochem. J.* **1954**, *58*, 593–600. [CrossRef]
147. Bhide, S.V.; Kale, N.R. Ligand-Induced Structural Changes in Amylose Partially Complexed with Iodine. *Biochim. Biophys. Acta (BBA)—Gen. Subj.* **1976**, *444*, 719–726. [CrossRef]
148. Benesi, H.A.; Hildebrand, J.H. A Spectrophotometric Investigation of the Interaction of Iodine with Aromatic Hydrocarbons. *J. Am. Chem. Soc.* **1949**, *71*, 2703–2707. [CrossRef]
149. Bernal-Uruchurtu, M.I.; Kerenskaya, G.; Janda, K.C. Structure, Spectroscopy and Dynamics of Halogen Molecules Interacting with Water. *Int. Rev. Phys. Chem.* **2009**, *28*, 223–265. [CrossRef]
150. Kireev, S.V.; Shnyrev, S.L. Study of Molecular Iodine, Iodate Ions, Iodide Ions, and Triiodide Ions Solutions Absorption in the UV and Visible Light Spectral Bands. *Laser Phys.* **2015**, *25*, 075602. [CrossRef]
151. Prasanna; Shrikanth, B.K.; Hegde, M.S. Formation and Structure of Iodine: Water (H_2O-I_2) Charge-Transfer Complex. *J. Chem. Sci.* **2021**, *133*, 51. [CrossRef]

Disclaimer/Publisher's Note: The statements, opinions and data contained in all publications are solely those of the individual author(s) and contributor(s) and not of MDPI and/or the editor(s). MDPI and/or the editor(s) disclaim responsibility for any injury to people or property resulting from any ideas, methods, instructions or products referred to in the content.

Review

Research Progress on Hypoglycemic Mechanisms of Resistant Starch: A Review

Jiameng Liu [1,†], Wei Lu [1,2,†], Yantian Liang [1], Lili Wang [1], Nuo Jin [1], Huining Zhao [1], Bei Fan [1,*] and Fengzhong Wang [1,*]

[1] Key Laboratory of Agro-Products Quality and Safety Control in Storage and Transport Process, Ministry of Agriculture and Rural Affairs, Institute of Food Science and Technology, Chinese Academy of Agricultural Sciences, Beijing 100193, China
[2] College of Food Science and Technology, Hebei Agricultural University, Baoding 071033, China
* Correspondence: caasBFan@163.com (B.F.); caasFZWang@163.com (F.W.)
† These authors contributed equally to this work.

Abstract: In recent years, the prevalence of diabetes is on the rise, globally. Resistant starch (RS) has been known as a kind of promising dietary fiber for the prevention or treatment of diabetes. Therefore, it has become a hot topic to explore the hypoglycemic mechanisms of RS. In this review, the mechanisms have been summarized, according to the relevant studies in the recent 15 years. In general, the blood glucose could be regulated by RS by regulating the intestinal microbiota disorder, resisting digestion, reducing inflammation, regulating the hypoglycemic related enzymes and some other mechanisms. Although the exact mechanisms of the beneficial effects of RS have not been fully verified, it is indicated that RS can be used as a daily dietary intervention to reduce the risk of diabetes in different ways. In addition, further research on hypoglycemic mechanisms of RS impacted by the RS categories, the different experimental animals and various dietary habits of human subjects, have also been discussed in this review.

Keywords: resistant starch; diabetes; hypoglycemic mechanisms; intestinal microbiota

1. Introduction

With changes in diet and lifestyle, the rising prevalence of diabetes has constituted one of the major threats to human health, globally. In the past thirty years, the number of people with diabetes has quadrupled, and diabetes has become the ninth leading cause of death [1]. It was estimated that the number of people with diabetes will reach 6.43 million worldwide by 2030. In addition, it cost at least USD 966 billion in global health expenditure in 2021, according to the International Diabetes Federation (IDF) [2], which showed an enormous global economic burden of diabetes. In the long term, lifestyle interventions, especially the changes in diet, have been suggested as the primary treatment for regulating the blood glucose level for patients. The consumption of easily digestible carbohydrates, such as sucrose and starch can affect the level of blood glucose directly, since they can be digested in the human gastrointestinal tract. RS is defined as a kind of starch that cannot be digested by amylases in the small intestine and eventually is fermented in the colon by microbiota [3]. There are relevant studies that show that consuming RS has a positive effect on regulating human blood glucose levels [4,5]. It has been confirmed that the glycemic response could be reduced by RS, compared with normal carbohydrates in an approved European Food Standards Agency claim (EFSA) [6].

As a special kind of dietary fiber, RS can't be digested in the small intestine [7]. It can be divided into five categories (i.e., RS1, RS2, RS3, RS4, RS5), based on its different physical and chemical structures [8]. The relevant studies have reported the positive effects of RS in the regulation of type 2 diabetes mellitus (T2DM) [9,10]. There are also studies which found that the postprandial glycemic response could be controlled [11,12], the insulin sensitivity

(IS) could be increased [13,14] and the expression of the inflammatory markers could be reduced by RS, as observed in clinical trials and in rodent models. However, the exact mechanism of how RS exerts its hypoglycemic effect has not been fully understood.

This review summarizes the current studies on RS in diabetes control and elaborates on the potential mechanisms (Figure 1, illustrating the structures of different types of, and the relevant hypoglycemic mechanisms of RS). At the same time, we summed up the limitations of the current research and provided a reference for future research perspectives.

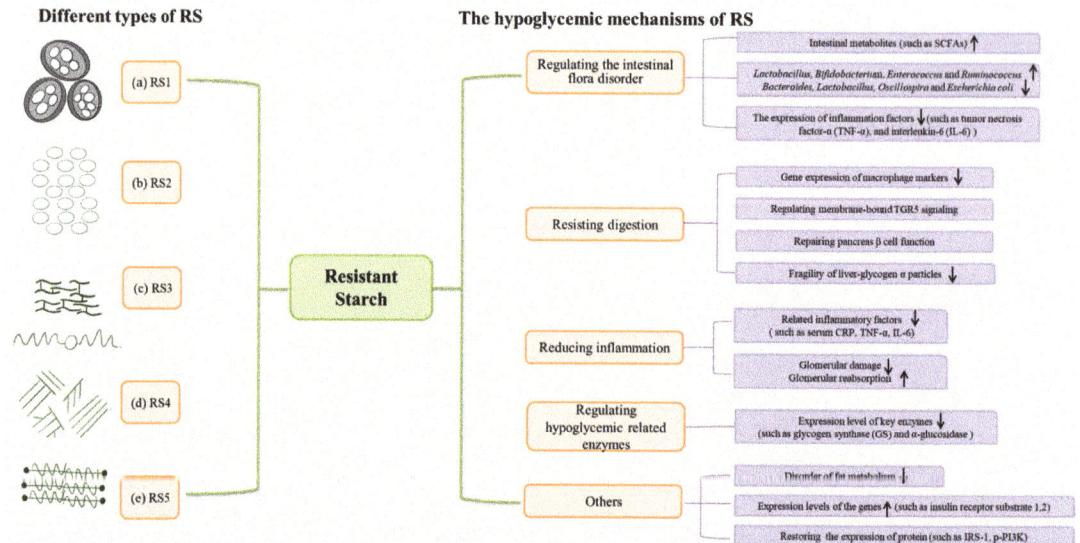

Figure 1. Different types of, and the potential hypoglycemic mechanisms of RS. (**a**) RS1, the most common RS found in all grains, is a kind of physically inaccessible starch.; (**b**) RS2, in raw potato or high-amylose maize, has a B- or C-type polymorph; (**c**) RS3, in cooked and cooled potatoes, a kind of retrograded starch; (**d**) RS4, chemically modified starch, through the addition of cross-linkages or chemical derivatives; (**e**) RS5, in processing or cooking oil, a lipid-modified starch. Adapted from Wong et al. [10].

2. Regulating the Intestinal Microbiota Disorder

The risk of development of T2DM could be prevented by RS through the gut microbiota, with the help of regulating the abundance of microbiota, to produce starch-degrading enzymes, and improving the intestinal barrier function [15]. Gut microbiota are composed of a variety of commensal microorganisms, including certain amounts of bacteria, fungi and viruses. They play an important role in regulating the metabolic, endocrine and immune functions [16].

Short chain fatty acids (SCFAs) are one of the important bio products of intestinal microbiota, mainly including acetic acid (C2), propionic acid (C3) and butyric acid (C4). These three SCFAs are also the most abundant SCFAs in the human body [16]. SCFAs are mainly produced from the fermentation of non-digestible carbohydrates (e.g., RS) [17]. They can improve the insulin resistance (IR) and T2DM, by regulating the related metabolic pathways. Different from the mechanisms that affect glucose homeostasis directly, SCFAs impact the host health at the cellular, tissue and organ levels [18].

SCFAs can promote the secretion of two important key intestinal hormones, namely the glucagon-like peptide-1 (GLP-1) and peptide YY (PYY). This secretion-boosting effect is able to increase the satiety by acting on the gut-brain axis. By this pathway, SCFAs can reduce appetite and food intake indirectly, which could prevent weight gain and thereby lower the risk of diabetes. SCFAs can also regulate the blood glucose concentrations by increasing the insulin secretion mediated by GLP-1 [19]. Nielsen et al. [20] found that, compared with the Western-style diet (WSD) group, there were 2- to 5-fold increases of butyrate pool size in the large intestinal digesta in the RS diet (RSD) and arabinoxylan diet (AXD) groups. They inferred that the result of stimulating the insulin secretion was caused by the promotion of the intestinal endocrinology of PYY, which inhibited the timing of the gastric and intestinal translocation. Then, the appetite was suppressed while the GLP-1 secretion was promoted. Hughes et al. [21] found that the fasting and peak concentration of peptide PYY3-36 increased while the peak concentration and AUC of the glucose-dependent insulinotropic peptide decreased after the healthy adult subjects ingested RS2-enriched wheat. Binou et al. [22] found that the bread rich in the β-glucans (βGB) groups and the bread rich in the RS (RSB) groups elicited a lower incremental area under the curve (AUC) for the glycemic response, compared with the control group (glucose solution, GS). At 15 min after the βGB and RSB intakes, a significant reduction in appetite and an increase in satiety were detected in the healthy adults, and this trend continued up to the 180th min. The result showed that the food containing RS could retard the absorption of glucose. Maziarz et al. [23] found that the total concentration of PYY in the high-amylose maize type 2 resistant starch (HAM-RS2) group was significantly higher than in the control group (p = 0.043) at 120 min. At the same time, the AUC glucose (p = 0.028) was decreased at the end of 6 weeks in the HAM-RS2 group, while this trend was not related to the changes in the subjects' physical composition or total energy intake. This result might be caused by the SCFAs that are produced from the fermentation of HAM-RS2 by the bacteria in the lower GI tract. At the same time, the relevant studies have suggested that HAM-RS2 might show its benefits by increasing the SCFAs in the blood to alter the free fatty acid and glycerol that are released by adipocytes, regulate the bile acid metabolism [24,25] or alter the intestinal microbiota profile [26].

Mohr et al. [27] have reported similar results as they found that the postprandial blood glucose and insulin levels could be reduced by the combination of RS and whey protein. Thus, they inferred that RS could enhance the variety of metabolites in the gut, e.g., the production of the SCFAs could be improved. Furthermore, Zhou et al. [28] found that the plasma GLP-1 and PYY levels were both increased at different time points within a 24 h period in mice fed with RS (53.7% RS2, 10 d), and this result was not related to diets, the different glycemic indexes or the time of the blood sample collections. Chen et al. [29] found that the glucose of diabetic mice in three (corn, mung bean and Pueraria) RSs were significantly lower than the diabetic groups, after 14 weeks. The GLP-1 content of diabetic mice, at 19 weeks, showed significant differences in corn RS groups and mung beans RS groups (p < 0.01). Chen et al. [30] found that the serum blood glucose level of mice in the high-dose multiple composite RS group was reduced by 59.71%, compared with the model control group, indicating that the blood glucose could be controlled by multiple compounds effectively, and the effect of the high-dose multiple composite RS on reducing the blood glucose was better. Boll et al. [31] found that arabinoxylan oligosaccharides (one of the dietary fibers, AXOS) showed the ability to improving the glucose tolerance in an overnight perspective. The possible mechanism was that IS and the gut fermentation could be improved by the breads containing an AXOS-rich wheat bran extract and RS, separately or combined, on the glucose tolerance and the intestinal markers in healthy subjects.

Many studies found that the concentration of SCFAs C2, C3, C4 could be promoted by RS, leading to the decrease of the pH in the intestine. The falling pH would promote the production of the beneficial bacteria and reduce the number of intestinal spoilage bacteria, achieving a balanced state of the intestinal microbiota [32,33]. Zhang et al. [34] found that the blood glucose could be reduced, the response to the IR and the glucose tolerance test could be ameliorated, and the pathological damage could all be relieved by RS3, in T2DM mice, from the *canna edulis* (Ce-RS3). In this study, 24 diabetic mice, induced by streptozotocin (STZ,) were randomly divided into a T2DM group (Model), a RS group (Ce-RS3) and a metformin group (Met). Eleven weeks later, they found the microbial and metabolic disorder of the mice in the RS and Met groups were significantly regulated. Among them, Ce-RS3 showed a better regulatory effect and an improved diversity of the intestinal microbiota, especially of the *Prevotella* genera. The SCFAs levels were significantly increased, since the abundance of the gut bacterial producing SCFAs was increased, such as *Phascolarctobacterium*, *Ruminococcaceae_NK4A214_group*, *Ruminococcaceae_UCG_014*, *Helicobacter* and *Ruminooccus*. Therefore, they inferred that the intestinal microbiota characteristics of the RS group were closely associated with the T2DM-related indicators. Zhou et al. [35] found that the intestinal flora microbiota abundance was regulated by the intake of BRS (Buckwheat-RS), which increased the abundance of the beneficial bacteria *Lactobacillus*, *Bifidobacterium* and *Enterococcus*, while the abundance of *Escherichia coli* decreased. Compared with the HFD (high-fat diet) group, the content of the SCFAs in the mice colons was increased in the BRS group. In this study, the male C57BL/6 mice were fed a normal diet (CON), HFD, and HFD supplemented with BRS (HFD + BRS) for 6 weeks, separately. The quantities of four common and major intestinal microbiota (*Bifidobacterium*, *Lactobacillus*, *Enterococcus*, *E. coli*) were analyzed by qPCR and the absolute quantification methods. It has been speculated that the changes in the intestinal microbiota caused by the BRS, might be related to its ability to regulate the intestinal redox status. Sánchez-Tapia et al. [36] found that the RS in black beans could improve the glucose response, because the gut microbiota, such as *Clostridia*, could be mediated by the black beans' RS. Zhu et al. [37] found that the gut microbiota composition in the T2DM mice, changed obviously. The abundance of the genus *Clostridium* and *Butyricoccus* could be increased, while the genus *Bacteroides*, *Lactobacillus*, *Oscillospira* and *Ruminococcus* could be decreased by the ORS (oat RS). In addition, the Pearson correlation analysis showed that the genus *Bacteroides*, *Butyricoccus*, *Parabacteroides*, *Lactobacillus*, *Oscillospira*, *Ruminococcus* and *Bifidobacterium* were positively correlated with the occurrence of diabetes and inflammation ($p < 0.05$), while genus *Clostridium* and *Faecalibacterium* showed a negative correlation ($p < 0.05$). The result indicated that the anti-diabetic effects of the ORS was achieved by altering the gut microbiota.

Additionally, the metabolites of intestinal microbiota can improve the intestinal barrier, reduce the IR and the expression of the related inflammatory factors [38]. Jiang et al. [39] found that, compared with the NC group (normal control, normal mice on a basal diet) and the MC group (model control, diabetic mice on a basal diet), *Firmicutes* and *Bacteroidetes* were the dominant bacterial phyla in the IG group (intervention group, the diabetic mice fed with of Ganoderma lucidum spores encapsulated within the RS (EGLS)). The abundance of *Proteobacteria* (mostly identified as pathogenic bacteria) in diabetic mice was the highest. The elevated level of *Proteobacteria* might indicate the intestinal inflammation in the MC group, which might be related to the occurrence of T2DM. However, compared with the MC group, the proportion of *Proteobacteria* in the IG group was significantly reduced. Therefore, they speculated that the blood glucose in mice was decreased since the fecal microbial community abundance associated with promoting the anti-inflammatory responses were modulated by EGLS. Kingbeil et al. [40] found that, compared with the low-fat chow (LF, 13% fat) and the HF (45% fat) intervention, the isocaloric HF supplemented with a 12% potato RS (HFRS) intervention in the HF-fed mice, would lead to changes in the composition of the gut microbes. They found this result correlated with the improved inflammatory status and the vagal signaling by the potato RS. Beyond that, they found that the energy consumed by the HFRS-fed mice was significantly less, compared with the HF-fed mice. Additionally, the systemic inflammation and the glucose homeostasis were improved in the HFRS group, compared to the HF group. Another study [41] showed that the improved intestinal barrier function in the potato RS treated mice was associated with the reduced systemic inflammation and the improved glucose homeostasis. For the HFD mice, the intestinal barrier function was decreased and the inflammation responses were initiated because of the gut microbiota dysbiosis. However, the RS supplementation could increase the SCFAs production that might decrease the effects of the HFD by enhancing the gut barrier function, reducing the levels of the systemic lipopolysaccharide (LPS) and increasing GLP-1 levels. In addition, the IR could be sufficiently promoted by the chronic elevation of the circulating LPS.

Keenan et al. [8] found that, in human subjects, the IS was increased after consuming the RS. However, only one of several studies reported an increase in the serum GLP-1 associated with RS added to the diet. This means that RS might reduce the blood glucose through other mechanisms, such as the increased intestinal gluconeogenesis, which might be associated with the promotion of the improved IR. Indeed, there were several studies suggesting that the SCFAs could decrease the hepatic glycolysis and gluconeogenesis but increase the glycogen synthesis [42–44].

According to the above studies, it can be found that the mechanisms of RS, based on the intestinal hypoglycemia, could be divided into three categories. Firstly, the RS could be fermented into intestinal metabolites related to the regulation of the blood glucose, such as the SCFAs. Secondly, the abundance of the beneficial bacteria, such as *Lactobacillus*, *Bifidobacterium*, *Enterococcus* and *Ruminococcus* could increase, while the abundance of the harmful bacteria, such as *Bacteroides*, *Lactobacillus*, *Oscillospira* and *Escherichia coli* would decrease by the RS to regulate the metabolic pathway of the blood glucose. Thirdly, RS could decrease the expression of the inflammation factors, such as tumor necrosis factor-α (TNF-α), and interleukin-6 (IL-6). Additionally, the hepatic glycolysis and gluconeogenesis could be decreased by the SCFAs fermented by the gut microbes. Table 1 illustrates the reported hypoglycemic mechanisms of RS by regulating the intestinal microbiota disorder.

Table 1. The reported mechanisms of regulating the intestinal microbiota disorder.

Type of RS and Its Source	Model	Dosage/Duration	Intestinal Hormone/Intestinal Microbiota	Inferences	Ref.
76% HAM-RS2 (high-amylose maize) +24% raw potato starch	Thirty female pigs (BW 63.1 ± 4.4 kg)	RSD and AXD diets: 2.7% of average BW (75 kg); WSD diet: 2.44% of BW/3 weeks	Increase PYY	PYY promoted intestinal secretion, promotes GLP-1 secretion and stimulates insulin secretion	[19]
Bread enriched with resistant starch (RSB) (15% of total starch)	Ten apparently healthy subjects (mean 27 years; SD 3.9) with a normal body mass index (mean 24.5 kg m^{-2}; SD 2.8)	An amount corresponding to 50 g of available carbohydrates or a solution containing 50 g of glucose diluted in 250 mL of water/Test sessions, total 4 weeks.	Increase GLP-1 and PYY	The food contains RS and could retard the absorption of glucose	[20]
HAM-RS2 (high-amylose maize type 2 resistant starch)	Eighteen overweight, healthy adults	Either muffins enriched with 30 g HAM-RS2 ($n = 11$) or 0 g HAM-RS2 (control; $n = 7$) daily/6 weeks	Increase PYY	The consumption of HAM-RS2 can improve glucose homeostasis, lower leptin concentrations, and increase fasting PYY	[21]
Pancake with RS	Eight healthy, adult man, middle-aged (51.4 ± 11.5 years), normal- and over-weight (BMI = 29.84 ± 7.77 kg/m^2; percent body fat = 26.42 ± 11.62%)	Consumed together with water (180 mL) within 12 min	Increase SCFA production	Combination of the RS and WP might enhance the gut SCFA production and reduce the blood glucose	[25]

Table 1. Cont.

Type of RS and Its Source	Model	Dosage/Duration	Intestinal Hormone/Intestinal Microbiota	Inferences	Ref.
RS (Hi-Maize 260)	One hundred adult male Sprague–Dawley rats	On the basis of the amount of Hi-Maize (56% RS) used/10 days	Increase GLP-1 and PYY	The plasma GLP-1 and PYY levels that regulate blood glucose were increased	[26]
Corn, mung bean and Pueraria RS	Fifteen diabetic rats induced with STZ	19 weeks	Increase GLP-1	The GLP-1 show a different content, the level of it might be related to the level the blood glucose	[27]
Ce-RS3 (RS3 from canna edulis)	Twenty-four diabetic mice induced with STZ	2 g/kg/11 weeks	Improve *Phascolarctobacterium*, *Ruminococ-caceae_NK4A214_group*, *Ruminococ-caceae_UCG_014*, *Helicobacter* and *Ruminooccu*; Decrease *Streptococcus* and *Bacillus* genus	The gut microbial properties of the RS group were tightly associated with the T2DM-related indexes	[32]
BRS (Buckwheat-RS)	Twenty-seven male 4-week-old C57BL/6 mice	6 weeks	Increase *Lactobacillus*, *Bifidobacterium* and *Enterococcus*; Decrease *Escherichia coli*	The gut microbiota change caused by BRS might be associated with the capacity of regulating the gut redox status	[33]
ORS (oat RS)	Fifty male Sprague–Dawley rats (4 weeks old, WD 105 ± 10 g)	6 weeks	Increase *Clostridium* and *Butyricoccus*; Decrease *Bacteroides*, *Lactobacillus*, *Oscillospira* and *Ruminococcus*	The anti-diabetic effects of the ORS were achieved by altering the gut microbiota	[35]
RS (Hi-maize TM)	Twenty-four healthy Sprague–Dawley rats (male, 190 ± 10 g weight)	10.5 g/kg bw/day/ 28 days	Increase *Proteobacteria*	The reduction in the blood glucose might be related to the changes in the fecal microbial community which promoted an anti-inflammatory response	[37]

3. Resisting Digestion

It has been shown that RS could regulate the levels of glucose and insulin in vivo and be beneficial to maintain the homeostasis of glucose. Due to its metabolic characteristics of slow absorption, RS plays a significant role in controlling and intervening in the condition of diabetes by reducing fasting and the postprandial blood glucose, as well as increasing the IS [45].

Bindels et al. [46] have shown that the increase of the insulin level mediated by RS also occurred in the absence of the relevant microbiota, through parallel experiments on RS fed conventional mice and sterile mice. The cecal concentrations of several bile acids (BAs) were changed, and the gene expression of the macrophage markers was reduced in the adipose tissue, of which the polarization phenotypes was implicated in the control of IS in both mice groups. The result showed that both the IS and the glucose homeostasis could be regulated by the BAs via the nuclear farnesoid X receptor (FXR) and the membrane-bound TGR5 signaling.

Wang et al. [47] found that the average blood glucose and the postprandial blood glucose could be reduced significantly in T2DM patients, with the blood glucose fluctuations decreasing after the RS diet treatment and the oral administration of glucose. The results were preliminarily inferred to be related to the anti-digestion characteristics of RS. Strozyk et al. [48] found that, compared with the fresh rice (NR) group, the peak of the postprandial blood glucose in type 1 diabetes was lower in the cooling and reheated rice (CR) group. A shorter time of the glycemic peak has also been observed in the CR group, suggesting a beneficial effect to the glycemic control, as the delayed glycemic peak could improve the activity of the short-acting insulin analogues. Yadav et al. [49]] have also found that the content of RS was increased in starch products with multiple heating/cooling cycles, while the content of digestible carbohydrates was reduced. Haini et al. [50] found that, compared with the control group, the 2-h postprandial glucose of healthy female subjects was lower in the high-amylose maize starch 30 (HM30) group. In the HM group, 30% wheat flour has been replaced by HM in a Chinese steamed bun (CSB), which decreased the content of the digestible starch and the digestion speed of the starch. Therefore, the glycemic response and the increase in the postprandial blood glucose of healthy adult subjects have been delayed. Djurle et al. [51] found similar results, a slower rise of the postprandial glucose in healthy adult subjects was observed in the RS bread group. In this group, the breads were made with refined flour containing RS. Maki et al. [52]. have assessed the effects of the two doses of HAM-RS2 intake on the IS participants with different waist circumferences. The participants were randomized to receive 0 (control starch), 15, or 30 g/d (double-blind) of HAM-RS2 for four weeks with washout intervals of three weeks. At the end of each period, the minimal model IS had been evaluated by using an insulin-modified intravenous glucose tolerance test. The present results suggested that the intake of HAM-RS2 at 15–30 g/d could improve IS in obese men whereas no significant change in IS was observed in women for reasons that remain to be determined. Zeng et al. [53]. found that the type 3 resistant starch (RS3) couldn't be degraded into glucose by the digestive enzymes in the human intestine, which could reduce the amount of the glucose conversion by the human body. The RS3 could also reduce the glycemic index that helped to reduce the postprandial blood glucose. At the same time, Wang's study [54] has shown that RS3 could stabilize the human blood glucose by repairing the pancreas β cell function, as well as improving the IS and IR of the peripheral tissues. Gourineni et al. [55] have completed a study on type 4 resistant starch (RS4). In this study, a nutritional bar containing a control (2 g), medium (21 g) and high (30 g) fiber, were consumed by healthy adults (n = 38). Venous glucose, insulin, and the capillary glucose were measured at the end. They found that the concentrations of the capillary glucose and venous insulin in the two fiber groups were significantly lower than those in the control group. At the same time, they found that the postprandial glucose and insulin responses were significantly reduced in the generally healthy adults who consumed the bar containing the potato RS4 fiber.

There are also several other studies about RS4. Stewart et al. [56] have proved that substituting RS4 for a digestible carbohydrate in scones significantly lowered the blood glucose levels in healthy adults. Likewise, Mah et al. [57] have replaced the digestible starch with cassava RS4, to reduce the available carbohydrates and they found that the postprandial blood glucose and insulin concentrations decreased significantly in the healthy subjects. Other studies also found similar results by using RS4 (25 g of VERSAFIBE™ 1490 (Ingredion Incorporated, Bridgewater, NJ, USA)) to replace the normal starch in cookies [58]. In general, there was a study that showed that glycemia could be reduced by replacing the rapidly digestible starch with RS4. This result might be caused by the incomplete release of glucose and the anti-digestibility of the starch [59]. Wang et al. [60] have postulated that the diabetes-related liver glycogen fragility could also be attenuated by RS. They found that both the diabetic group and the non-diabetic group of mice, fed with two types of high-amylose RS, contained less hepatic glycogen than those fed with normal corn starch (NCS). In addition, the molecular size and the chain-length distributions of the liver glycogen were characterized to detect the fragility of the liver glycogen before and after the dimethyl sulfoxide (DMSO) treatment. The result showed that the high-amylose RS diet could prevent the fragility of the liver-glycogen α particles, which were consistent with the hypothesis that hyperglycemia was related to the glycogen fragility. They postulated the reason was that the high-amylose RS was eventually fermented in the large intestine rather than in the small intestine, which elicited beneficial effects on the glycemic response and T2DM.

Through the above studies, it is not hard to find that the IS could be affected by RS through the reducing gene expression of the macrophage markers in the adipose tissue, regulating the membrane-bound TGR5 signaling, repairing the pancreas β cell function and preventing the fragility of the liver-glycogen α particles. Meanwhile, the RS shows less effect on the blood glucose, since it cannot be degraded by the digestive enzymes in the small intestine but only be fermented in the large intestine that reduces the absorption of glucose.

4. Reducing Inflammation

Studies have shown that the damaged pancreatic β cells could be repaired, while the expression of the binding genes, such as the C-reactive protein (CRP), TNF-α and interleukin, could be down-regulated by the RS to show the hypoglycemic effect [61].

Gargari et al. [62] found that the glycated hemoglobin (HbA1c) (−0.3%, −3.2%) and TNF-α (−3.4 pg/mL, −18.8%) could be decreased by the RS2, compared with the placebo groups. In this study, 28 females with diabetes took RS (intervention group) and 32 took a placebo (placebo group) at 10 g/d for 8 weeks. The fasting blood sugar (FBS), HbA1c, lipid profile, high-sensitive CRP (HS-CRP), IL-6 and TNF-α were measured at the end of the trial. The results suggested that the glycemic status and the inflammatory markers in the women with T2DM could be improved. Based on the results, they speculated that the improvement in the glycemic status was due to the reduction of the TNF-α levels. And Tayebi Khosroshahi et al. [63] came to similar conclusions through their research. They found that the IR level and the body's IS could be improved by RS. In the study, a 20–25 g high linear chain RS and wheat flour, daily, were used to treat hemodialysis patients for 8 weeks, respectively. The results showed that the serum IL-6 and TNF-α levels in the RS group were significantly decreased.

Xu et al. [64] found that the blood glucose of obese mice could be reduced efficiently by RS. The obese mice were placed into four groups: NC, HF, URS (intervention group with RS from untreated lentil starch) and ARS (intervention group with RS from autoclaved lentil starch). The mice in the ARS and URS groups were administrated intragastrically with the ARS and URS (400 mg/kg·BW) suspension, once daily. Furthermore, the histological analysis and the gut microbiota analysis suggested the results above might be achieved, based on the improvement of the inflammatory state and the changes of the microbial components related to vagal signals. Yuan et al. [65] have reported the similar results. Compared with the normal rice (NR)-treated diabetes mice, the levels of the related inflammation factor, such as the serum CRP, TNF-α, IL-6, nuclear factor-k-gene binding (NF-κB) and leptin (LEP), were lower while the Adiponutrin (ADPN) level was higher in the selenium-enriched rice with a high RS content (SRRS) treated mice and the normal rice with the high RS content (NRRS) treated mice. The results suggested that the hypoglycemic effects might be achieved by the high RS rice treatment because of the improvement of the chronic inflammation.

It is not hard to see that the levels of the related inflammatory factors, such as CRP, IL-6, TNF-α and NF-κB, were lowered by the RS. The reduction of glomerular damage and the enhancement of the glomerular reabsorption alleviated the development of diabetes.

5. Regulating Hypoglycemic Related Enzymes

The level of the blood glucose could be regulated by some metabolic enzymes, such as glycogen synthase (GS), phosphoenolpyruvate carboxy kinase (PEPCK) and α-glucosidase. The activity of these enzymes could be regulated by the RS to achieve a hypoglycemic effect.

Zhou et al. [66] found that the blood glucose level in the RS administration group diabetic mice was lower than that in the control group. Moreover, the expression of the insulin-induced genes Insig-1 and Insig-2, that were related to the glycolipid metabolism, were also significantly up-regulated after the RS administration in mice. The blood glucose level in the diabetic mice fed with RS was regulated by promoting glycogen synthesis and the inhibiting gluconeogenesis. Further studies suggested that the expression level of isoform 1 of the glucose-6-phosphatase (G6PC1) catalytic subunit, was lower in the RS group than it was in the MC group. In addition, this study found that the expression of the glycogen synthesis genes, the GS and glycogenin1 (GYG1) increased more than twofold after the RS intake, which suggested a progressive stimulation of the hepatic glycogen synthesis in the liver. These results suggested that the inhibition of the gluconeogenesis and the promotion of the glycogen synthesis may be one of the main ways for RS to decrease the blood glucose. This study demonstrated that the mRNA encoding enzymes involved in the gluconeogenesis could be reduced by the RS to alleviate the glucose metabolic disorders in diabetic mice. Zhu et al. [67] found that after the intervention with a kind of RS in banana powder, the glucose uptake in the liver, the glycogen synthesis, the IS and IR of the db/db diabetes mice, were improved, while the mRNA expression of the key enzyme PEPCK, the carbohydrate response element binding protein (ChREBP) of the gluconeogenesis and the GSK-3 of the glycogen synthesis, were all significantly down regulated by the RS. Hao et al. [68] found that the green banana powder was rich in RS2 and made biscuits from it, which verified its feasibility. Xiao et al. [69] found that the blood glucose of T2DM Kunming (KM) mice was increased by 10.9% in the control group, while it was decreased by 14.7% in the RS group. The inhibition rate of α-glucosidase that related to the blood glucose peak in the T2DM mice, was measured. In RS group, the inhibition rate was 23.13%, showing a certain inhibitory effect. Since the activity of α-glucosidase could be inhibited by the RS, the consumption of the liver glycogen would be reduced and the trend of weight loss would be alleviated as well.

Above of all, it's not difficult to find that the present studies of the related enzymes were all carried out in mice. In addition, to reduce the blood glucose, the expression level of the key enzymes, such as GS, G6PC1, PEPCK, ChREBP, GK and α-glucosidase, could be lowered by the RS treatment, leading to the reduction of IS and IR.

6. Other Mechanisms of the Hypoglycemic Action

In addition to the above mechanisms, there are some other pathways of RS that play a role in hypoglycemia.

Li et al. [70] found that the blood glucose of the model group was significantly increased, compared with the control group ($p < 0.05$). In this study, the hyperglycemia mice induced by HFD, were treated with a dioscorea alata L. high RS (HRS) for 4 weeks. The blood glucose increased since the ability of converting the glucose into lipids was weakened because of the disorder of the fat metabolism. Li et al. [71] found that the blood glucose of the mice with diabetes was regulated after the intake of the biscuit with the added RS3 from *Purple Disocorea Alata*. L. Wang et al. [72] found the colonic proglucagon expression and the adiponectin levels in visceral fat could be increased by HAMRS2, which indicated that the IS in the visceral fat has been improved. Sun et al. [73] found that the glucose tolerance, the insulin content and glucose metabolism in diabetic mice were regulated by the RS2 treatment. In the study, they treated the T2DM mice with a high-glucose-fat diet and a low-dose STZ with low, medium, and high doses of RS2 (100, 150, and 200 g/kg) for 28 days. Furthermore, the western blot and real-time polymerase chain reaction (RT-PCR) results showed that the expression levels of the insulin receptor substrate 1 and the insulin receptor substrate 2, were enhanced in the pancreas. Based on the above results, the blood glucose in the diabetic mice can be regulated by the RS by altering the expression level of the genes related to the glucose metabolism and improving the pancreatic dysfunction. Wang et al. [74] found that the blood glucose level was reduced by 16.0–33.6% and the serum insulin level was recovered by 25.0–39.0% in T2DM mice fed on a lotus seed RS (LSRS). They elucidated the molecular basis of the hypoglycemic effect by supplying different doses of the LSRS on the T2DM mice. Through the relevant analysis of genes, they have suggested that the protective effect of the LSRS was most likely achieved by modulating the expression levels of the various key factors involved in the insulin secretion, insulin signal transmission, cell apoptosis, antioxidant activity and p53 signaling pathways.

MacNeil et al. [75] found that, as an effective substitute for the available carbohydrate (CHO) in baked food, RS could lower the T2DM diabetes' blood glucose excursion by using a randomized crossover design. Furthermore, the GIP-insulin axis was influenced after ingesting more RS because of the hyperglycemic effect of the RS. In this study, 12 patients with T2DM underwent four different bagel treatments. Abby et al. [76] have found that the glucose, insulin and glucagon-like peptide-1 have been reduced significantly in the fasted subjects. In the study, the fasted subjects (n = 20) consumed either a low-fiber control breakfast or one of four breakfasts that contained a 25 g soluble corn fiber (SCF) or RS, alone or in combination with pullulan (SCF + P and RS + P). The results suggested that the satiety or the energy intake would not be influenced by the fiber treatments, compared to the control. Though the definite mechanism of the results haven't been described, it may be related to the secretion of GLP-1, and the aging-related decline in the glucose tolerance could be recuperated by it [77]. Song et al. [78] found the value of the fasting blood glucose of T2DM mice was decreased by investigating the effects of the Kudzu RS on the IR, the gut physical barrier and the gut microbiota. The expression of IRS-1, p-PI3K, p-Akt, and Glut4 were restored by the study of the relevant expression of the protein, which led to the improvements of the insulin synthesis efficiency and the glucose sensitivity in the T2DM mice.

In conclusion, other mechanisms of the RS hypoglycemic actions could be divided into three categories. Firstly, the disorder of the fat metabolism could be decreased by the RS. The tryptophan related to the gut microbiota function and the IS in the visceral fat could be regulated by the HRS. Secondly, the expression levels of the genes related to glucose, such as the insulin receptor substrate 1 and the insulin receptor substrate 2, could be enhanced in the pancreas by the RS. The expression levels of the various key factors involved in the insulin secretion, insulin signal transmission, cell apoptosis, antioxidant activity and p53 signaling pathways could also be modulated by the RS. Thirdly, the expression of protein, such as IRS-1, p-PI3K, p-Akt, and Glut4 could be restored by the RS. All in all, the relevant studies still have some limitations and need further study.

7. Conclusions

According to numerous previous studies, RS has been confirmed as a kind of dietary fiber to prevent diabetes. In this review, several mechanisms of the glycemic control with a RS consumption were summarized, mainly including regulating the intestinal microbiota disorder, resisting digestion, reducing the inflammation and regulating the hypoglycemic related enzymes. Several specific intestinal microbiota, signaling pathways, gene targets and relevant enzymes of those mechanisms have been clarified in the above studies. Therefore, RS seems to hold great promise in the prevention and treatment of diabetes. Based on the above research, we have concluded the studies on the prevention of T2DM by RS (as shown in Table 2).

Table 2. Studies on the prevention of T2DM by RS.

Kind of RS	Results	Conclusion	Ref.
RS2	The glycemic status and inflammatory markers in women with T2DM could be improved.	The improvement in glycemic status was due to the reduction of the TNF-α levels.	[62]
RS2	The expression levels of the insulin receptor substrate 1 and the insulin receptor substrate 2 were enhanced in the T2DM mice.	RS could regulate the expression level of the genes related to the glucose metabolism and improving the pancreatic dysfunction.	[73]
Lotus seed RS (LSRS)	The blood glucose level was reduced by 16.0–33.6% and the serum insulin level was recovered by 25.0–39.0% in the T2DM mice.	The LSRS achieved the hypoglycemic effect by modulating the expression levels of the various key factors.	[74]
Kudzu RS	The value of the fasting blood glucose of the T2DM mice was decreased.	Kudzu RS restored the expression of the relevant protein and it led to the improvements of the insulin synthesis efficiency and the glucose sensitivity in the T2DM mice.	[78]
Bagel with high-amylose maize RS (RS2)	The fasting IS of the RS bagel treatment is lower than the control bagel treatment.	The amount of insulin required to manage the postprandial glucose were reduced by the high-HAM-RS2 bagel through the improvement of the glycemic efficiency, while improving the fasting IS in adults at an increased risk of T2DM.	[79]
Indica rice resistant starch (IR-RS) prepared by modification	The blood glucose of the rats with T2DM was lower than those in the control group.	The IR-RS digestibility was affected as well as the blood glucose levels of the diabetic mice	[80]
White wheat flour bread (WWB) enriched RS	The glucose tolerance and GLP-1 were improved, compared with that without WWB.	The consumption of RS might affect the glycemic excursions through a mechanism involving colonic fermentation.	[81]
Banana starch (NBS) with a high resistant starch (RS)	The 24 h mean blood glucose (24 h MBG) value of the T2DM patients was lower in the NBS treatment but not significant.	The result might be influenced by different baseline microbiota, an underlying dietary variability, or other environmental factors.	[82]

However, previous studies are incomplete, since most of the studies have been focused on animal experiments, rather than on human subjects. To our knowledge, the relevant literatures on animal experiments are mainly focused on mice, but a few studies are on large animals. Although there are a couple of relevant studies on pigs, the change of the PYY levels showed different results, compared with the mice. It is indicating that the role of PYY on the RS in different animals is controversial. Thus, further investigations are needed.

In addition, even the reports on the human subjects are not comprehensive. There are several factors leading to inaccurate conclusions. Firstly, the RS has been classified into five main categories, according to the causes of indigestion. They process the different molecular structures and amylase binding sites. Meanwhile, the RS molecules are linked together by different glycosidic bonds that cause diverse effects on the blood glucose. Therefore, it is necessary to clarify the mechanisms about how the blood glucose is influenced by the different RS structures for further studies, especially in the changes of the different enzymes related to the different RS structures. Secondly, the current conclusions obtained in mice are not necessarily applicable to human subjects, since the metabolic pathways and targets are different. Therefore, more thorough studies on humans should be conducted. In order to clarify the specific metabolites or the proteins that are related to the blood glucose in the human body, such studies should not only observe the level of blood glucose, but also analyze the metabolites in the blood and urine by metaomics, such as metatranscriptomics and metaproteomics. With the help of the statistical analysis, the main pathways, key enzymes and target genes related to the hypoglycemic activity in humans will be screened out. Thirdly, the various dietary habits may also cause the relevant genetic changes in the same ethnic groups, which may lead to the different responses to the same RS. There are certain differences in the intestinal microbiota and metabolic pathways among people with different dietary habits. For example, the abundance of some harmful intestinal microbiota, such as *Bacteroides* and *Lactobacillus*, will be increased in people with a high animal protein or high animal fat diet, while the abundance of several intestinal microbiota that produce SCFAs, such as *Helicobacter*, *Akermanniella* and *Bifidobacterium* will be increased in people with a high dietary fiber diet. In addition, there are also studies that suggest that the postprandial cardiovascular and metabolic indexes are different in the equal-energy diets with different proportions of macronutrients, such as fiber, fat and protein, as well as influenced by body weight and exercise. To date, data from the intervention studies that systematically assess the effects of the different diet habits on glucose and metabolomics, are particularly lacking. Therefore, to avoid the interference of the results on the RS hypoglycemic activity, a long-term analysis of the dietary habits, should be carried out in future studies.

In conclusion, the mechanisms of the RS hypoglycemic activity in the human body need to be further studied from the aspects discussed above. Those studies can help us to understand the mechanisms comprehensively, as well as to provide the theoretical basis for people to choose a specific type of RS to control their blood glucose.

Author Contributions: Writing—original draft preparation, J.L. and W.L.; writing—review and editing, Y.L., L.W., N.J. and H.Z.; supervision, B.F. and F.W. All authors have read and agreed to the published version of the manuscript.

Funding: This research was funded by the project of Science and Technology Department of Qinghai province (2021-NK-A3); Agricultural Science and Technology Innovation Program of Institute of Food Science and Technology, Chinese Academy of Agricultural Sciences (CAAS-ASTIP-G2022-IFST-06) and the project of Science and Technology Department of Yunnan province (2003AD150016).

Institutional Review Board Statement: Not applicable.

Informed Consent Statement: Not applicable.

Data Availability Statement: Not applicable.

Acknowledgments: We acknowledge the partial support from the Science and Technology Department of Qinghai province with the project number 2021-NK-A3 and the Institute of Food Science and Technology with the project number CAAS-ASTIP-G2022-IFST-06. We also appreciate the financial research support of the Chinese Academy of Agricultural Sciences, Science and Technology Department of Yunnan province. All the individuals included in this section have consented to the acknowledgement.

Conflicts of Interest: The authors declare no conflict of interest.

References

1. Zheng, Y.; Ley, S.H.; Hu, F.B. Global Aetiology and Epidemiology of Type 2 Diabetes Mellitus and Its Complications. *Nat. Rev. Endocrinol.* **2018**, *14*, 88–98. [CrossRef] [PubMed]
2. International Diabetes Federation. *IDF Diabetes Atlas*, 10th ed.; International Diabetes Federation: Brussels, Belgium, 2021; ISBN 978-2-930229-98-0.
3. Englyst, H.; Kingman, S.; Cummings, J. Classification and Measurement of Nutritionally Important Starch Fractions. *Eur. J. Clin. Nutr.* **1992**, *46* (Suppl. S2), S33–S50.
4. Birt, D.F.; Boylston, T.; Hendrich, S.; Jane, J.-L.; Hollis, J.; Li, L.; McClelland, J.; Moore, S.; Phillips, G.J.; Rowling, M.; et al. Resistant Starch: Promise for Improving Human Health. *Adv. Nutr.* **2013**, *4*, 587–601. [CrossRef] [PubMed]
5. Robertson, M.D. Dietary-Resistant Starch and Glucose Metabolism. *Curr. Opin. Clin. Nutr. Metab. Care* **2012**, *15*, 362–367. [CrossRef]
6. EFSA Panel on Dietetic Products; Nutrition and Allergies (NDA) Scientific Opinion on the Substantiation of Health Claims Related to Resistant Starch and Reduction of Post-Prandial Glycaemic Responses (ID 681), "Digestive Health Benefits" (ID 682) and "Favours a Normal Colon Metabolism" (ID 783) Pursuant to Article 13(1) of Regulation (EC) No 1924/2006. *EFSA J.* **2011**, *9*, 2024. [CrossRef]

7. Englyst, H.N.; Trowell, H.; Southgate, D.A.; Cummings, J.H. Dietary Fiber and Resistant Starch. *Am. J. Clin. Nutr.* **1987**, *46*, 873–874. [CrossRef]
8. Keenan, M.J.; Zhou, J.; Hegsted, M.; Pelkman, C.; Durham, H.A.; Coulon, D.B.; Martin, R.J. Role of Resistant Starch in Improving Gut Health, Adiposity, and Insulin Resistance. *Adv. Nutr.* **2015**, *6*, 198–205. [CrossRef] [PubMed]
9. Bodinham, C.L.; Smith, L.; Thomas, E.L.; Bell, J.D.; Swann, J.R.; Costabile, A.; Russell-Jones, D.; Umpleby, A.M.; Robertson, M.D. Efficacy of Increased Resistant Starch Consumption in Human Type 2 Diabetes. *Endocr. Connect.* **2014**, *3*, 75–84. [CrossRef] [PubMed]
10. Maki, K.C.; Phillips, A.K. Dietary Substitutions for Refined Carbohydrate That Show Promise for Reducing Risk of Type 2 Diabetes in Men and Women. *J. Nutr.* **2015**, *145*, 159S–163S. [CrossRef] [PubMed]
11. Wong, T.; Louie, J. The Relationship between Resistant Starch and Glycemic Control: A Review on Current Evidence and Possible Mechanisms. *Starch Stärke* **2016**, *69*, 1600205. [CrossRef]
12. Shen, L.; Keenan, M.J.; Raggio, A.; Williams, C.; Martin, R.J. Dietary-Resistant Starch Improves Maternal Glycemic Control in Goto-Kakizaki Rat. *Mol. Nutr. Food Res.* **2011**, *55*, 1499–1508. [CrossRef] [PubMed]
13. Robertson, M.D.; Currie, J.M.; Morgan, L.M.; Jewell, D.P.; Frayn, K.N. Prior Short-Term Consumption of Resistant Starch Enhances Postprandial Insulin Sensitivity in Healthy Subjects. *Diabetologia* **2003**, *46*, 659–665. [CrossRef]
14. Johnston, K.L.; Thomas, E.L.; Bell, J.D.; Frost, G.S.; Robertson, M.D. Resistant Starch Improves Insulin Sensitivity in Metabolic Syndrome. *Diabet. Med.* **2010**, *27*, 391–397. [CrossRef] [PubMed]
15. Liu, H.; Zhang, M.; Ma, Q.; Tian, B.; Nie, C.; Chen, Z.; Li, J. Health Beneficial Effects of Resistant Starch on Diabetes and Obesity via Regulation of Gut Microbiota: A Review. *Food Funct.* **2020**, *11*, 5749–5767. [CrossRef] [PubMed]
16. Portincasa, P.; Bonfrate, L.; Vacca, M.; De Angelis, M.; Farella, I.; Lanza, E.; Khalil, M.; Wang, D.Q.-H.; Sperandio, M.; Di Ciaula, A. Gut Microbiota and Short Chain Fatty Acids: Implications in Glucose Homeostasis. *Int. J. Mol. Sci.* **2022**, *23*, 1105. [CrossRef] [PubMed]
17. Morrison, D.J.; Preston, T. Formation of Short Chain Fatty Acids by the Gut Microbiota and Their Impact on Human Metabolism. *Gut Microbes* **2016**, *7*, 189–200. [CrossRef] [PubMed]
18. Chambers, E.S.; Preston, T.; Frost, G.; Morrison, D.J. Role of Gut Microbiota-Generated Short-Chain Fatty Acids in Metabolic and Cardiovascular Health. *Curr. Nutr. Rep.* **2018**, *7*, 198–206. [CrossRef]
19. Canfora, E.E.; Jocken, J.W.; Blaak, E.E. Short-Chain Fatty Acids in Control of Body Weight and Insulin Sensitivity. *Nat. Rev. Endocrinol.* **2015**, *11*, 577–591. [CrossRef]
20. Nielsen, T.S.; Theil, P.K.; Purup, S.; Nørskov, N.P.; Bach Knudsen, K.E. Effects of Resistant Starch and Arabinoxylan on Parameters Related to Large Intestinal and Metabolic Health in Pigs Fed Fat-Rich Diets. *J. Agric. Food Chem.* **2015**, *63*, 10418–10430. [CrossRef] [PubMed]
21. Hughes, R.L.; Horn, W.F.; Wen, A.; Rust, B.; Woodhouse, L.R.; Newman, J.W.; Keim, N.L. Resistant Starch Wheat Increases PYY and Decreases GIP but Has No Effect on Self-Reported Perceptions of Satiety. *Appetite* **2022**, *168*, 105802. [CrossRef]
22. Binou, P.; Yanni, A.E.; Stergiou, A.; Karavasilis, K.; Konstantopoulos, P.; Perrea, D.; Tentolouris, N.; Karathanos, V.T. Enrichment of Bread with Beta-Glucans or Resistant Starch Induces Similar Glucose, Insulin and Appetite Hormone Responses in Healthy Adults. *Eur. J. Nutr.* **2021**, *60*, 455–464. [CrossRef]
23. Maziarz, M.P.; Preisendanz, S.; Juma, S.; Imrhan, V.; Prasad, C.; Vijayagopal, P. Resistant Starch Lowers Postprandial Glucose and Leptin in Overweight Adults Consuming a Moderate-to-High-Fat Diet: A Randomized-Controlled Trial. *Nutr. J.* **2017**, *16*, 14. [CrossRef] [PubMed]
24. Higgins, J.A.; Higbee, D.R.; Donahoo, W.T.; Brown, I.L.; Bell, M.L.; Bessesen, D.H. Resistant Starch Consumption Promotes Lipid Oxidation. *Nutr. Metab.* **2004**, *1*, 8. [CrossRef] [PubMed]
25. Ebihara, K.; Shiraishi, R.; Okuma, K. Hydroxypropyl-Modified Potato Starch Increases Fecal Bile Acid Excretion in Rats. *J. Nutr.* **1998**, *128*, 848–854. [CrossRef]
26. Venkataraman, A.; Sieber, J.R.; Schmidt, A.W.; Waldron, C.; Theis, K.R.; Schmidt, T.M. Variable Responses of Human Microbiomes to Dietary Supplementation with Resistant Starch. *Microbiome* **2016**, *4*, 33. [CrossRef] [PubMed]
27. Mohr, A.E.; Minicucci, O.; Long, D.J.; Miller, V.J.; Keller, A.; Sheridan, C.; O'brien, G.; Ward, E.; Schuler, B.; Connelly, S.; et al. Resistant Starch Combined with Whey Protein Increases Postprandial Metabolism and Lowers Glucose and Insulin Responses in Healthy Adult Men. *Foods* **2021**, *10*, 537. [CrossRef] [PubMed]
28. Zhou, J.; Martin, R.J.; Tulley, R.T.; Raggio, A.M.; McCutcheon, K.L.; Shen, L.; Danna, S.C.; Tripathy, S.; Hegsted, M.; Keenan, M.J. Dietary Resistant Starch Upregulates Total GLP-1 and PYY in a Sustained Day-Long Manner through Fermentation in Rodents. *Am. J. Physiol. Endocrinol. Metab.* **2008**, *295*, E1160–E1166. [CrossRef]
29. Chen, Y.C.; Cui, W.W.; Huang, G.H. Long Term Effect on Diabetic Rats of Feeding Resistant Starch. *Sci. Technol. Food Ind.* **2014**, *35*, 374–378, 382.
30. Chen, Y.C.; Liu, J.H.; Zhang, X.; Cui, J.A.; Chen, X.P. Metabolic Regulation and Mechanism of Multi-Component Resistant Starch on High-Sugar and High-Fat Model Mice. *Sci. Technol. Food Ind.* **2021**, *42*, 357–362.
31. Boll, E.V.J.; Ekström, L.M.N.K.; Courtin, C.M.; Delcour, J.A.; Nilsson, A.C.; Björck, I.M.E.; Östman, E.M. Effects of Wheat Bran Extract Rich in Arabinoxylan Oligosaccharides and Resistant Starch on Overnight Glucose Tolerance and Markers of Gut Fermentation in Healthy Young Adults. *Eur. J. Nutr.* **2016**, *55*, 1661–1670. [CrossRef] [PubMed]

32. Zhang, W.Q.; Zhang, Y.M. Effects of Resistant Starch on Insulin Resistance in Diabetes Mellitus Rats. *Acta Nutr. Sin.* **2008**, *30*, 257–261.
33. Haub, M.D.; Hubach, K.L.; Al-Tamimi, E.K.; Ornelas, S.; Seib, P.A. Different Types of Resistant Starch Elicit Different Glucose Reponses in Humans. *J. Nutr. Metab.* **2010**, *2010*, 230501. [CrossRef]
34. Zhang, C.; Ma, S.; Wu, J.; Luo, L.; Qiao, S.; Li, R.; Xu, W.; Wang, N.; Zhao, B.; Wang, X.; et al. A Specific Gut Microbiota and Metabolomic Profiles Shifts Related to Antidiabetic Action: The Similar and Complementary Antidiabetic Properties of Type 3 Resistant Starch from Canna Edulis and Metformin. *Pharmacol. Res.* **2020**, *159*, 104985. [CrossRef] [PubMed]
35. Zhou, Y.; Zhao, S.; Jiang, Y.; Wei, Y.; Zhou, X. Regulatory Function of Buckwheat-Resistant Starch Supplementation on Lipid Profile and Gut Microbiota in Mice Fed with a High-Fat Diet. *J. Food. Sci.* **2019**, *84*, 2674–2681. [CrossRef] [PubMed]
36. Sánchez-Tapia, M.; Hernández-Velázquez, I.; Pichardo-Ontiveros, E.; Granados-Portillo, O.; Gálvez, A.; Tovar, A.R.; Torres, N. Consumption of Cooked Black Beans Stimulates a Cluster of Some Clostridia Class Bacteria Decreasing Inflammatory Response and Improving Insulin Sensitivity. *Nutrients* **2020**, *12*, 1182. [CrossRef] [PubMed]
37. Zhu, Y.; Dong, L.; Huang, L.; Shi, Z.; Shen, R. Effects of Oat β-Glucan, Oat Resistant Starch, and the Whole Oat Flour on Insulin Resistance, Inflammation, and Gut Microbiota in High-Fat-Diet-Induced Type 2 Diabetic Rats. *J. Funct. Foods* **2020**, *69*, 103939. [CrossRef]
38. Zhao, L.; Zhang, F.; Ding, X.; Wu, G.; Lam, Y.Y.; Wang, X.; Fu, H.; Xue, X.; Lu, C.; Ma, J.; et al. Gut Bacteria Selectively Promoted by Dietary Fibers Alleviate Type 2 Diabetes. *Science* **2018**, *359*, 1151–1156. [CrossRef]
39. Jiang, Y.; Zhang, N.; Zhou, Y.; Zhou, Z.; Bai, Y.; Strappe, P.; Blanchard, C. Manipulations of Glucose/Lipid Metabolism and Gut Microbiota of Resistant Starch Encapsulated Ganoderma Lucidum Spores in T2DM Rats. *Food Sci. Biotechnol.* **2021**, *30*, 755–764. [CrossRef]
40. Klingbeil, E.A.; Cawthon, C.; Kirkland, R.; de La Serre, C.B. Potato-Resistant Starch Supplementation Improves Microbiota Dysbiosis, Inflammation, and Gut-Brain Signaling in High Fat-Fed Rats. *Nutrients* **2019**, *11*, 2710. [CrossRef]
41. De La Serre, C.B.; Ellis, C.L.; Lee, J.; Hartman, A.L.; Rutledge, J.C.; Raybould, H.E. Propensity to High-Fat Diet-Induced Obesity in Rats Is Associated with Changes in the Gut Microbiota and Gut Inflammation. *Am. J. Physiol. Gastrointest. Liver Physiol.* **2010**, *299*, G440–G448. [CrossRef] [PubMed]
42. He, J.; Zhang, P.; Shen, L.; Niu, L.; Tan, Y.; Chen, L.; Zhao, Y.; Bai, J.; Hao, X.; Li, X.; et al. Short-Chain Fatty Acids and Their Association with Signalling Pathways in Inflammation, Glucose and Lipid Metabolism. *Int. J. Mol. Sci.* **2020**, *21*, 6356. [CrossRef]
43. Li, X.; Chen, H.; Guan, Y.; Li, X.; Lei, L.; Liu, J.; Yin, L.; Liu, G.; Wang, Z. Acetic Acid Activates the AMP-Activated Protein Kinase Signaling Pathway to Regulate Lipid Metabolism in Bovine Hepatocytes. *PLoS ONE* **2013**, *8*, e67000. [CrossRef]
44. Li, H.; Gao, Z.; Zhang, J.; Ye, X.; Xu, A.; Ye, J.; Jia, W. Sodium Butyrate Stimulates Expression of Fibroblast Growth Factor 21 in Liver by Inhibition of Histone Deacetylase 3. *Diabetes* **2012**, *61*, 797–806. [CrossRef] [PubMed]
45. Hu, Z.Z.; Hao, M.Z.; Meng, Y.; Wang, Q.; Xu, H.Z.; Wei, Y.L.; Cheng, L.Y. Preparation, Efficacy and Application of Resistant Starch. *Food Nutr. China* **2021**, *27*, 30–35.
46. Bindels, L.B.; Segura Munoz, R.R.; Gomes-Neto, J.C.; Mutemberezi, V.; Martínez, I.; Salazar, N.; Cody, E.A.; Quintero-Villegas, M.I.; Kittana, H.; de Los Reyes-Gavilán, C.G.; et al. Resistant Starch Can Improve Insulin Sensitivity Independently of the Gut Microbiome. *Microbiome* **2017**, *5*, 12. [CrossRef]
47. Wang, B.; Liu, J.; Zhou, Z.H. Effect of Resistant Starch on Blood Glucose in Patients with Type 2 Diabetes. *J. Tongji Univ.* **2014**, *35*, 56–60.
48. Strozyk, S.; Rogowicz-Frontczak, A.; Pilacinski, S.; LeThanh-Blicharz, J.; Koperska, A.; Zozulinska-Ziolkiewicz, D. Influence of Resistant Starch Resulting from the Cooling of Rice on Postprandial Glycemia in Type 1 Diabetes. *Nutr. Diabetes* **2022**, *12*, 21. [CrossRef] [PubMed]
49. Yadav, B.S.; Sharma, A.; Yadav, R.B. Studies on Effect of Multiple Heating/Cooling Cycles on the Resistant Starch Formation in Cereals, Legumes and Tubers. *Int. J. Food Sci. Nutr.* **2009**, *60* (Suppl. S4), 258–272. [CrossRef]
50. Haini, N.; Jau-Shya, L.; Mohd Rosli, R.G.; Mamat, H. Effects of High-Amylose Maize Starch on the Glycemic Index of Chinese Steamed Buns (CSB). *Heliyon* **2022**, *8*, e09375. [CrossRef] [PubMed]
51. Djurle, S.; Andersson, A.A.M.; Andersson, R. Effects of Baking on Dietary Fibre, with Emphasis on β-Glucan and Resistant Starch, in Barley Breads. *J. Cereal Sci.* **2018**, *79*, 449–455. [CrossRef]
52. Maki, K.C.; Pelkman, C.L.; Finocchiaro, E.T.; Kelley, K.M.; Lawless, A.L.; Schild, A.L.; Rains, T.M. Resistant Starch from High-Amylose Maize Increases Insulin Sensitivity in Overweight and Obese Men. *J. Nutr.* **2012**, *142*, 717–723. [CrossRef] [PubMed]
53. Zeng, C.; Liu, Y.L.; Xiao, M.F.; Liu, B.; Zeng, F. Research Progress of RS3 Resistant Starch. *Food Ind. Sci. Technol.* **2020**, *41*, 338–344.
54. Wang, Q. Study on the Hypolipidemic Effect of Lotus Seed Resistant Starch and Its Mechanism. Ph.D. thesis, Fujian Agriculture and Forestry University, Fujian, China, 2018.
55. Gourineni, V.; Stewart, M.L.; Wilcox, M.L.; Maki, K.C. Nutritional Bar with Potato-Based Resistant Starch Attenuated Post-Prandial Glucose and Insulin Response in Healthy Adults. *Foods* **2020**, *9*, 1679. [CrossRef] [PubMed]
56. Stewart, M.L.; Wilcox, M.L.; Bell, M.; Buggia, M.A.; Maki, K.C. Type-4 Resistant Starch in Substitution for Available Carbohydrate Reduces Postprandial Glycemic Response and Hunger in Acute, Randomized, Double-Blind, Controlled Study. *Nutrients* **2018**, *10*, 129. [CrossRef]

57. Mah, E.; Garcia-Campayo, V.; Liska, D. Substitution of Corn Starch with Resistant Starch Type 4 in a Breakfast Bar Decreases Postprandial Glucose and Insulin Responses: A Randomized, Controlled, Crossover Study. *Curr. Dev. Nutr.* **2018**, *2*, nzy066. [CrossRef] [PubMed]
58. Stewart, M.L.; Zimmer, J.P. A High Fiber Cookie Made with Resistant Starch Type 4 Reduces Post-Prandial Glucose and Insulin Responses in Healthy Adults. *Nutrients* **2017**, *9*, 237. [CrossRef]
59. Du, Y.; Wu, Y.; Xiao, D.; Guzman, G.; Stewart, M.L.; Gourineni, V.; Burton-Freeman, B.; Edirisinghe, I. Food Prototype Containing Resistant Starch Type 4 on Postprandial Glycemic Response in Healthy Adults. *Food Funct.* **2020**, *11*, 2231–2237. [CrossRef]
60. Wang, Z.; Hu, Z.; Deng, B.; Gilbert, R.G.; Sullivan, M.A. The Effect of High-Amylose Resistant Starch on the Glycogen Structure of Diabetic Mice. *Int. J. Biol. Macromol.* **2022**, *200*, 124–131. [CrossRef] [PubMed]
61. Yan, G.S.; Zheng, H.Y.; Sun, M.Y.; Zhang, Z.H.; Xu, H.; Chen, H. Recent Progress in Physiological Functions and Mechanism of Action of Resistant Starch. *Food Sci.* **2020**, *41*, 330–337.
62. Gargari, B.P.; Namazi, N.; Khalili, M.; Sarmadi, B.; Jafarabadi, M.A.; Dehghan, P. Is There Any Place for Resistant Starch, as Alimentary Prebiotic, for Patients with Type 2 Diabetes? *Complement. Ther. Med.* **2015**, *23*, 810–815. [CrossRef] [PubMed]
63. Tayebi Khosroshahi, H.; Vaziri, N.D.; Abedi, B.; Asl, B.H.; Ghojazadeh, M.; Jing, W.; Vatankhah, A.M. Effect of High Amylose Resistant Starch (HAM-RS2) Supplementation on Biomarkers of Inflammation and Oxidative Stress in Hemodialysis Patients: A Randomized Clinical Trial. *Hemodial. Int.* **2018**, *22*, 492–500. [CrossRef] [PubMed]
64. Xu, J.; Ma, Z.; Li, X.; Liu, L.; Hu, X. A More Pronounced Effect of Type III Resistant Starch vs. Type II Resistant Starch on Ameliorating Hyperlipidemia in High Fat Diet-Fed Mice Is Associated with Its Supramolecular Structural Characteristics. *Food Funct.* **2020**, *11*, 1982–1995. [CrossRef]
65. Yuan, H.; Wang, W.; Chen, D.; Zhu, X.; Meng, L. Effects of a Treatment with Se-Rich Rice Flour High in Resistant Starch on Enteric Dysbiosis and Chronic Inflammation in Diabetic ICR Mice. *J. Sci. Food Agric.* **2017**, *97*, 2068–2074. [CrossRef]
66. Zhou, Z.; Wang, F.; Ren, X.; Wang, Y.; Blanchard, C. Resistant Starch Manipulated Hyperglycemia/Hyperlipidemia and Related Genes Expression in Diabetic Rats. *Int. J. Biol. Macromol.* **2015**, *75*, 316–321. [CrossRef]
67. Zhu, X.H.; Yang, G.M.; Chen, A.M. Research the Improvement and Mechanism on Insulin Resistance of Raw Banana Powder in Type II Diabetic Mellitus. Ph.D. Thesis, South China Agricultural University, Guangzhou, China, 2016.
68. Hao, X.; Chen, F.; Wang, J. Study on Processing Technology of Banana Resistant Starch Biscuit. *Food Technol.* **2019**, *44*, 223–227.
69. Xiao, B.; Deng, D.W. Resistant Starch to Diabetic Mice the Regulation of Blood Sugar and the Effect of Short Chain Fatty Acid. Master's Thesis, Nanchang University, Nanchang, China, 2018.
70. Li, T.; Teng, H.; An, F.; Huang, Q.; Chen, L.; Song, H. The Beneficial Effects of Purple Yam (*Dioscorea Alata* L.) Resistant Starch on Hyperlipidemia in High-Fat-Fed Hamsters. *Food Funct.* **2019**, *10*, 2642–2650. [CrossRef] [PubMed]
71. Li, W.Q. Preparation of Resistant Starch Type 3 from Purple Disocorea Alata L and Its Application on Biscuit. Master's thesis, Nanchang University, Nanchang, China, 2015.
72. Wang, Y.; Perfetti, R.; Greig, N.H.; Holloway, H.W.; DeOre, K.A.; Montrose-Rafizadeh, C.; Elahi, D.; Egan, J.M. Glucagon-like Peptide-1 Can Reverse the Age-Related Decline in Glucose Tolerance in Rats. *J. Clin. Investig.* **1997**, *99*, 2883–2889. [CrossRef]
73. Sun, H.; Ma, X.; Zhang, S.; Zhao, D.; Liu, X. Resistant Starch Produces Antidiabetic Effects by Enhancing Glucose Metabolism and Ameliorating Pancreatic Dysfunction in Type 2 Diabetic Rats. *Int. J. Biol. Macromol.* **2018**, *110*, 276–284. [CrossRef]
74. Wang, Q.; Zheng, Y.; Zhuang, W.; Lu, X.; Luo, X.; Zheng, B. Genome-Wide Transcriptional Changes in Type 2 Diabetic Mice Supplemented with Lotus Seed Resistant Starch. *Food Chem.* **2018**, *264*, 427–434. [CrossRef]
75. MacNeil, S.; Rebry, R.M.; Tetlow, I.J.; Emes, M.J.; McKeown, B.; Graham, T.E. Resistant Starch Intake at Breakfast Affects Postprandial Responses in Type 2 Diabetics and Enhances the Glucose-Dependent Insulinotropic Polypeptide—Insulin Relationship Following a Second Meal. *Appl. Physiol. Nutr. Metab.* **2013**, *38*, 1187–1195. [CrossRef]
76. Klosterbuer, A.S.; Thomas, W.; Slavin, J.L. Resistant Starch and Pullulan Reduce Postprandial Glucose, Insulin, and GLP-1, but Have No Effect on Satiety in Healthy Humans. *J. Agric. Food Chem.* **2012**, *60*, 11928–11934. [CrossRef] [PubMed]
77. Warman, D.J.; Jia, H.; Kato, H. The Potential Roles of Probiotics, Resistant Starch, and Resistant Proteins in Ameliorating Inflammation during Aging (Inflammaging). *Nutrients* **2022**, *14*, 747. [CrossRef] [PubMed]
78. Song, X.; Dong, H.; Zang, Z.; Wu, W.; Zhu, H.; Zhang, H.; Guan, Y. Kudzu Resistant Starch: An Effective Regulator of Type 2 Diabetes Mellitus. *Oxid. Med. Cell. Longev.* **2021**, *2021*, 4448048. [CrossRef]
79. Dainty, S.A.; Klingel, S.L.; Pilkey, S.E.; McDonald, E.; McKeown, B.; Emes, M.J.; Duncan, A.M. Resistant Starch Bagels Reduce Fasting and Postprandial Insulin in Adults at Risk of Type 2 Diabetes. *J. Nutr.* **2016**, *146*, 2252–2259. [CrossRef] [PubMed]
80. Zhou, Y.; Meng, S.; Chen, D.; Zhu, X.; Yuan, H. Structure Characterization and Hypoglycemic Effects of Dual Modified Resistant Starch from Indica Rice Starch. *Carbohydr. Polym.* **2014**, *103*, 81–86. [CrossRef]
81. Nilsson, A.C.; Ostman, E.M.; Holst, J.J.; Björck, I.M.E. Including Indigestible Carbohydrates in the Evening Meal of Healthy Subjects Improves Glucose Tolerance, Lowers Inflammatory Markers, and Increases Satiety after a Subsequent Standardized Breakfast. *J. Nutr.* **2008**, *138*, 732–739. [CrossRef]
82. Arias-Córdova, Y.; Ble-Castillo, J.L.; García-Vázquez, C.; Olvera-Hernández, V.; Ramos-García, M.; Navarrete-Cortes, A.; Jiménez-Domínguez, G.; Juárez-Rojop, I.E.; Tovilla-Zárate, C.A.; Martínez-López, M.C.; et al. Resistant Starch Consumption Effects on Glycemic Control and Glycemic Variability in Patients with Type 2 Diabetes: A Randomized Crossover Study. *Nutrients* **2021**, *13*, 52. [CrossRef]

MDPI
St. Alban-Anlage 66
4052 Basel
Switzerland
www.mdpi.com

Molecules Editorial Office
E-mail: molecules@mdpi.com
www.mdpi.com/journal/molecules

Disclaimer/Publisher's Note: The statements, opinions and data contained in all publications are solely those of the individual author(s) and contributor(s) and not of MDPI and/or the editor(s). MDPI and/or the editor(s) disclaim responsibility for any injury to people or property resulting from any ideas, methods, instructions or products referred to in the content.

www.ingramcontent.com/pod-product-compliance
Lightning Source LLC
LaVergne TN
LVHW070153120526
838202LV00013BA/1053